D1765936

Prestressed Concrete Bridges

Design and Construction

Second edition

ice | publishing
Institution of Civil Engineers

Prestressed Concrete Bridges

Design and Construction

Second edition

Nigel Hewson
Hewson Consulting Engineers Ltd

Published by ICE Publishing, 40 Marsh Wall, London E14 9TP.

Full details of ICE Publishing sales representatives and distributors can be found at:
www.icevirtuallibrary.com/info/printbooksales

First edition published 2003

Also available from ICE Publishing

Stress Ribbon and Cable-supported Pedestrian Bridges, Second edition.
J. Strasky. ISBN 978-0-7277-4146-2
ICE Manual of Bridge Engineering, Second edition.
G. Parke (ed.). ISBN 978-0-7277-3452-5
Designers' Guide to Eurocode 1: Actions on bridges.
J.-A. Calgaro. ISBN 978-0-7277-3158-6
Designers' Guide to EN 1992-2. Eurocode 2: Design of concrete structures.
Part 2: Concrete bridges.
C.R. Hendy. ISBN 978-0-7277-3159-3

Associate Commissioning Editor: Victoria Thompson
Production Editor: Imran Mirza
Market Development Executive: Catherine de Gatacre

www.icevirtuallibrary.com

A catalogue record for this book is available from the British Library

ISBN: 978-0-7277-4113-4

© Thomas Telford Limited 2012

ICE Publishing is a division of Thomas Telford Ltd, a wholly-owned subsidiary of
the Institution of Civil Engineers (ICE).

Typeset by Academic + Technical, Bristol
Printed and bound in Great Britain by CPI Group (UK) Ltd, Croydon, CR0 4YY

Contents

Preface to the 2nd edition

It is eight years since the first edition of this book and much has changed since then on both a personal level and in the prestressed concrete bridge industry. On a personal level I established Hewson Consulting Engineers Ltd in 2005 and H&T Associates in Malaysia in 2009, and since then we have designed some of the largest bridges in the world, and been involved with the design and construction of many prestressed concrete bridges including some of the longest prestressed concrete viaduct projects.

In the prestressed concrete bridge industry the sophistication of design software has leapt forward, while Eurocodes are now fully implemented in Europe. In construction the basic components of prestressing remain unchanged but construction techniques have become more sophisticated and the use or precast segmental construction has flourished, establishing strong footholds in India and China.

This second edition incorporates the changes over the last eight years with reference to the projects I have worked on and the lessons learnt. I have also updated the text with reference to the Eurocodes where appropriate, although I have not tried to replicate the many guidance publications that give a much more comprehensive insight into the Eurocode requirements.

Preface to the 1st edition

This book is an expansion of the chapter on prestressed concrete bridges in the *Manual of Bridge Engineering*, and it seeks to give a wider coverage of the practical aspects involved in the design and construction of prestressed concrete bridge decks. Concrete remains the most common material for bridge construction around the world, and prestressed concrete is frequently the material of choice for bridge decks with spans greater than 25 m. As well as the more common highway and rail bridges, prestressed concrete has also been successfully used on some of the larger cable-stayed structures, major river crossings and urban viaducts.

Much has been learnt about prestressed concrete over the six decades since it was first used on bridgeworks and the current techniques employed in both their design and their construction have evolved greatly from those used by the early pioneers such as Freyssinet and Magnel. Higher-strength concrete and improvements in the prestressing steels coupled with sophisticated design tools have given prestressed concrete a greater versatility.

There are many different ways to design and build prestressed concrete bridges and it is true to say that every bridge is different in one way or another. All bridge designers have their own way of doing things and their own preferences in the design approach and the details to adopt, while individual contractors come up with a different solution to the same problem. No single publication can cover all the possible ways to design or build prestressed concrete bridges; however, this book presents the author's experiences, collected over 25 years in the industry.

Although there are several good publications covering general prestress concrete design, and many short articles and guidance notes on the different practical aspects of designing and constructing prestressed concrete bridges, there is little available literature bringing all this together. It is the aim of this book to combine all the aspects of prestressed concrete bridge decks into one volume.

Chapters 1 and 2 cover the general aspects of prestressing, its principles and the components that make up the prestressing systems. Chapters 3 and 4 consider durability issues, while Chapters 5 to 8 cover a range of general design issues. Chapters 9 to 17 discuss the design and construction of different deck forms and construction techniques. Chapter 18 looks briefly at some of the problems that have occurred in the past.

While reviewing the design and construction of the different types of prestressed concrete bridge decks and the prestressing systems used, this book assumes that the reader has a basic understanding of prestressed and reinforced concrete design, which can be applied to the specific application of bridges.

Dedication

To my wife Alison, who has visited many bridges over the years, and to my daughters Sarah and Laura, who may come to use them one day.

Acknowledgements

Since the first edition of this book I have been fortunate to have had the opportunity to work on many new prestressed concrete bridges, either in their design or their construction, and I wish to thank all those who I have worked with and who have helped me along the way; there are too many to mention here but they know who they are.

Special thanks to Andrew Hodgkinson, Jeremy Barnes, George Daoutis, Stuart Moore and all the others at Hewson Consulting Engineers Limited and to Teh Tzyy Wooi and colleagues at H&T Associates who have provided support, assistance and entertainment over recent years.

For their assistance with the first edition I would like to thank Graham Davenport, Keith Simm and Dick Thomas who gave me the initial encouragement and opportunity to become a bridge engineer. Special thanks also to Andrew Barbour, Louise Smith, Alan Major, Francis Kung, Peter Fox, Martin Morris, Bill Hard, Tom Williams, Roger Knight, Stephen Cardwell, Bernard Fortier, Flemming Pedersen, Robert Uthwatt, Gordon Clark, Paul Bottomley, Dr Brenni, Tony Dempsey, Ronald Yee, and Jean-Philippe Mathieu, and to VSL, Freyssinet, DYWIDAG, BBR and McCall's for the information on their systems and for their permission to publish extracts from their brochures, and to Sarah Hewson for help with some of the diagrams.

Prestressed Concrete Bridges, 2nd edition
ISBN: 978-0-7277-4113-4

ICE Publishing: All rights reserved
doi: 10.1680/pcb.41134.001

Chapter 1
Prestressed concrete in bridgeworks

Introduction

Developed over the last 80 years, prestressed concrete has provided the bridge engineer with the ability to build economic, durable and efficient structures, often using local resources and labour, combined with the latest design and construction technology to provide improved access and transportation links for the local communities. Prestressing of concrete is now a well-established technique in all countries, with a wide portfolio of bridge types and span lengths constructed, ranging from major sea crossings to urban viaducts, motorways, rail structures and footbridges. For all these structures, the principles behind the design and construction of prestressed concrete bridges remain the same, which is to combine the tensile strength of the prestressing with the compressive strength of the concrete to create a balanced enhanced structure.

Concrete is strong in compression but weak in tension; however, prestressing can be used to ensure that it remains within its tensile and compressive capacity under the range of loading applied. Prestressing in bridgeworks is normally applied as an external force to the concrete by the use of wires, strands or bars, and can greatly increase the strength of the concrete alone. This can result in longer or slender spans, which improves aesthetics while providing economy in the construction. There are many fine examples in the UK and around the world of bridges and viaducts with prestressed concrete decks, such as the Byker Viaduct (Figure 1.1), built utilising precast match-cast segments, the Ceiriog Viaduct box girder (Figure 1.2), which was launched into position, and the River Dee crossing (Figure 1.3), with a cable-stayed deck. On motorways and other highway networks, numerous overbridges and underbridges have also utilised prestressing with either in situ or precast concrete beams such as the Killarney Overbridge (Figure 1.4), or with an in situ box girder deck such as the viaduct under construction in Dubai (Figure 1.5).

Development of prestressed concrete bridges has given the bridge engineer increased flexibility in his selection of bridge form and in the construction techniques available. As a result, prestressed concrete is frequently the material of choice for bridge decks. Spans range from less than 25 m, using precast beams, to more than 500 m for cable-stayed bridges.

Prestressed concrete bridges include a wide variety of different forms: from cast in situ to precast; from beams to box girders; and from simply supported to cable-stayed. Their functions range from the carrying of pedestrians and cyclists to provision for road or rail traffic and they make up a significant proportion of the bridge stock in existence around the world today.

The design of prestressed concrete bridges both greatly influences and is dependent on the construction process envisaged. The construction sequence and practical considerations in positioning the tendons influence the prestress layout much more than the desire to achieve a concordant profile. The deck

1

Figure 1.1 Byker Viaduct, Newcastle, England

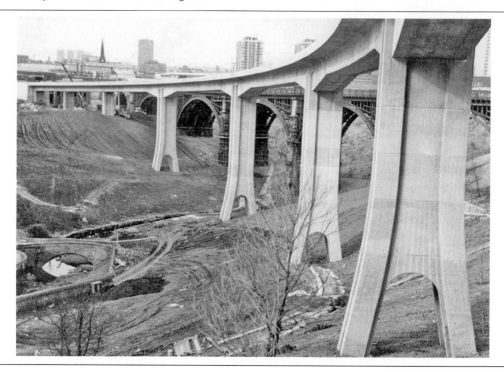

Figure 1.2 Ceiriog Viaduct, Wales (reproduced courtesy of Tony Gee and Partners LLP, copyright reserved)

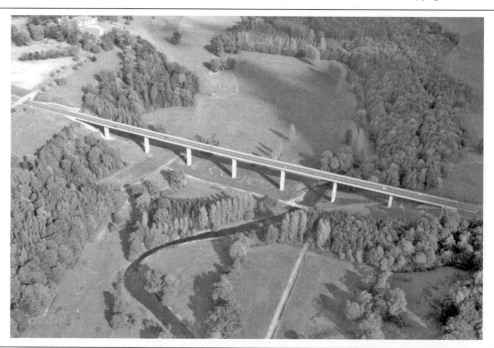

Figure 1.3 River Dee crossing, Wales (reproduced courtesy of Gifford, part of Ramboll, copyright reserved)

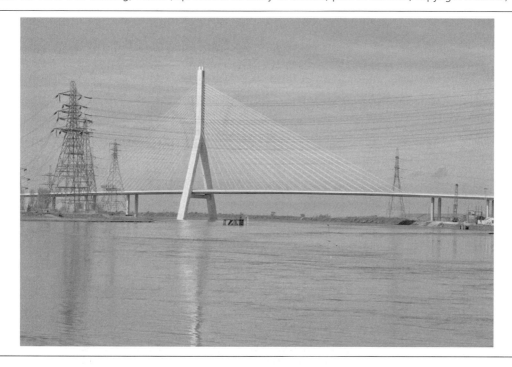

Figure 1.4 Killarney Overbridge, Ireland (reproduced courtesy of Roughan & O'Donovan, Dublin, copyright reserved)

Figure 1.5 Dubai interchange viaduct

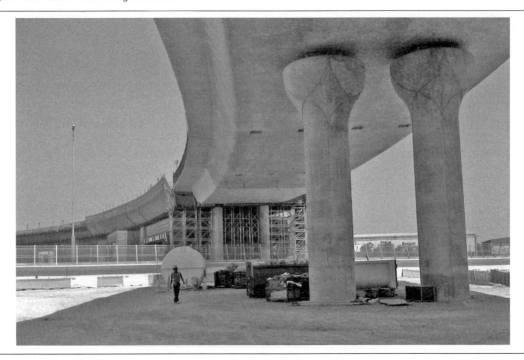

section and concrete shape is often dictated by the placement of the prestress tendons and their anchorages, while the need for rapid construction or difficulties with access may dictate the type of structure and construction methodology adopted. On the other hand, the assumptions made in the analysis and design may govern the construction method and sequence adopted. Solutions using precast segments (see Chapter 13) are used over difficult ground or where rapid construction is needed. The deck may be launched (see Chapter 15) over a deep valley to reduce the need for falsework and to minimise the disruption and environmental impact on the landscape below. Precast beams (see Chapter 10) are used where access to the ground underneath is restricted, and stay cables (see Chapter 16) allow longer spans and reduce the number of piers. In all cases, the design and the construction methodology must be developed together.

Durability and maintenance requirements feature prominently in any design produced today as lessons learnt from problems with existing bridges are fed back into the industry. Aesthetics are another important consideration, as designers strive to improve the appearance of bridges to fit into the environment. Today's bridge designer has to take into account all these factors in order to develop the optimum solution for any particular bridge project undertaken.

After deciding the type of bridge structure to be used and the construction approach to be adopted, the design of the prestressed concrete deck is carried out. This usually involves the use of specialist software, although hand calculations or spreadsheets are valid approaches that are still widely used by many designers. The analysis must accurately model the effect of the prestress and applied load on the structure, to ensure that all the strength and serviceability requirements are met. This often includes 3D finite-element modelling, as well as stage-by-stage analysis with the creep and shrinkage behaviour included.

The many aspects that contribute to the design and construction of prestressed concrete bridge decks, including the materials and equipment used, the design requirements and procedures, and the construction techniques adopted, are covered in subsequent chapters. In this first chapter only the basic principles of prestressing are discussed, and a brief history of prestressing and prestressed concrete bridges presented.

Principles of prestressing

Force transfer between the prestress tendon and the concrete achieves prestressing of the concrete member. Tendons are pulled and stretched and then anchored against the concrete, with the tension in the tendon being balanced by compression in the concrete. In this way an external compressive force is applied to the concrete and is used to counter the tensile stresses generated under the bending moments and shear forces present.

The tendons are placed either within the concrete member as internal tendons, or alongside the concrete as external tendons, and can be unbonded or bonded to the concrete. They can be pre-tensioned or post-tensioned and consist of wires, strands or bars; however, their effect on the concrete, and the basic principles of design, are the same in all cases.

The effect of prestressing can be demonstrated by attempting to pick up a row of children's building blocks, as depicted in Figure 1.6. By pushing the ends together, and applying a 'prestressing' force to the blocks, it is possible to pick them up; however, if no force were applied to the ends then the row of bricks would fall apart. The joints between the blocks are unable to resist tension and, without the 'prestressing' force, they open up under their self-weight. With the 'prestressing' force applied the joints are kept in compression over their full depth.

Figure 1.6 Prestressing building blocks

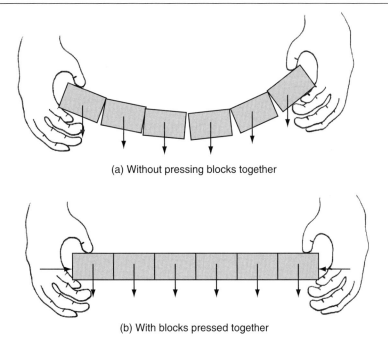

(a) Without pressing blocks together

(b) With blocks pressed together

Figure 1.7 Change to stresses in beam

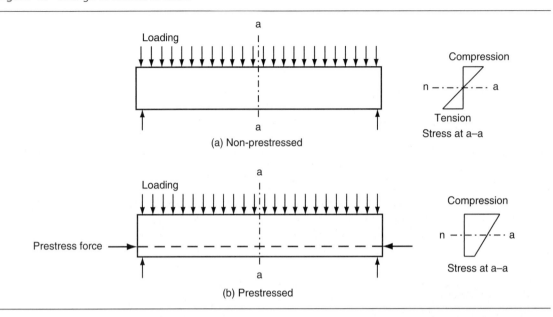

The same principles apply to bridge decks where, in Figure 1.7(a), the non-prestressed concrete beam would be subjected to tension along the bottom fibre of the span due to the bending moments generated by the applied load. With a 'prestress' force applied as indicated in Figure 1.7(b), the tension can be eliminated and the concrete kept in compression over the full section.

The concrete can be either fully prestressed, which ensures that the longitudinal stresses are always in compression, or partially prestressed, which allows some small tension to occur within the concrete under certain loading conditions.

Pre-tensioning
For pre-tensioning, the tendon is stressed by jacking against an anchor frame before the concrete is placed. The force is released from the anchor frame into the concrete when the concrete has obtained sufficient strength. Used primarily in precast beams, the tendons usually consist of 7-wire strands or individual wires.

The precasting beds can be hundreds of metres long and provide space to cast several beams in one go, or they can be arranged to cast one beam at a time. The beds have soffit and side shutters that are arranged around the prestressing tendons, which are tensioned before fixing the reinforcement and placing the concrete. Figure 1.8 shows a typical precasting bed arrangement, with the beam side shutters being prepared in the centre of the picture and the jacking frame at the lower right corner. At the far end of the bed, another anchor frame is needed to hold the strands when they are stressed, and this is shown in the inset in Figure 1.8.

After the concrete in the beams has reached the required transfer strength, the strands or wires are released from the jacking frames. The force in the strands transfers to the beams by the bond between the strand and the concrete over the anchorage length at the beam ends.

Figure 1.8 Precasting yard for the Bosporus Crossing project, Turkey

Post-tensioning

For post-tensioning, the tendon is pulled and stretched using a jack, and the resulting force is trans-ferred directly on to the previously cast and hardened concrete through the tendon anchor. The tendons consist of bars, single strands or multi-strands that can be arranged with a varying vertical and horizontal profile along the bridge deck, which allows for the most efficient arrangements of prestressing to be adopted. Post-tensioning can be used on many different types of structure, including precast beams, in situ or precast box girders and cable-stayed structures.

Internal tendons are placed inside a duct cast into the concrete and installed either before or after the concrete is placed, while external tendons are installed in ducts placed outside the concrete section after the concreting has been completed. Figure 1.9 shows internal prestressing ducts within the reinforce-ment cage installed in a deck web prior to concreting, while Figure 1.10 shows external tendons running inside a box girder adjacent to internal tendons anchored on blisters.

The force is generated in a post-tensioned tendon by the use of a stressing jack, as shown in Figure 1.11 for a multi-strand tendon. The jack pulls the strands and transfers the force from the tendon on to the anchor plate and into the concrete. Reinforcement in the surrounding concrete resists the local tensile stresses around the anchor and assists in transferring the force into the deck.

Brief history of prestressed concrete bridges

Prestressing of concrete dates back to the early 1900s, when several engineers experimented with the technique, but it was Eugène Freyssinet (1879–1962) who first applied for a patent in 1928 covering

Figure 1.9 Internal ducts in reinforcement prior to concreting

Figure 1.10 External tendons inside box girder deck

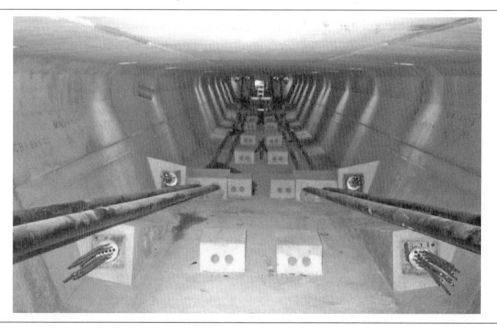

Figure 1.11 Stressing of post-tensioned tendon (reproduced courtesy of Freyssinet International, copyright reserved)

the principle. Freyssinet is reported to have built an experimental arch in 1908, which incorporated prestressing tendons, and in 1930 he utilised prestressing during the construction of the Plougastel Bridge in France. In Germany, Dyckerhof and Widmann used post-tensioned bars during the construction of the arch bridge at Alsleben in 1927.

During this time, much was still being learnt about prestressed concrete and the nature of time-dependent effects. Engineers were experimenting with ways of using prestressed concrete in building and other applications. The first prestressed concrete bridges were built in the mid-1930s with the Oued Fodda Bridge in Algeria (1936), Aux Bridge in Saxony (1936) and the Oella Bridge in Germany (1938) leading the way, but it was not until after the Second World War that prestressed concrete became firmly established in bridge building.

Eugène Freyssinet is generally considered the father of modern-day prestressing, and it was his six bridges across the Marne River in France, built between 1945 and 1950, that established the technique. The first of these, the Luzancy Bridge, is a 55 m-long structure built as a 'frame' and integral with the abutments. The remaining five are each 73 m long and again built as a frame with the deck gently curving up towards the centre, where the deck depth is only 1 m. One of these, the Annet Bridge, is shown in Figure 1.12. The decks are either a concrete box or concrete 'I' beam arrangement, with prestressing being used vertically in the webs as well as horizontally along the deck.

After the Second World War, with a shortage of steel in Europe the use of prestressed concrete became popular in the reconstruction of bridges across the continent, and the 1950s saw prestressed concrete being used widely for bridgeworks. In this period, Eugène Freyssinet continued to design many new prestressed concrete bridges, while Gustave Magnel developed the technique on several notable

Figure 1.12 Annet Bridge, France (reproduced courtesy of Freyssinet International, copyright reserved)

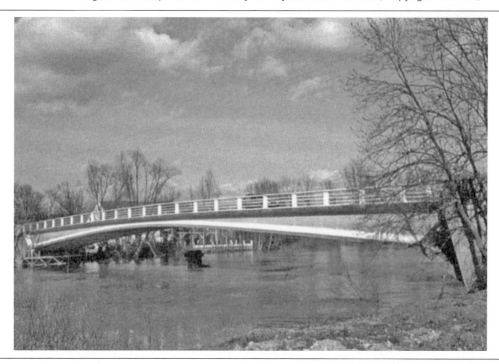

structures in Belgium. In the UK, LG Mouchel and Partners, in collaboration with the Prestressed Concrete Company, used prestressed concrete for a number of projects, including bridges. In Germany, Fritz Leonhardt was a leading exponent of prestressed concrete. The new material quickly caught on and its application was developed by many notable engineers of that time.

In 1939, Freyssinet developed and patented the first conical friction anchor. After 1945 other systems, such as those produced by Magnel-Blaton, BBRV and Lee-McCall, began to appear, as prestressing of concrete became popular. The early prestressing systems used were comprised of wires usually of 5 mm or 7 mm diameter, tensioned and anchored by a gripping device at the ends that transferred the load to the concrete. With the BBRV system, the wires were fitted with button heads at the ends to hold the individual wires against the anchor head, a system that is still used by BBR today. Towards the late 1950s, more systems using bars were developed; while the use of the wire systems evolved, with larger tendons employing wire-by-wire stressing and wedge anchors.

The very early attempts at prestressing concrete used normal steel as 'tie rods', and were not very successful, as the low level of prestress in the steel was lost due to the shrinkage and creep in the concrete. By the time that prestressing was beginning to be used more widely in bridgeworks, higher strength steel wires were available with ultimate strengths of up to 1725 N/mm^2 and a yield limit of 1240 N/mm^2. Low-relaxation grades of steel and higher-strength wires, with ultimate strengths of up to 1860 N/mm^2, have now become standard in the industry.

By the early 1960s, the wires were being assembled into strands and anchored by wedges onto an anchor cone cast into the concrete. This led to the development of the standard '7-wire' strand most

commonly used today. Also at this time, large capacity jacks were developed that could tension the large multi-strand cables in one operation, and this type of tendon began to dominate, although bars and wires were still used for more specialist applications.

From the early 1950s and over the next 40 years, Jean Muller carried on with Freyssinet's work in developing the art of prestressed concrete bridge design and construction. He introduced the technique of match-cast precast segmental concrete deck construction on the Shelton Road Bridge, USA, in 1952. In 1962, Muller used the technique on a series of motorway overbridges around Paris, the first of which was the Choisy-le-Roi Bridge over the River Seine. He went on to use this approach on numerous viaducts and major bridges worldwide, with one of the most notable bridges being the Linn Cove Viaduct, shown in Figure 1.13. This was built by a 'top down' approach to minimise the disruption to the environmentally sensitive area beneath. The deck was cantilevered out using precast segments and when the pier position was reached, the precast substructure was placed from the deck and then used to support the deck as the construction continued.

In 1956, the 38 km-long Lake Pontchartrain Bridge, in the USA, was constructed with precast prestressed girders, each with a span length of 17 m, and is still one of the longest bridges in the world today.

The UK's first use of prestressed concrete for bridges involved precast prestressed beams manufactured and stockpiled at the beginning of the Second World War and used for the reconstruction of two bridges carrying roads over railways. In 1948, Nunns Bridge in Lincolnshire became the first in situ

Figure 1.14 St James's Park Footbridge, England

post-tensioned concrete road bridge in the UK, with a span of 22.5 m. This was followed by several other short-span bridges formed using precast beams, and in 1954 the Northam Bridge in South-ampton became the first major prestressed concrete bridge in the UK, with a total length of 148 m and spans up to 32 m. Figure 1.14 shows the footbridge in St James's Park, London, which was built in 1957 as a slender beam with a span of 21 m and mid-span depth of 0.36 m and remains as a fine example of an early prestressed concrete structure. A detailed history of prestressed concrete road bridges in the UK is described in the book by Sutherland *et al.* (2001) and a paper by Sriskandan (1989).

In 1960, the Mangfall Bridge in Germany was the first prestressed concrete truss bridge to be built; it has a span of 108 m. In more recent years several truss decks using prestressed concrete have been built for multi-span viaducts, although this has not become a common form of structure for concrete bridges.

The first balanced cantilever bridge was built in reinforced concrete across the Rio de Peixe in Brazil in 1930, but it was not until 1950 that this form of construction was first used with prestressed concrete on the Lahn Bridge at Balduinstein, Germany. When the Medway Bridge in the UK, shown in Figure 1.15, was opened in 1963 it was at the time the longest span in the world constructed as an in situ concrete segmental box girder using the balanced cantilever technique. The central span was 150 m long and included a 30 m-long suspended precast beam section in the centre. The approach spans of this 1000 m-long river crossing were constructed using precast prestressed concrete beams.

The Hammersmith Flyover, as described by Rawlinson and Stott (1962), is a landmark structure in the UK. Completed in 1961, it was the first modern elevated urban motorway. The continuous concrete box girder was built using precast segments erected on falsework and joined with 75 mm in situ stitches

Figure 1.15 Medway Bridge, UK (reproduced courtesy of Hyder Consulting (UK) Ltd, copyright reserved)

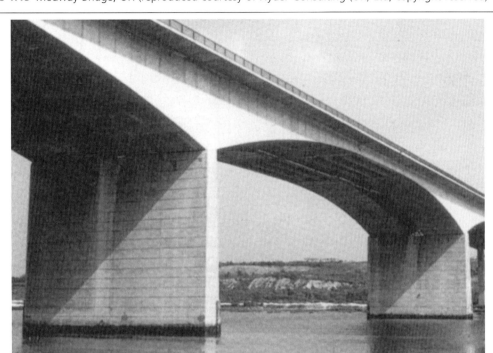

between the segments before being prestressed. The centre spine for the deck was erected first and precast cantilevers added to give a total width of 18.6 m, with a maximum span of 42.7 m and an overall length of 623 m.

Throughout the UK in the 1960s, the development of the motorway and bypass network led to the widespread use of precast beams for both overbridge and underpass construction. The production of families of precast beams in factory conditions led to economic structures with the minimum of formwork required and rapid construction. The inverted 'T' beams, 'I' beams, box-beams, 'M' beams and pseudo-box beams were all popular and competed against each other, with a typical arrangement shown in Figure 1.16.

The technique of incrementally launching a post-tensioned concrete box girder was first used on the bridge over the Rio Caroni in Venezuela in 1962, with the system commonly seen today first used on the Inn Bridge, Kufstein, Austria, in 1965. Both these bridges had the basic design carried out by Fritz Leonhardt and his partner Willi Baur. This technique has since been used on many bridges including several notable viaducts on the motorway systems in Germany, France and Spain, and is especially well suited for crossing over deep valleys or difficult ground. Dornoch Firth crossing in Scotland, shown in Figure 1.17, was opened in 1991 and was incrementally launched over the estuary during a ten-month period. With 20 spans it has a total length of 890 m.

Prestressed concrete was used in major cable-stayed bridges from the 1960s, with several notable structures built including the Rafael Urdaneta Bridge in Venezuela, with a 235 m main span. Refinements in the design and construction of concrete cable-stayed bridges have led to their common use for spans of

Figure 1.16 Precast beam flyover in Hong Kong

Figure 1.17 Dornoch Firth crossing, Scotland (reproduced courtesy of Tony Gee and Partners LLP, copyright reserved)

up to 500 m. Another early example is Brotonne Bridge in France, shown in Figure 1.18. Built in 1976, it has a 320 m-long main span formed using a prestressed concrete box girder deck and a single plane of stays, which produced an elegant and striking appearance.

From the mid-1970s, the rapid growth of many major cities around the world, and the increase in the volume of highway traffic, led towards the development of long urban viaducts, with a need to minimise disruption during their construction, and to build them quickly and economically. Prestressed concrete in the form of precast beams or precast segmental box girders dominated this type of construction. In 1992, the 60 km of viaducts for the Bangkok Second Expressway System, Sectors A and C, were completed utilising match-cast precast segmental box girders over poor ground and through congested streets. With 12 erection gantries, and a casting yard producing an average of 36 segments a day, the viaducts were completed within a two-year construction period. More recently, the Hung Hom Bypass in Hong Kong, shown in Figure 1.19, utilised precast match-cast segments to construct a viaduct crossing over several major highways, a mainline railway and part of the harbour.

Crossing the Tagus River, the Vasco da Gama Bridge (Figure 1.20) in Portugal opened in 1998. It is an excellent example of modern prestressed concrete bridge design, with a wide range of construction techniques used on the 18 km-long crossing. Spanning the main navigational channel, the Main Bridge has an 830 m-long cable-stayed deck with a main span of 420 m, utilising in situ prestressed concrete edge beams and concrete deck slab. The approach structures are divided into a number of different construction types. The Expo Viaduct and the 130 m navigation spans on the Central Viaduct were constructed using match-cast precast segmental box girder decks erected as balanced cantilevers, while the remainder of the Central Viaduct used precast full-length span elements weighing 2200 tonnes, lifted into place by a floating crane. The other major elements of the crossing are the Northern Viaduct, which is an in situ ribbed slab cast on full-height falsework, and the Southern Viaduct, which is a twin-beam and deck slab arrangement cast in situ using formwork supported on a travelling gantry.

Figure 1.19 Hung Hom Bypass, Hong Kong (reproduced courtesy of Hyder Consulting (UK) Ltd, copyright reserved)

Figure 1.20 Vasco da Gama Bridge, Portugal (reproduced courtesy of Yee Associates, copyright reserved)

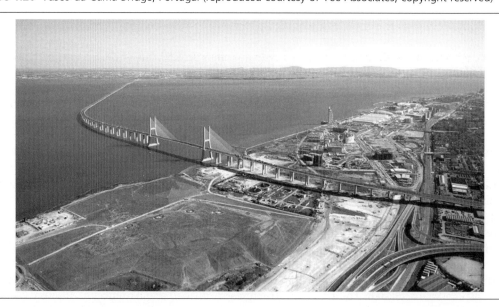

A variation to the cable-stayed bridge, the 'extradosed' arrangement, has short pylons and can be economical in the 200 m to 300 m span ranges. The Sunniberg Bridge in Switzerland, shown in Figure 1.21, opened in 1998 and is an outstanding example of this form of structure. Extending the 'prestressing' tendons above the deck gives an efficient structural arrangement which can be used with either a box girder or a beam and slab form of deck.

Figure 1.21 Sunniberg Bridge, Switzerland (reproduced courtesy of Tiefbauamt Graubünden, copyright reserved)

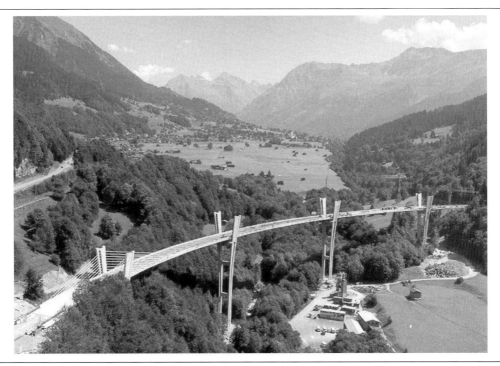

The development of the different forms of prestressed concrete bridges is illustrated in Figure 1.22, which shows the increase in span length achieved over the last 80 years. As concrete quality and strengths have increased, and prestressing systems improved, construction techniques have evolved to fully utilise the advantages inherent in prestressed concrete, thus allowing longer and more economic spans to be built.

The first designs for prestressed concrete bridges were based on experience gained from experiments and application of the technique in buildings and other structures, as no codified rules existed. In the early days, Freyssinet and other prominent engineers published articles and gave lectures, forming the main source of exchanging information. In 1949, Abeles published his book, one of the first on the principles of prestressed concrete, followed by Magnel (1950) and Guyun (1951) a few years later. Design rules were published in France in 1953, followed in 1959 by the UK standard, CP 115 published by the BSI (British Standards Institution). In the USA, the AASHTO (American Association of State Highway and Transportation Officials) Standard Specifications (AASHTO, 1931–1999) developed in the 1950s to include prestressed bridges. Design codes evolved significantly as more knowledge was gained, especially in relation to the creep, shrinkage and temperature effects. Between the early 1980s and 1990s the comprehensive modern codes BS 5400 (BSI, 1978) and AASHTO LRFD (load and resistance factor design) (AASHTO, 1994) became established. More recently the Eurocodes (BSI, 2004–2005) have been introduced as the common design standard for the European Union, and are being implemented in other countries around the world. Design codes are becoming more sophisticated in terms of their definition of the loading and performance of structures, although they only cover a small part of the knowledge and expertise required for the successful design and

Figure 1.22 Longest prestressed concrete bridge span length plotted against year

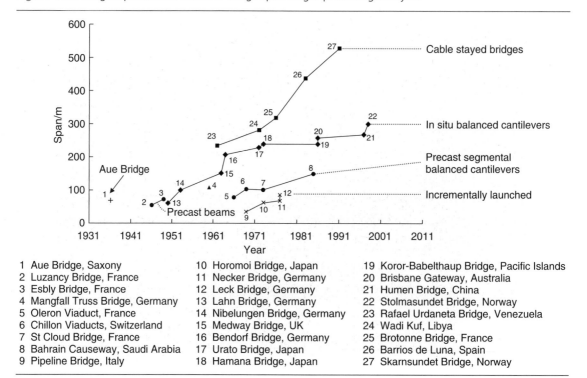

1 Aue Bridge, Saxony	10 Horomoi Bridge, Japan	19 Koror-Babelthaup Bridge, Pacific Islands
2 Luzancy Bridge, France	11 Necker Bridge, Germany	20 Brisbane Gateway, Australia
3 Esbly Bridge, France	12 Leck Bridge, Germany	21 Humen Bridge, China
4 Mangfall Truss Bridge, Germany	13 Lahn Bridge, Germany	22 Stolmasundet Bridge, Norway
5 Oleron Viaduct, France	14 Nibelungen Bridge, Germany	23 Rafael Urdaneta Bridge, Venezuela
6 Chillon Viaducts, Switzerland	15 Medway Bridge, UK	24 Wadi Kuf, Libya
7 St Cloud Bridge, France	16 Bendorf Bridge, Germany	25 Brotonne Bridge, France
8 Bahrain Causeway, Saudi Arabia	17 Urato Bridge, Japan	26 Barrios de Luna, Spain
9 Pipeline Bridge, Italy	18 Hamana Bridge, Japan	27 Skarnsundet Bridge, Norway

construction of prestressed concrete bridges. They should always be treated as guidelines around which the experienced bridge engineer can develop his designs.

The first prestressed concrete bridges tended to use post-tensioned tendons placed externally to the concrete. Prestressed wires were either left bare and open to the elements, or wrapped in some form of protection, or encased in concrete after stressing. Most of these forms of protection resulted in the external tendons having corrosion problems, which for many years led to post-tensioned designs usually adopting tendons placed internally.

There was some experimenting with external tendons in the 1960s and 1970s, but in the UK problems with durability occurred which effectively resulted in a UK-wide ban on the use of external tendons for several years. In France, during the early 1970s, the need to strengthen several bridges led to the development of external tendon arrangements, while the 'prestressing' strand was beginning to be used as stays on cable-stayed bridges. These external tendons and stay arrangements adopted a more robust protection system, developed using high-density polyethylene ducts filled with cement grout or grease, which led to increased confidence in external tendons.

In the UK in 1985, a collapse of the Ynys-y-Gwas Bridge in Wales due to corroded internal tendons initiated a series of inspections and investigations. By 1992 this led to a ban on internal tendons and all UK designs at that time had to use external tendons. This ban was partially lifted in 1996 when, with good detailing and construction practice, internal tendons were again allowed, except for precast segmental construction where concerns over the water-tightness of the joints remained. For the

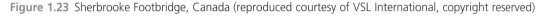

Figure 1.23 Sherbrooke Footbridge, Canada (reproduced courtesy of VSL International, copyright reserved)

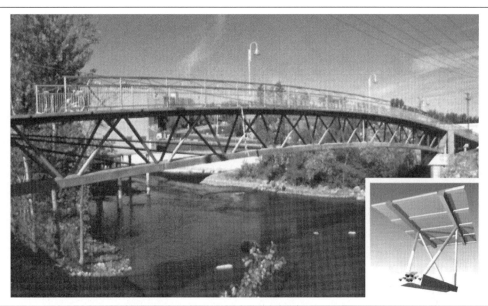

future, prestressed concrete is likely to continue to dominate as the material of choice for bridge spans in the range of 30 m to 250 m, while its use on longer-span cable-stayed bridge decks should continue to grow. High-strength concretes continue to be developed and used in building works; they are also becoming more common for bridgeworks which, when combined with prestressing, will enable longer and lighter structures to be adopted. Two recent bridges using high-strength concrete with prestressing are the Sherbrooke Footbridge in Canada and the Peace Footbridge in Korea.

The Sherbrooke Footbridge, completed in 1997, uses a special high-strength concrete product in its composition developed by Bouygues's research department. The Reactive Powder Concrete (RPC) is made using small particles to give a dense mixture. Concrete used on the Sherbrooke Footbridge achieved a compressive strength of 200 N/mm^2 and tensile strengths of 7 N/mm^2 and 40 N/mm^2 under direct load and bending respectively. Shown in Figure 1.23, the 3.5 m-deep truss deck spans 60 m with a RPC top slab and bottom beam. The steel tube diagonal members are also filled with RPC concrete. The top slab is transversely prestressed, with external prestressing tendons used longitudinally.

Similar high-strength concrete has also been used on the Peace Footbridge, Korea, completed in 2001. The fibre-reinforced high-strength concrete is prestressed both longitudinally and transversely for the beam and slab section that makes up the 120 m-span arch seen in Figure 1.24. The specially developed concrete enabled very thin sections to be used, reducing the total weight of the deck while the dense concrete is expected to be highly durable. Non-metallic prestressing tendons using carbon fibre reinforced polymers have been developed and tested in recent years. These have the advantage of having very high strength, and are not susceptible to corrosion; however, they are currently much more expensive than steel tendons and have proved difficult to anchor effectively. As this technology develops their use is likely to become more common.

Figure 1.24 Peace Footbridge, Seoul, Korea (reproduced courtesy of VSL International, copyright reserved)

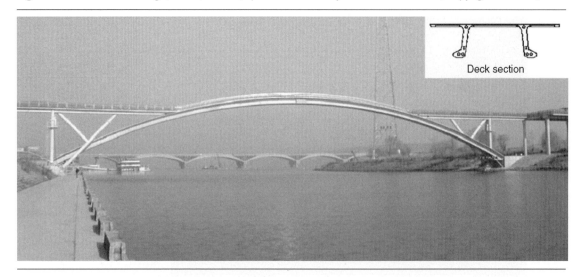

Deck section

Figure 1.25 Tampines Viaduct, Singapore (reproduced courtesy of Projalma, copyright reserved)

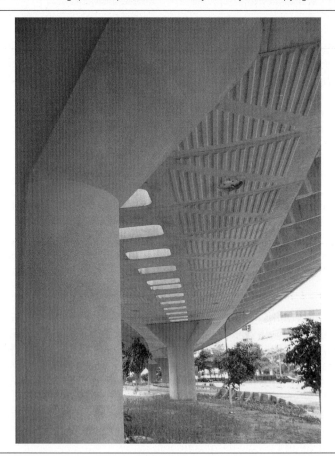

Figure 1.26 Dubai LRT, UAE

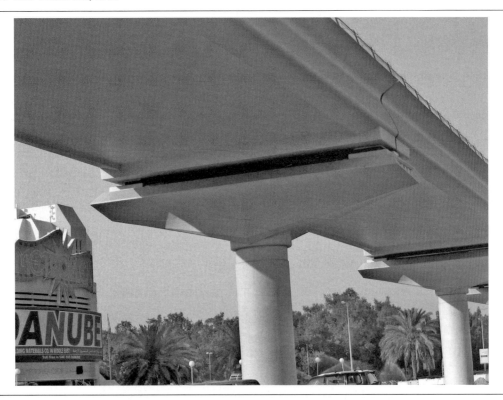

In recent years, the importance of aesthetics for bridges has been more widely recognised by bridge owners, consultants and contractors as well as the general public. A bridge is constructed with a design life of over 100 years, and all being well provides a useful service for much longer. The impact that the structure's appearance has is paramount to its surrounding environment. The extra cost involved with designing and constructing a bridge with a more pleasing appearance is being recognised as a good investment for the future and more bridge owners are insisting on improving the aesthetics of their bridge stock. The versatility of form, shape and texture of concrete has enabled more architectural style structures to be adopted, such as the Tampines Viaduct shown in Figure 1.25 and the Dubai LRT (Light Rail Transit) shown in Figure 1.26.

REFERENCES
AASHTO (1931–1999) Standard Specification for Highway Bridges. AASHTO, Washington, DC.
AASHTO (1994) LRFD Specification for Design of Highway Bridges. AASHTO, Washington, DC.
Abeles PW (1949) *The Principles and Practice of Prestressed Concrete*. Crosby Lockwood, London.
BSI (1959) CP 115: The structural use of prestressed concrete in building. BSI, London.
BSI (1978) BS 5400: Steel, concrete and composite bridges, part 4. Code of practice for the design of concrete bridges. BSI, London.
BSI (2004–2005) BS EN 1992: Eurocode 2: Design of concrete structures. BSI, London.
Guyun Y (1951) *Prestress Concrete*. Edition Eyrolles, Paris.
Magnel G (1950) *Prestressed Concrete*. Concrete Publications, London.

Rawlinson J and Stott PF (1962) The Hammersmith Flyover. *Proceedings of the Institution of Civil Engineers* **23:** December, 565–624, doi:10.1680/iicep.1962.10813.

Sriskandan K (1989) Prestressed concrete road bridges in Great Britain: A historical survey. *Proceedings of the Institution of Civil Engineers* **86:** April, 269–303, doi:10.1680/iicep.1990.4665.

Sutherland J, Humm D and Chrimes M (eds) (2001) *Historic Concrete – The Background to Appraisal*. Thomas Telford Publishing, London.

Prestressed Concrete Bridges, 2nd edition
ISBN: 978-0-7277-4113-4

ICE Publishing: All rights reserved
doi: 10.1680/pcb.41134.023

Chapter 2
Prestressing components and equipment

Introduction

Prestressing systems are made up of a number of standard components. The basic prestressing element is the wire, strand or bar, while with post-tensioning the anchors, couplers and ducts form the embedded items. Specialist equipment for placing, stressing and grouting the tendon are required, and proprietary prestressing systems have developed and become established worldwide. This chapter looks at the different components and equipment that make up the prestressing system. More details on the proprietary systems from individual suppliers are given in Appendix D, while grouting equipment is included in Chapter 4.

Proprietary systems

The specialist nature of prestressing concrete has led to a number of proprietary prestressing systems being widely used throughout the industry; several of the principal prestressing companies are mentioned below. Most multi-strand prestressing systems utilise similar types of strand, either 13 mm or 15 mm diameter, while BBR also provide a parallel wire system and there are several bar systems available. When designing prestressed concrete members, the choice of which type of prestressing to be used is usually made by the designer, while the contractor chooses which particular system to adopt. However, the designer must ensure that the design and detailing adequately cater for the actual equipment and components to be used.

The VSL post-tensioning systems have been used throughout the world since 1956 and are protected by patents. They are used in every branch of prestressed concrete construction and although used primarily for bridges and buildings, they are used for numerous other applications as well, including soil and rock anchors and for lifting and sliding heavy loads. VSL originated in Switzerland and rapidly expanded, with their system being adopted internationally. In 1988, the group restructured under a new name, VSL International Ltd, to reflect their worldwide business, and today have subsidiaries or licensees in most parts of the world. For bridgeworks, VSL provide both multi-strand and stressbar systems.

Eugène Freyssinet was one of the pioneers of prestressed concrete: developing the materials and equipment, as well as the design philosophy. He took out several patents and worked closely with contractors in France to establish the techniques. In 1943, Edme Campenon established a company, Société Technique pour l'Utilisation de la Préconstrainte (STUP), to develop Freyssinet's ideas. STUP were involved with many early prestressed concrete structures and became established worldwide in both the design and the construction of prestressed concrete bridges, as well as for other uses of the technique. In 1976, STUP changed its name to Freyssinet International (STUP), in order to honour Eugène Freyssinet and to develop the company on an international scale. Today, Freyssinet International provides multi-strand and bar stressing systems.

Founded in 1865, DYWIDAG is the abbreviation of its mother company, Dyckerhoff and Widmann AG, which is one of the oldest construction companies in Germany. The first ever structure built with a prototype DYWIDAG Post-Tensioning System using bars was the arch-bridge at Alsleben (Germany) in 1927, followed by the three-span Aue Bridge (Germany) in 1936, where unbonded post-tensioned tendons were used. DYWIDAG began licensing their system in 1950 on a worldwide basis and in 1979 formed the DSI group to co-ordinate their international business. DYWIDAG provide both multi-strand and prestressing bar systems for bridgeworks.

Bureau BBR Ltd, established in 1944, was originally a partnership formed by three Swiss civil engineers: M. Birkenmaier, A. Brandestini and M.R. Ro. Today it is a limited company registered in Zurich, Switzerland, and specialises in prestressing technology and related construction work. BBR operates on an international scale with numerous licensees, consultants, contractors and suppliers. BBR provides multi-strand and multi-wire prestressing systems for use in bridgeworks.

Macalloy is a registered trademark of McCalls Special Products Ltd, based in the UK, who produce high-tensile alloy-steel bars for prestressing concrete as well as a range of other applications.

Details of some of the above proprietary systems are given in Appendix D; however, each proprietary system may have different details when used in different countries and the systems develop over time, so it is important to check the specifications with local prestressing suppliers for any particular project.

Wires

Individual wires are sometimes used in pre-tensioned beams but have become less common in favour of strand, which has better bond characteristics. The wire is cold-drawn from hot-rolled rods of high-carbon steel, and is stress-relieved to give the required properties. Wire diameters are typically between 5 mm and 7 mm, with a minimum tensile strength of between 1570 N/mm^2 and 1860 N/mm^2 (newton per square millimetre) carrying forces up to 45 kN (kilonewtons) each. Material properties for pre-stressing wire are specified in the British Standard 5896 (1980). Typical details and properties for wires are given in Table 2.1.

Galvanised wire is available, although ungalvanised wire is used for most standard prestressing applications. Stainless steel wire can also be obtained, but this is not usually used for the prestressing of concrete.

Strands and multi-strand tendons

The most common form of prestressing is 7-wire strand, which is made up of individual cold-drawn wires with six outer wires twisted around an inner core wire. An example of a multi-strand tendon and its anchorage partly cut away is shown in Figure 2.1. The strand is stress-relieved and is usually of a low-relaxation grade. Strands can be galvanised should greater protection be required, although this is unusual for normal prestressing applications. Epoxy-coated strands are also available and have been used for prestressing bridge decks, but again this is not a common practice. The concern over the use of galvanised or epoxy-coated strands is in the effectiveness of this form of protection at critical locations such as the anchorages, where the wedges can penetrate the coatings applied.

For post-tensioning, 13 mm or 15 mm diameter, 7-wire strand is used, either singly for pre-tensioning or in bundles to form multi-strand tendons. The most common post-tensioned tendon sizes utilise 7, 12, 19 or 27 strands to suit the standard anchor blocks available, although systems are available for tendons incorporating up to 55 strands when necessary. Stressing the strands to the common limit

Table 2.1 Wire, strand and bar properties

		Nominal diameter: mm	Nominal area: mm²	Nominal mass: kg/m	Yield strength: N/mm²	Tensile strength: N/mm²	Minimum breaking load: kN	Modulus of elasticity: kN/mm²	Relaxation[1]: class 2 or low relaxation
7-wire strand low-relaxation									
13 mm (0.5")	Euronorm 138–79, or BS 5896: 1980, Super	12.9	100	0.785	1580	1860	186	195	2.5%
	Euronorm 138–79, or BS 5896: 1980, Standard	12.5	93	0.73	1500	1770	164	195	2.5%
	Euronorm 138–79, or BS 5896: 1980, Drawn	12.7	112	0.89	1580	1860	209	195	2.5%
	ASTM A416-85, Grade 270	12.7	98.7	0.775	1670	1860	183.7	195	2.5%
15 mm (0.6")	Euronorm 138–79, or BS 5896: 1980, Super	15.7	150	1.18	1500	1770	265	195	2.5%
	Euronorm 138–79, or BS 5896: 1980, Standard	15.2	139	1.09	1420	1670	232	195	2.5%
	Euronorm 138–79, or BS 5896: 1980, Drawn	15.2	165	1.295	1550	1820	300	195	2.5%
	ASTM A416-85, Grade 270	15.2	140	1.10	1670	1860	260.7	195	2.5%
Stress bars									
20 mm	BS 4486: 1980	20	314	2.39	835	1030	323	170/205	3.5%
25 mm	BS 4486: 1980	25	491	3.9	835	1030	505	170/205	3.5%
32 mm	BS 4486: 1980	32	804	6.66	835	1030	828	170/205	3.5%
40 mm	BS 4486: 1980	40	1257	10	835	1030	1300	170/205	3.5%
50 mm	BS 4486: 1980	50	1963	16.02	835	1030	2022	170/205	3.5%
Cold-drawn wire									
7 mm	BS 5896: 1980	7	38.5	302	1300	1570	60.4	205	2.5%
	BS 5896: 1980				1390	1670	64.3	205	2.5%
5 mm	BS 5896: 1980	5	19.6	154	1390	1670	32.7	205	2.5%
	BS 5896: 1980				1470	1770	34.7	205	2.5%

Notes
Reference should be made to the manufacturers' literature for the properties of individual wires, strands or bars being used.
[1] Relaxation after 1000 hrs, at 20°C and 70% UTS.

Figure 2.1 Multi-strand tendon (reproduced courtesy of DYWIDAG-Systems International GmbH, copyright reserved)

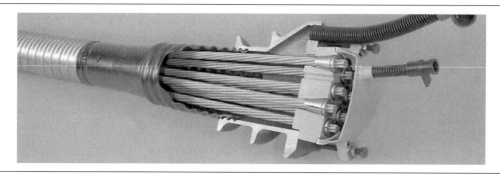

of 75% of UTS gives typical jacking forces of 140 kN or 199 kN for the 13 mm or 15 mm diameter strands respectively. The larger multi-strand tendons can carry forces up to 10 000 kN. Material properties for prestressing strand are specified in the British Standard 5896 (1980) and by ASTM A416M-99 (1999). Table 2.1 gives typical details and properties of the strand used, while Table 2.2 summarises the most common tendon sizes and the associated minimum breaking loads for 'super strand' and ASTM grade 270 steel.

Table 2.2 Multi-strand tendon sizes

	13 mm (0.5″) strand				15 mm (0.6″) strand			
	Minimum breaking load: kN		Typical duct sizes internal/external diameter		Minimum breaking load: kN		Typical duct sizes internal/external diameter	
No. of strands	Euronorm 138–79, or BS 5896: 1980, Super	ASTM A416-85, Grade 270	Steel ducts: mm	Plastic ducts: mm	Euronorm 138–79, or BS 5896: 1980, Super	ASTM A416-85, Grade 270	Steel ducts: mm	Plastic ducts: mm
1	186	184	25/30		265	261	25/30	
2	372	367	40/45		530	521	40/45	
3	558	551	40/45		795	782	40/45	
4	744	735	45/50		1060	1043	50/55	
6	1116	1102	50/55		1590	1564	60/67	
7	1302	1286	55/60		1855	1825	60/67	59/73
12	2232	2204	65/72	59/73	3180	3128	80/87	76/91
18	3348	3307	80/87		4770	4693	95/102	
19	3534	3490	80/87	76/91	5035	4953	95/102	100/116
22	4092	4041	85/92		5830	5735	110/117	100/116
31	5766	5695	100/107	100/116	8215	8082	130/137	130/146
37	6882	6797	120/127	130/146	9805	9646	140/150	130/146
43	7998	7899	130/137		11 395	11 210	150/160	
55	10 230	10 104	140/150	130/146	14 575	14 339	170/180	

Figure 2.2 Prestressing bar and anchor

At the ends of tendons the strands are anchored either by splaying out the wires and encasing them in the concrete as a dead-end anchorage, or by passing them through an anchor arrangement and fixing them into an anchor block for a live-end anchorage.

Bars

Prestressing bars, as shown in Figure 2.2, are available in different diameters from 15 mm up to 75 mm, and are used in post-tensioned construction as well as a number of temporary works situations for bridge construction. They typically have a minimum ultimate characteristic tensile strength of between $1000 \, \text{N/mm}^2$ and $1080 \, \text{N/mm}^2$, although a higher-strength steel grade is available from some manufacturers. Jacking forces range from 135 kN to over 3000 kN. Bars are formed in straight lengths by the hot rolling of steel rods and are either smooth or deformed on the surface. Galvanised bars are available, but are not commonly used in standard post-tensioned concrete applications. Material properties for prestressing bars are specified in the British Standard 4486 (1980) and in ASTM A722M-98 (1998). Typical details for bar post-tensioning are given in Table 2.1.

The bars are generally placed into ducts cast into the concrete between two anchor blocks, located on the concrete surface. Bars are stressed by pulling from one end, using a stressing jack placed against the anchorage arrangement, and then held in place by a nut assembly.

Anchorages

At each end of a tendon the force is transferred into the concrete by an anchorage system. For pre-tensioned strands the anchorage is by bond and friction of the bare strand cast into the concrete, while for post-tensioned tendons anchorage is achieved by using anchor blocks or an encased dead-end anchor.

Figure 2.3 Multi-strand tendon live-end anchor (reproduced courtesy of Freyssinet International, copyright reserved)

In Figure 2.2, a typical anchor arrangement is shown for a post-tensioned bar within a pocket formed in the concrete. The bar is held in place by the threaded nut and the force from the bar is transferred through the threads and the nut and on to the anchor block that is cast into the concrete. Holes are left through the anchor block to assist with the grouting of the duct and the anchorage arrangement after stressing has been completed.

Stressing of multi-strand tendons is undertaken using jacks placed over the anchorage and tendon. The jack grips each strand and pulls the tendon until the required force is generated. Wedges are then pushed into place around the strand and seated into the anchor block, so that on the release of the force by the jack the wedges grip the strand and transfer the force onto the anchorage and into the concrete. Figure 2.3 shows a typical live-end anchorage for a multi-strand tendon after the strands have been stressed and trimmed back to the wedges. The strands and wedges are seated in holes formed in the anchor block, which rests against the bearing plate and trumpet, cast into the concrete. The hole at the top of the anchor is for the grout to flow through, to ensure that the duct and anchor are fully filled during the grouting operation.

Strand tendons may also be anchored at the non-stressing end with a cast-in dead-end anchorage arrangement, as shown in Figure 2.4, where the strands are spread out after emerging from the duct. In this arrangement, the tendon is installed before concreting, but is not stressed until after the concrete has attained the required transfer strength. As the strands emerge from the end of the ducting, a seal prevents the ingress of concrete into the duct and, where the strands spread out, a tension ring resists the splitting forces caused by the change in angle. At the end of the strands the wires are opened out to create a 'bulb' that, when encased within the concrete, ensures a good anchorage.

External, post-tensioned tendons should be removable and replaceable, and the detail of the anchorage is arranged to allow for this. Where cement grout is used, a lining is provided in the central hole of the anchor arrangement to ensure that the grout around the tendon does not bond to the anchor or adjacent concrete. This facilitates the cutting and pulling out of the tendon, if necessary. Alternatively, where grease or wax filler is used, the tendon can be de-stressed or re-stressed using a jack. To provide

Figure 2.4 Strand dead-end anchorage (reproduced courtesy of VSL International, copyright reserved)

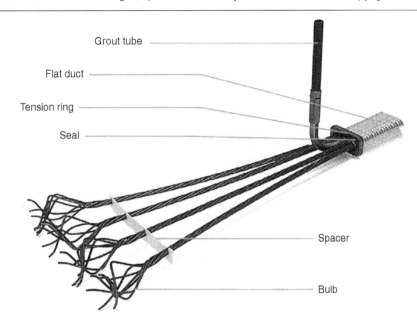

Grout tube

Flat duct

Tension ring

Seal

Spacer

Bulb

for this the strands must protrude far enough behind the anchor block to allow the jack to grip them. A special extended capping assembly is used to protect the strand, which is filled with the grease or wax to prevent corrosion.

All bonded post-tensioned tendon anchor blocks should be detailed to allow them to fully fill up with grout and prevent any air being trapped during the grouting operation. This may be achieved by providing vents and holes through the anchor blocks, thus allowing the grout to flow freely.

Tendon couplers
Tendons can be coupled to extend their length during stage-by-stage construction. Figure 2.5 shows a coupler system for multi-strand tendons. A special anchor block and coupler arrangement is used to enable the tendon to be extended after it has been stressed. The standard bearing plate, shown on the right-hand side of the figure, is cast into the first stage of concreting, with the tendon being placed and stressed as the construction progresses. During construction of the next stage and before concreting, the next length of tendon is positioned with the ends of the strands fixed into the coupling head, as shown in the centre of the figure. After concreting the next stage, the new length of tendon is stressed, with the force being transferred through the coupler to the first tendon. This arrangement can simplify the tendon layout and save the cost of a separate anchor.

Bars can also be joined together by using a simple threaded coupler, as shown in Figure 2.6. The coupler is threaded on to the ends of the bars to transfer the load across the connection. Couplers enable long lengths of bars to be assembled and allow previously stressed bars to be extended into subsequent concrete pours.

Where the couplers are located, the concrete section has to be thick enough to adequately surround the arrangement, with sufficient room for the reinforcement and the concrete cover. If the duct runs

Figure 2.5 Multi-strand tendon coupler (reproduced courtesy of VSL International, copyright reserved)

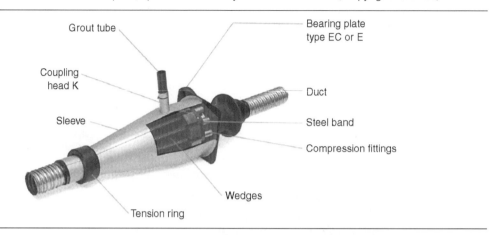

through a thin web or slab, then a thickened section is often needed at the location of the coupler. The thickening required can be significant on large multi-strand tendons, where the size of the coupler is much greater than the tendon duct.

Where couplers are used to extend cables or bars in stage-by-stage construction, the design must consider that although the 'anchor' side at the coupler may not carry much load in the final condition, it carries the full tendon load during the stage stressing. Therefore, the anchor part of the coupler and surrounding concrete must be designed to cater for this load in the temporary condition.

Figure 2.6 Prestress bar coupler

Ducting

With internal post-tensioning a duct is used to form a void through the concrete into which the tendon is placed. Traditionally, ducts are made from corrugated steel, often from strips rolled on site into a helix and crimped together to form the duct. The requirements for steel ducts are covered by British Standard EN 523 (2003). Recent concern over the durability of prestressing tendons has led to the use of corrugated plastic ducts, which are made watertight along their complete length. Figure 2.7 shows typical steel ducts installed within a reinforcement cage, while Figure 2.8 shows a typical plastic duct. Steel ducts usually have a wall thickness of 0.25 mm, while corrugated plastic ducts are normally 2.5 mm to 3 mm thick. Indicative duct sizes for different tendon systems are given in Table 2.2, with details of some of the proprietary systems given in Appendix D.

Figure 2.7 Steel ducts inside reinforcement cage

Figure 2.8 Corrugated plastic duct and coupler (reproduced courtesy of VSL International, copyright reserved)

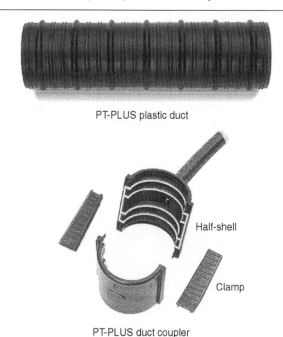

PT-PLUS plastic duct

Half-shell

Clamp

PT-PLUS duct coupler

External tendons are usually placed inside high-density polyethylene (HDPE) ducts, shown in Figure 2.9. These ducts have to be strong enough to withstand the abrasion from the strand as it is threaded and stressed, pressure from the strand as it goes around any curves in the tendon alignment, and the pressure from the grouting. As a result, the HDPE duct thickness is normally at least 6 mm.

Ducting is manufactured in convenient handling lengths, normally between 4 m and 6 m long and joined in situ to make up the full length required. Steel ducts are joined with a short coupler that goes around the outside of the ducts on both sides of a joint to provide continuity in the ducting system. The coupler is usually a slightly larger diameter duct that fits tightly over the corrugations and is 'twisted' into position. Corrugated plastic ducts are made continuous and sealed with couplers, which provide a watertight joint. A typical plastic coupler fitting is shown in Figure 2.8.

HDPE ducts are joined together by either hot welding or sleeving. During hot welding, the carefully prepared ends of the ducts are positioned in a frame and a hot knife or plate used to soften the end faces of the HDPE. The duct ends are then brought together and held until the material has hardened again and a seal formed. When using sleeves, a short length of tight-fitting prefabricated HDPE duct is slid over the joint and sealed into position.

Grout inlets, outlets, vents and drainage tubes are needed along the ducts to remove any water that may accumulate inside and to facilitate the grouting operation. For steel ducting, vents are formed by creating a hole in the duct and fixing a plastic saddle, onto which a plastic tube is connected leading to the concrete surface. Special adapters are used for corrugated plastic ducts, while for HDPE ducts the inlet or vent tube is 'welded' to the duct around a suitably sized hole. Examples of the different vent arrangements are given in Chapter 4.

Figure 2.9 HDPE duct

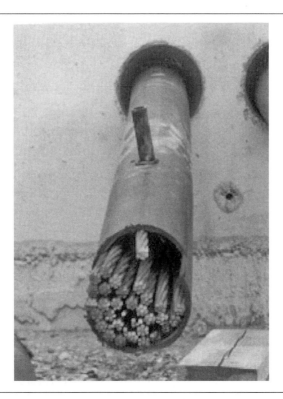

Equipment for placing tendons

Bar tendons are usually placed by hand, by pushing the bars through the ducts. Hand placing can also be used for the shorter strand tendons and where a simple duct profile exists. For the larger and longer multi-strand tendons, special equipment is needed to position the strand. Coils of 7-wire strand arrive on site usually weighing up to 3 tonnes. The coils are placed in a frame, as seen in Figure 2.10, to enable the strand to be pulled out and made into the tendon. In a multi-strand tendon the strands are positioned by either pushing or pulling the strand into place. The most common method of placing the strand is by the push-through method, either using a machine or by hand. The push-through method as illustrated in Figure 2.10 involves pushing individual strands into the duct one at a time until the full number are in position. A push-through machine, also shown in Figure 2.10, grips the strand as it is pulled out of the coil and feeds it into the duct. The pull-through method is illustrated in Figure 2.11. Strands are first cut to length and bundled into a complete tendon at one end of the duct. A steel pulling rope is threaded through the duct and connected to a winch at one end. At the other end, the rope is fixed to a cable sock, usually made from a wire mesh, which is put over the leading end of the tendon. When pulled, the sock grips the strands and pulls them all through the duct together.

Stressing jacks

Typical jacks for single-strand and multi-strand tendons are shown in Figures 2.12 and 2.13 respectively. Figure 2.14 shows a typical jack for bar tendons. These jacks are hydraulically operated, with oil pumped into the piston to apply load to the tendon. The larger jacks can generate a pulling

Figure 2.10 Push-through placing of strand (reproduced courtesy of VSL International, copyright reserved)

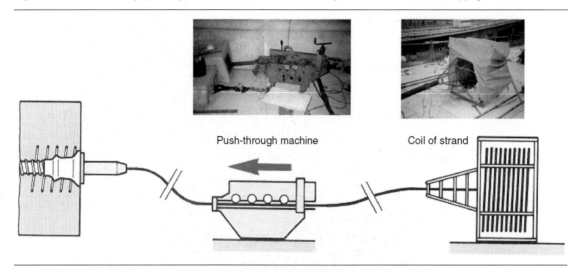

Push-through machine   Coil of strand

Figure 2.11 Pull-through placing of strand (reproduced courtesy of VSL International, copyright reserved)

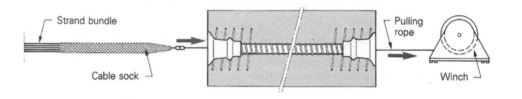

Strand bundle

Cable sock

Pulling rope

Winch

Figure 2.12 Jack for stressing single strand (reproduced courtesy of VSL International, copyright reserved)

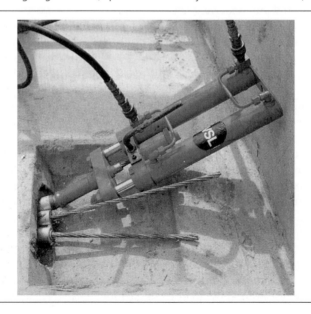

Figure 2.13 Jack for multi-strand tendon

force in excess of 1200 tonnes. The same jacks are also used to de-stress tendons where this is necessary. For de-stressing, special stools are used to move the jack back and to give access to the anchor wedge or nut, allowing them to be loosened or removed.

Jacks for single strands and small tendons, or smaller bars, typically weigh up to 250 kg, and can be easily handled and manoeuvred into position. For the larger tendons special lifting equipment, lifting frames or cranes, are required to move the jacks, which can weigh up to 2000 kg. Figure 2.15 shows typical jack-lifting equipment being used inside a box girder. The design and detailing of the bridge deck should always take into account the access needed to set up and operate the jacks and associated equipment.

Health and safety

Installing and stressing the tendons can present safety risks for the operatives, especially with the very high forces involved during the prestressing operation. The installation of the tendons can involve moving heavy material and equipment in confined spacing, and detailed method statements should be developed and risk assessments carried out to ensure that all operations can be carried out in a safe manner.

During stressing it is not unusual for wires or strands to break, or anchors to fail, causing the jacks to jump forward or sideways. There have been a number of cases where the jacks have jumped back many metres if a tendon or one of its components fails, while individual strands have also been known to shoot out of the ducts. During stressing it is important that only essential personnel are in the vicinity of the stressing operation, and these people should be fully trained to operate the prestressing system in a safe manner. While the jacks are in operation, no one should walk behind the jack or the dead-end anchorage, and it is usually prudent to install protection barriers at vulnerable locations.

Figure 2.14 Jack for prestressing bar

Figure 2.15 Placing of large prestressing jack onto tendon

The health and safety aspects should also be considered during the design process. The bridge decks should be arranged and detailed so that safe access is possible for the installation and stressing of the tendons, and for any future inspections and maintenance.

REFERENCES

ASTM (1999) ASTM A416/A416M-10 Standard Specification for Steel Strand, Uncoated Seven-Wire for Prestressed Concrete. ASTM, Pennsylvania.

ASTM (2003) ASTM A722/A722M-98 Standard Specification for Uncoated High Strength Steel Bar for Prestressing Concrete. ASTM, Pennsylvania.

BSI (1973) BS:4447 Specification for the performance of prestressing anchorages for post-tensioned construction. BSI, London.

BSI (1980) BS 5896:1980. Specification for high tensile steel wire and strand for the prestressing of concrete. BSI, London.

BSI (1980) BS 4486:1980. Specification for hot rolled and hot rolled and processed high tensile alloy steel bars for the prestressing of concrete. BSI, London.

BSI (2003) BS EN 523:2003. Steel strip sheaths for prestressing tendons. Terminology, requirements, quality control. BSI, London.

Prestressed Concrete Bridges, 2nd edition
ISBN: 978-0-7277-4113-4

ICE Publishing: All rights reserved
doi: 10.1680/pcb.41134.039

Chapter 3
Durability and detailing

Introduction

Discovery of corroded tendons in several bridges in the UK and internationally during the 1980s and early 1990s resulted in concern over the long-term durability of prestressed concrete bridges. Subsequent inspections and testing programmes have led to a better understanding of the factors that caused the corrosion and also of the influence that good design and construction of these bridge decks has. In the UK, a review of current practice and 'state of the art' research led to a report issued by the Concrete Society (1996), with a second edition (2002) giving guidance on some of the requirements for achieving durable post-tensioned concrete bridges.

The prestressing elements in a bridge deck are a primary structural component and they need to be protected from risk of damage or deterioration. This can be achieved through appropriate design and detailing. Good access to inspect as much of the prestressing tendons and other elements of the deck as possible is required for long-term maintenance, while, for external tendons, provisions for their re-stressing or replacement need to be incorporated.

With well-designed details and good construction practices, prestressed concrete bridge decks provide durable and low-maintenance structures. This is clearly demonstrated by the large number of existing bridges that have performed well and are still in a good condition after many years in service, both in the UK and around the world. However, this cannot be taken for granted and good practices must be adhered to for the delivery of durable structures in the future.

This chapter discusses some of the aspects concerning durability and associated topics in more detail. The successful grouting of post-tensioned tendons is paramount to achieving good protection to the stressed components and is covered separately in Chapter 4.

Recent history of durability issues in the UK

Durability issues became the subject of much discussion and extensive research and development following the collapse of the Ynys-y-Gwas Bridge in December 1985. The key features of the Ynys-y-Gwas Bridge collapse, as described by Woodward and Williams (1998), were as follows.

(*a*) The bridge collapsed without warning.
(*b*) The bridge was a highway structure built in 1953 and subjected to de-icing salts.
(*c*) The ducts were predominantly unlined, except at segment joints where they passed through ineffective short lengths of cardboard or steel cylinders, such that tendon protection was limited.
(*d*) The segment joints were 25 mm wide and packed with mortar, resulting in permeable areas where water could easily find a path to a tendon.

(*e*) The tendons were generally well grouted, but, where grouting was inadequate, corrosion travelled along the length of the tendon instead of remaining focused at the segment joints, as was the case elsewhere.

The main lesson that has been learnt from this collapse was that the integrity of a post-tensioned structure relies on preventing corrosion of the tendons. This is best achieved by the application of multi-layer protection systems, allied to robust design details at construction joints, expansion joints, vents, anchorages and other key areas, combined with a high-quality construction regime.

Before the Ynys-y-Gwas Bridge collapse, problems with post-tensioned concrete bridges had come to light in the UK. In the 1960s, two footbridge collapses were put down to corrosion of post-tensioned anchorages and in the early 1980s several cases of corroded tendons were discovered. However, it was the sudden collapse of the Ynys-y-Gwas Bridge without any warning that caused most concern.

Following this collapse and the identification of defects in other prestressed concrete bridges in the UK, as outlined in the report by the Highways Agency and TRL (1999), a review of key design details, specifications and construction techniques was launched. A moratorium was introduced by the UK's Department of Transport in September 1992 on the new design of post-tensioned bridges using internal tendons, to allow the issues to be fully researched. The moratorium was partially lifted in 1996 after the publication of the Concrete Society's report, TR47 (1996), which specified improved detailing and grouting procedures. For precast segmental structures with match-cast joints, the ban was maintained due to concerns over the effectiveness of the protection to the tendons as they pass across the joints, where traditionally the ducts are not continuous. Subsequently, the 1990s witnessed a revival in the use of external tendons.

External tendons had become unpopular in the UK after a moratorium was introduced in October 1977, following the discovery of corroded wires in the Braidley Road Bridge in Bournemouth, built in 1970. The industry had become wary of external tendons due to corrosion and anchorage problems caused by poor detailing and inadequately protected strands. The use of grease and wax for filling external tendon ducts was unreliable at that time as in several cases it leaked out, leaving tendons vulnerable to corrosion. Of the seven externally post-tensioned bridges built in the UK during the first phase of external tendon construction in the 1970s, at least two that had strands individually sheathed in PVC had suffered from wire corrosion of such severity that the tendons have now been replaced. The second phase of externally post-tensioned structures in the 1990s saw improved detailing and better protection systems, including the use of HDPE ducts with either cement grout or grease to protect the tendons.

The consequences of corroded tendons are so serious that the reaction of the industry has been very vigorous. The number of bridges affected over the last 50 years has only been a small percentage of the total bridge stock, but even this was considered unacceptable. The numerous specifications, technical guidance, learned papers and other publications that now exist on the topics of grouting, tendon protection, ducting systems and best practice bear testament to the importance of this subject. The multi-layer protection philosophy, along with the many refined design details developed, should result in more durable post-tensioned concrete structures in the future.

Guidance on the requirements to achieve durable post-tensioned concrete structures can be found in a number of publications, including the design manual published by the Highways Agency (2001a, 2001b) and the book by the Comité Euro-International du Béton (1992).

Corrosion of post-tensioning outside the UK

The UK has not been the only country to suffer problems with their post-tensioned concrete bridge stock. In the USA, 11 bridges were found with corroded tendons in the late 1990s and early 2000s. The Niles Channel Bridge in Florida was found to have a failed external tendon during a routine inspection in 1999. This was due to corrosion of the strands caused by wind-borne salt water accumulating at the anchorages near an expansion joint. Another bridge in Florida, Mid-Bay Bridge, was found in 2000 to have several failed external tendons just seven years after opening. In one case, the corrosion was caused by failure of the duct, letting moisture in along the length of the tendon, while in another the anchorages had again been subjected to corrosion. In all 11 bridges, tendons were found heavily corroded and needed replacing.

In Europe, corrosion of tendons has also been reported in Switzerland, Denmark, France and Belgium. In 1992, the Schelde River Bridge in Belgium collapsed resulting from corrosion of post-tensioning through a joint. Asian countries have also suffered corrosion of their post-tensioned bridges, most notably with Japan reporting a number of cases.

Corrosion protection and ducting

Wires, bars and strand are generally used uncoated; however, to protect the tendons during storage or to reduce friction losses during stressing, they can be coated with soluble oil which is washed off before grouting the duct. Galvanised and epoxy-coated bars and strand are also available, although neither has been commonly adopted in normal bridge prestressing works. This is partly due to the higher costs but also due to concerns over the effectiveness of these protection coatings, which can become damaged during handling on site and when installing the tendon, and when the wedges grip the strands at the anchorages. Stainless steel bars are also available, and are often used as stays or hangers on cable-supported bridges, but again they have not found a market in the prestressing of concrete, most probably due to the much higher costs involved.

Pre-tensioned strands and wires in precast elements is fully encased in the concrete and is well protected from corrosion. The ends of the strands can be painted over where they finish on the concrete surface to provide a water-tight seal and prevent the ingress of moisture. As long as the concrete remains uncracked and sound, there is no direct path for any moisture to reach the strand, although long-term deterioration of the concrete can expose the strands to risk of corrosion, and this requires monitoring.

For post-tensioned arrangements, the different protection systems used for both internal and external post-tensioned tendons are illustrated in Figure 3.1. Post-tensioned tendons are placed inside ducts to allow them to be stressed after the concrete has hardened and to provide protection to the tendons. The ducts for internal tendons are traditionally manufactured using galvanised mild-steel strips which are wound into a tube and crimped together, making a duct that is fairly rigid but not watertight. In recent years, plastic ducting systems have been adopted to provide a watertight barrier around the tendon as further protection against corrosion. Details of the ducting used are given in Chapter 2, 'Ducting'. Plastic or PVC ducting can be air-pressure tested before concreting to confirm its integrity and allow repairs if necessary.

After stressing internal post-tensioned tendons, the duct is filled with cement grout that flows into the voids, expelling any water and air while providing a benign environment that discourages corrosion in the pre-stressing steel. Achieving full grouting of a duct can be difficult and requires experienced personnel and the adoption of good site practice. This aspect is discussed in more detail in Chapter 4.

Figure 3.1 Multi-strand tendon protection systems

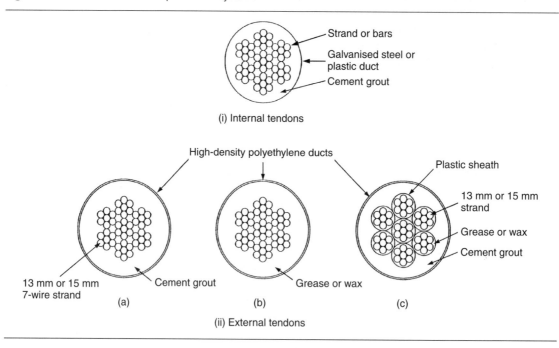

HDPE ducts are used with external tendons. The HDPE ducts need to be strong enough to prevent deformation when pressurised during grouting and to resist the strand punching through at deviated positions during threading and stressing.

There are several different protection systems used with external tendons, as illustrated in Figure 3.1(ii). The simplest is with the bare strand placed inside the HDPE duct and a cement grout used, as detailed in (a). Alternatively, grease or wax filler can be used instead of the cement grout to fill the void, as in (b). A third option (c) is for the individual strands to be placed inside a plastic tube with grease or wax filler injected at the factory. The strand and tubes are then placed inside the HDPE ducts and surrounded with cement grout on site. HDPE provides long-term protection to the tendon, while the advantage of using grease as a filler is that it allows the tendons to be more easily de-stressed and re-stressed or replaced. This is an important feature for external tendons. Where cement grout is used to fill the duct, re-stressing of the tendon is not possible, and removal of the external tendon involves cutting it up into short lengths and pulling it out of the deviators and anchor blocks. This requires careful detailing of the duct arrangements.

Concrete

To provide adequate protection to the prestressing steel the surrounding concrete must be well-compacted and free from defects. There must also be sufficient cover to prevent any deterioration in the concrete from penetrating to the level of the tendons during the design life of the structure. Durable concrete is largely achieved by good detailing and construction practices. The factors controlling the durability of concrete and the mechanisms of deterioration are well documented in the publication by the Comité Euro-International du Béton (1992).

Consideration should be given to the risk of cracks forming in the concrete next to the tendons, and the potential widths and depths of possible cracks should be investigated. Where appropriate, additional reinforcement is provided to reduce the risk of cracks forming, or the tendons moved away from potential problem areas. The risk of cracking occurring within the concrete perpendicular to the prestress tendons can be reduced by applying some of the prestress to the concrete at an early age, in advance of any significant shrinkage or thermal strains occurring.

When designing and detailing the prestress layout and reinforcement arrangement there must always be provision for sufficient space to be available for placing and compacting the concrete. In congested areas, it may be necessary to use concrete with small aggregate, and to detail the reinforcement with sufficient gaps for placing and vibrating the concrete. The use of self-compacting concrete with excellent flow characteristics should be considered for those areas where reinforcement congestion is high and access difficult.

The area around tendon anchorages can be highly congested, with the reinforcement and concrete subjected to high stresses making the need for dense and defect-free concrete paramount. Figure 3.2 shows a pair of multi-strand tendons where the reinforcement is tightly detailed, making concreting difficult. In these situations, special provisions such as temporary access windows in the shutters should be incorporated to allow easy inspection of the concreting. The windows can also be used to insert pokers to vibrate the concrete as it is placed, ensuring that a well-compacted material is achieved, while the reinforcement should be adjusted locally to provide clear paths for the concrete to flow and the vibrators to be inserted.

Figure 3.2 Congested reinforcement at tendon anchor

Detailing

Good detailing is essential to achieving a durable prestressed concrete structure. The quality, and hence eventual durability, of construction is improved by the use of section shapes and sizes that permit easy concreting. Careful detailing of the reinforcement and prestress tendons to simplify fixing, while maintaining sufficient concrete cover, contributes to long-term durability.

A primary requirement in the detailing is the provision of an adequate drainage system for the efficient removal of any rainwater. This minimises the risk of water getting below the waterproofing layers or under the deck, and causing deterioration of the concrete or reinforcement, or prestressing steel. Potential water paths to the prestressing tendons should be identified and then detailed to reduce the risk of water ingress. This may involve the provision of additional waterproofing or other protection systems.

For box sections, drainage holes, as illustrated in Figure 3.3, should be provided through the bottom slab at all low points and next to all obstructions within the box, such as diaphragms and deviators. This enables any water entering the box to drain away, preventing ponding on the bottom slab. Drainage holes are particularly important if surface drainage or water mains are carried inside the box and they should be regularly inspected to ensure that they do not become blocked.

Potential leakage points, such as manholes or other penetrations through the deck slab, should be avoided if possible. Where drainage gullies are needed, the detailing of the interface between the gully and concrete deck slab should be such as to make the arrangement as watertight as possible. It is preferable for the surface-water drainage system not to penetrate the deck slab. For short- and medium-length bridge decks, side drains in the verges can be used to carry the water directly off the deck to be collected behind the abutments. This is not practical for long bridge decks, where instead drainage gullies are placed along the deck edges, feeding into a carrier pipe which is located either under the deck cantilevers, as illustrated in Figure 3.4, or within the box section.

Tendon anchorages should be well protected and detailed to minimise the risk of water collecting around them. It is preferable to anchor tendons beneath the deck slab or inside a box section where they are protected, rather than in exposed areas near the top surface of a section where moisture could seep through the construction joints in the concrete.

Figure 3.3 Soffit drainage hole

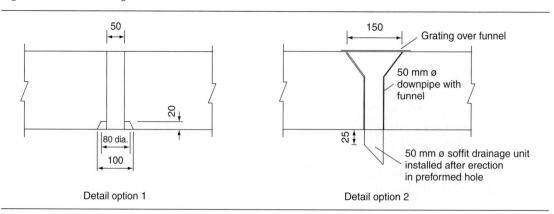

Detail option 1   Detail option 2

Figure 3.4 Deck drainage

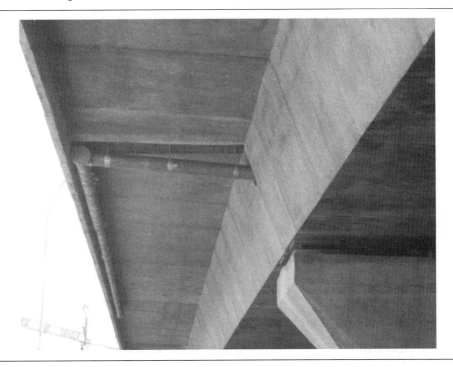

The ends of a deck at an expansion joint are potentially vulnerable to the ingress of water. This poses a risk for any prestressing tendons anchored there, especially with post-tensioned arrangements. A drainage system should be provided under the expansion joint units to collect any water seepage. If possible, the anchors should be recessed away from the expansion joint and extra waterproofing layers applied to protect them.

Small drip channels should be used along the edges of all soffits potentially exposed to rainwater. These prevent the water from running across the soffit, protect the concrete and fixtures, and prevent unsightly staining of the surface.

Advice on detailing for all aspects of concrete bridges can be found in the report by the Concrete Society (2002) and in the Ciria *Bridge Detailing Guide* (2001).

Access

All bridges need good access for regular inspections and maintenance, and this is of particular importance for critical items such as bearings, expansion joints and any exposed prestress tendons and anchorages. Access to the outer surfaces of a deck can often be from the ground, using movable scaffolding or working platforms. Where access from the ground is difficult, bridge inspection units as shown in Figure 3.5 can give good access from the top of the deck. Using this type of plant gives ready access to the outside of the webs and to the soffits of the deck slabs and box, although getting good access between beams or boxes can still be a problem with the larger structures. Box girder bridges need access into the voids. This is usually achieved at the abutments via the bearing shelf into the end of the deck and along the length of the deck via soffit access holes, as illustrated in

Figure 3.5 Bridge inspection unit (reproduced courtesy of MOOG GmbH, copyright reserved)

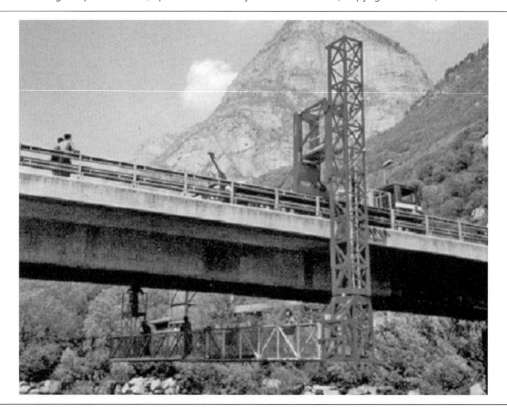

Figure 3.6. The soffit access holes are usually located towards the ends of a span and at locations where there is easy access from the ground. They are typically spaced between 150 m and 200 m apart, provided there is good access through any intermediate diaphragms. Lockable covers are used to give a secure arrangement.

At non-integral abutments for box-type decks, inspection galleries are provided between the end of the deck and the abutment backwall, as illustrated in Figure 3.7. Where expansion joints are located in the deck slab, the arrangement should allow easy access to all parts of the joint. At these locations there are often post-tensioned tendon anchorages which should also be accessible. Entrance to the inspection gallery is usually via a secure door system built into the abutment side wall, with a path constructed to provide access for transporting any equipment needed. The end diaphragm in the deck is detailed with a large hole through it to allow access into the box.

For cellular or box girder decks, access holes are provided through the internal diaphragms, as seen in Figure 3.8. Multi-cell boxes usually have access holes through the internal webs so that all the voids can be inspected and maintenance carried out. Although these holes are normally small to minimise their impact on the design, they should be made as large as possible to provide good access.

Stray current protection

For bridges supporting or near electrified railway systems, power lines or other sources of electrical currents, the prestress tendons and passive reinforcement should be earthed to protect them from the

Figure 3.6 Soffit access hole

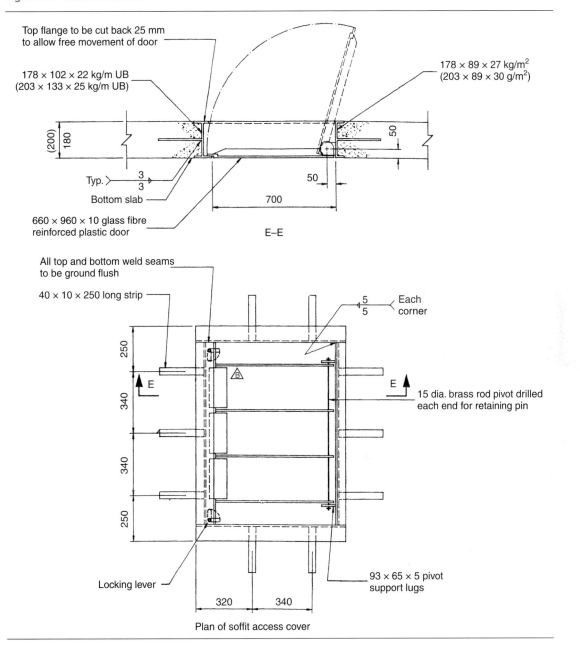

Top flange to be cut back 25 mm
to allow free movement of door

178 × 102 × 22 kg/m UB
(203 × 133 × 25 kg/m UB)

178 × 89 × 27 kg/m^2
(203 × 89 × 30 g/m^2)

(200) 180

50

Typ. $\frac{3}{3}$

50

Bottom slab

700

660 × 960 × 10 glass fibre
reinforced plastic door

E–E

All top and bottom weld seams
to be ground flush

40 × 10 × 250 long strip

$\frac{5}{5}$ Each corner

250

E

340

E

15 dia. brass rod pivot drilled
each end for retaining pin

340

250

Locking lever

93 × 65 × 5 pivot
support lugs

320

340

Plan of soffit access cover

stray currents that can occur, which cause corrosion in the steel. DC-powered railway systems present the greatest corrosion risk, although it is also usual to protect structures where AC power supplies occur.

All the prestressing tendons and reinforcement bars should be connected to provide electrical continuity and to provide a path for the stray current to travel through the structure and down to the ground. Prestress tendons are good conductors of electricity in themselves and are normally connected

Figure 3.7 Abutment inspection gallery

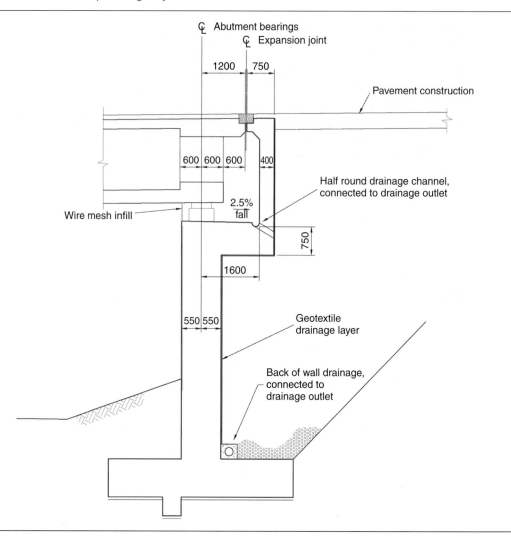

to the reinforcement cage by a copper strip, or similar conducting material, that is fixed between the anchor block and the adjacent reinforcement. The reinforcement cage should have additional mild steel bars incorporated that are welded to the cages at appropriate locations to ensure good continuity in the connection. Alternatively, a special earthing cable can be provided along the complete length of deck and the tendons and reinforcement connected to it. At appropriate piers, or at the abutment, the deck reinforcement is connected to the substructure reinforcement by a copper strip spanning across the bearing gap. The stray current is then taken down to ground level and connected to earthing rods, formed from copper bars installed into the ground to provide a good earthing system.

External tendon replacement

One of the advantages with external tendons is the ability to re-stress or remove and replace them in the future if required. To enable this to be possible the detailing of the tendon at the anchorages and deviators has to allow replacement to take place.

Figure 3.8 Access through box girder diaphragms

The anchorage arrangement for post-tensioned bars is readily suited to re-fixing the jack and de-tensioning where grease or wax filler has been used in the ducts instead of cement grout.

Each end of a multi-strand external tendon is anchored using a standard arrangement, similar to that shown in Figure 2.3, but with special provisions incorporated to allow the tendon to be de-stressed and replaced without damaging the anchor components. If the duct is to be cement-filled this involves providing a double-sleeve arrangement inside the trumpet to allow the tendon to be pulled out, with typical arrangements illustrated in Appendix D. The strands at the stressing end of a tendon with wax-filled ducts are left with sufficient length extending behind the anchor block to allow a jack to be fitted to enable de-tensioning or re-stressing of the strands.

Sufficient space is required around the anchorages of external tendons, and along their length, to give access to remove the old tendon and to install a new one. Access to the structure also needs to be adequate for the stressing and grouting equipment to be handled and positioned. This involves using larger access holes and entrances compared with those needed for routine maintenance. Good access should be provided at each abutment, with full-size doors into the abutment access chamber and large openings through the deck diaphragms. Where soffit access holes are provided they should provide a clear space of at least 1.0 m × 0.8 m.

Removal of the tendons first requires them to be de-stressed. Tendons in grease or wax-filled ducts are de-stressed using a stressing jack to release the force at the anchor. The tendon is then pulled out of the duct and the grease or wax collected. Tendons in cement grout-filled ducts have to be cut into sections to allow them to be removed. In this case, the ducting and grout are removed locally to expose the tendons in the area to be cut and the strands severed by cutting or burning, thus releasing the force. Another technique employed is to burn out the wedges at the anchor blocks, which releases the strands.

Special precautions are needed to prevent a sudden release of the large forces that are in the tendons and to protect the operatives in the vicinity. De-stressing of tendons should be carried out by specialists with experience in this type of operation.

REFERENCES

Comité Euro-International du Béton (1992) *Durable Concrete Structures*, 2nd edition. Thomas Telford, London.

Concrete Society (1996) *Technical report no. 47: Durable bonded post-tensioned concrete bridges.* Concrete Society, Slough.

Concrete Society (2002) *Technical report no. 47. Durable bonded post-tensioned concrete bridges*, 2nd edition. Concrete Society, Slough.

Highways Agency and TRL (1999) Post-tensioned concrete bridges Anglo-French liaison report. Thomas Telford, London.

Highways Agency (2001a) (MDRB) Departmental Standard, BD 57/01: Design for durability. HMSO, Norwich.

Highways Agency (2001b) (MDRB) Advice Note, BA 57/01: Design for Durability. HMSO, Norwich.

Ciria (2001) *Bridge Detailing Guide*. Ciria, London.

Woodward RJ and Williams FW (1998) Collapse of Ynys-y-Gwas Bridge, West Glamorgan. *Proceedings of the Institution of Civil Engineers, Part 1* **85:** 635–669.

Prestressed Concrete Bridges, 2nd edition
ISBN: 978-0-7277-4113-4

ICE Publishing: All rights reserved
doi: 10.1680/pcb.41134.051

Chapter 4
Grouting post-tensioned tendons

Introduction

The process of introducing grout into the duct of a post-tensioned tendon is a simple and effective way to improve the overall durability of the structure. The grout surrounds the strand or bar and acts as a protective layer to prevent corrosion. Cement-based grouts are the most common, although grease or wax fillers are sometimes used with external unbonded tendons. Figure 4.1 shows a grouted duct and tendon from a trial section which has been subsequently cut through to demonstrate the effectiveness of the grouting on the project.

The grouting process involves pumping the fluid grout into the duct at an inlet, which is at the lowest point and often part of an anchorage arrangement. As the grout flows through the duct it comes out at intermediate vents along the length of the tendon, until it finally emerges at the outlet at the far end of the tendon, having filled the duct. Cement-based grouts are pumped into the ducts on site after completion of the stressing operation, while grease or wax fillers are often injected into pre-assembled tendons at the factory.

The complicating factors include such aspects as inadequate design details, insufficiently robust ducting systems, material variability, time constraints when injecting the grout, equipment problems, meaningful testing, duct blockages and sealing segment joints.

This chapter reviews the problems encountered in the 1980s and 1990s, and primarily discusses the process of grouting with cement-based grout. The procedures needed to achieve fully grouted ducts are described. Grease and wax grouts are briefly covered towards the end of this chapter.

Recent history of grouting

Since the early 1990s, the UK has seen a series of refinement in the requirements for the grouting of tendons, including more detailed and exacting specifications. The collapse of the Ynys-y-Gwas Bridge, in December 1985, highlighted historical deficiencies in the grouting process and related design and specification issues. It also emphasised the importance of effective grouting to prevent corrosion. Subsequent extensive research led to the publication of the Concrete Society's report TR47 (1996) and resulted in development of the existing specifications and procedures used for grouting. In the past, the effectiveness of grouting varied from project to project, as the general guidelines available did not fully address all the issues which were often left to the contractor to resolve. Consequently, the quality achieved was largely a function of the experience and expertise of the individual operatives, site supervisors and design engineers.

The extensive programme of inspections for all prestressed concrete bridges in the UK during the 1990s revealed that many early post-tensioned decks had well-grouted ducts; however, grouting defects and

Figure 4.1 Grouted duct and tendon

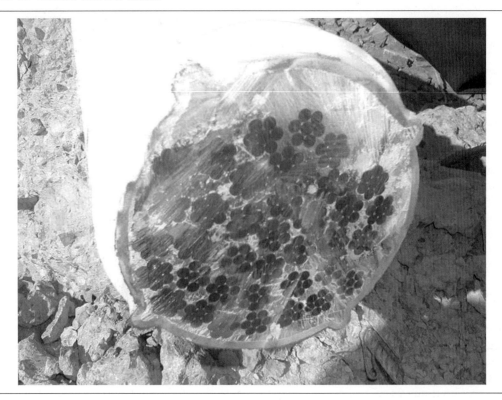

tendon corrosion were found in a significant number of cases, as indicated in Figure 4.2. A small percentage of these bridges were found to have unacceptably severe tendon corrosion. As the corrosion of the tendons is often hidden from view and can result in catastrophic failure of a deck, even a low percentage was considered an unacceptable risk. Consequently, new standards were adopted for all new projects in

Figure 4.2 UK post-tensioned bridge tendon survey results

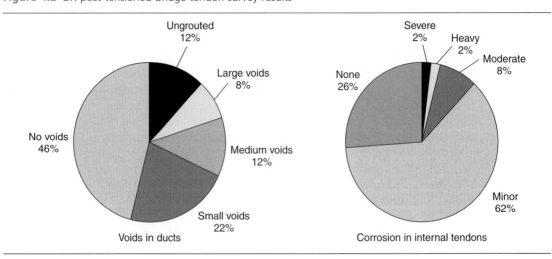

the UK. These standards have not been widely adopted outside the UK, although countries that adopt British Standards often also adopt the new grouting specifications and requirements. Other countries, including the USA, Japan and some other European countries, have also identified problems with grouted post-tensioned tendons and have introduced more effective grouting procedures.

The current UK standard specifications for bridges is the Highways Agency's *Manual of Contract Documents for Highway Works* (2009), which gives the specifications for the grout material and procedures to be adopted. It can be adapted for individual projects to take into account specific requirements and latest developments in the industry. Research since 1996 has resulted in the publication of the second edition of the Concrete Society's report TR47 (2002). This introduced further tests for grout that are intended to improve the assessment of how the grout will perform in a duct under the particular conditions experienced in practice. These tests are adopted in the Highway Agency's specifications (2009).

The European Standard BS EN 447 (BSI, 2007c) covers the basic requirements for grouting of prestressing tendons, while BS EN 446 (BSI, 2007b) covers the procedures for grouting and BS EN 445 (2007) (BSI, 2007a) covers the test methods. These are similar to the requirements of the UK's Highways Agency specifications (2009).

Grout material

Grout for filling post-tensioning ducts is traditionally a mix of water, cement and admixtures. Mixes are developed to limit certain detrimental properties, and designed to pass certain limiting tests such as those specified in the Highways Agency's *Manual of Contract Documents for Highway Works* (2009). The 'design' or selection of such a mix on this basis could be inappropriate given the inadequate nature of the commonly specified tests that do not always fully reflect the performance of the grout in practical situations. Grout-mix designs need to be developed to suit the individual project requirements and to give the required end product of a fully grouted tendon.

Uniformity in the quality and properties of the basic ingredient materials is critically important to the consistent production of a grout. Due to the variability of the performance of grout made from ordinary cements, the UK industry has moved towards the use of special grouts, made from pre-bagged materials with only the water measured and added on site. It has also moved away from the traditionally specified grout material tests, with new performance tests being introduced as discussed in the following section.

The current European Standards for grout, BS EN 447 (BSI, 2007c), BS EN 446 (BSI, 2007b) and BS EN 445 (BSI, 2007a), cover the traditional approach to grouting and testing and should be used in conjunction with the guidance given in TR47 (Concrete Society, 2002). Grout should always be made with Portland cement, either cement type Class 42.5N complying with BS12 (BSI, 1996), or cement type CEM1 complying with BS EN 197-1 (BSI, 2000) or similar standards.

Admixtures are supplied in either a liquid or powder form and can have expanding or non-expanding properties. Proprietary admixtures may also include fluidifying, water-reducing and super-plasticising agents. As the use of special grouts becomes the norm, admixtures should become available pre-measured to the correct dosages.

Trials should be carried out to ensure that the mix performs in the intended way, fully filling the duct and encasing the tendon. Commonly, these take the form of small-scale 'suitability' trials that test the material properties, and large-scale grouting trials where tendons and duct details replicating those in the works are grouted following the proposed procedures. The effectiveness of the grouting is determined by cutting up the completed tendons in the large-scale trials and visually inspecting the grouted ducts. These trials are discussed in more detail later in this chapter.

During winter it may be necessary to heat the water for the grout as well as storing the constituent materials in controlled temperatures. In concrete box girders, the grouting location temperature can be maintained above 5°C utilising electric heaters inside the deck, and suspended sheeting to cordon off work sections.

In the summer, the hotter environment reduces the fluidity of the grout and shortens its working life, which can create problems during grouting. This can be countered by adding ice to the water supply or installing air-conditioning units in the material storage facilities. The normal specification requirement is to wash out grouting plant every three hours. This should be carried out more frequently in hot weather to prevent setting grout blocking critical areas of the grouting plant and tendon ducts.

From the provision of the basic ingredients to the mixing and injection of the grout, the quality control of the material is critical to the successful completion of the grouting process.

Figure 4.3 Flow-cone test

Grout material tests

Traditionally, the fluidity of the grout after mixing and before installation in the works is measured by using a flow cone of known dimensions, as illustrated in the Concrete Society's report TR47 2nd Edition (2002) and shown in Figure 4.3. Additional tests for volume change, homogeneousness, sedimentation and grout strength are also often carried out. The flow-cone test measures the time for a known quantity of grout to pass through the cone and is a gauge of the grout's behaviour; however, this does not necessarily indicate how the grout will perform when passed through the small spaces in a tendon duct or when pumped under pressure. Thixotropic grouts, used in special circumstances, flow when pumped, but come to an equilibrium state when not agitated and therefore cannot be tested by the flow-cone method.

The bleed properties of the grout used to be measured using traditional 'jam jars', where the bleed and the volume change are measured at fixed times against a height of grout standing in the test cylinder. The bleed is simply the emergence of water from the mixed grout. This test gives no overall picture of how the grout will perform under pressure, in a profiled duct or passing through restrictions. A trial of new tests, seen in Figure 4.4, was carried out on vertical and inclined tubes, and with and without prestressing strands inserted. These tests showed that the presence of the strands increased the bleed from the grout.

A 'no bleed/no shrinkage' grout is the perfect choice of material, but, when gauged against the traditional bleed test, no true picture of its properties can be seen. Bleed water emerges from the grout under

Figure 4.4 Inclined duct test (reproduced courtesy of VSL International, copyright reserved)

pressure, travelling along individual strands, which act like wicks, to high points. The resultant void, filled with air and water, provides a perfect environment for corrosion.

The second edition of TR47 (Concrete Society, 2002) and later BS EN 447 (BSI, 2007c) addressed this issue by introducing a new test for bleeding. The grout is poured into a 1.5 m-high vertical duct with prestressing strands inside, and the height of the grout and amount of bleed water measured over a period of time. 'Flowmeters', which measure the amount of air in a fluid grout, have also undergone trials, giving operatives the possibility of using such measuring devices to watch grout flow into a duct and to determine the percentage voids in the duct whilst the grout is still fluid.

The full range of tests to be carried out and the criteria against which they are judged are described in TR47 (Concrete Society, 2002), and in BS EN 447 (BSI, 2007c) with BS EN 445 (BSI, 2007a). Grout material tests should be carried out using the same material, mix design, plant and operatives, as proposed for the permanent works. The environmental conditions should also be similar to those expected, with the temperature and humidity controlled.

Grouting equipment

Grout mixing on site is carried out using special mixing pans, as shown in Figure 4.5. These are used to mix the cement, water and any admixtures required to give a homogeneous grout. For long tendons, which require large quantities of grout, several mixing pans are needed, or a larger holding tank for mixed grout can be used to store two or three batches of grout. This prevents delays during the grouting which can cause blockages if the grout is allowed to stand for too long.

Figure 4.5 Grout-mixing equipment

A grout pump, often combined with the mixing pans, takes the fluid grout from the pans and pushes it into the duct via the inlet. The grout is usually pumped with a pressure up to $1 \, \text{N/mm}^2$, although if grouting is carried out above the pump the additional head can increase the required working pressure up to $3 \, \text{N/mm}^2$. If the pressure is allowed to get too high it is liable to cause the grout to create blockages at valves or other restrictions in the equipment of tendon, and the grout pumps often have pressure relief valves to protect against this.

All grouting equipment should be duplicated on site to ensure that back-up is always available in case of breakdowns. The use of easily replaceable fittings allows quick resolution of problems encountered during the grouting operation.

Vents and other details

The ability of the grout to flow round the strand over the whole tendon length is influenced by certain duct and vent parameters. Guidance on duct diameter is normally recommended by the suppliers of the prestressing system, but is also given in other technical publications such as the one by the FIP (Fédération International de la Préconstraint) (1990) and clause 9.26.4 of AASHTO (1996–2002). The area of the duct should be at least twice the steel area of the tendon inside.

The ends of the tendons are usually treated in one of two ways. The most straightforward method is to trim the strand protruding from the anchorage to the minimum permissible length, usually a strand diameter, and then place a cap over the anchor head to restrain the grout, as shown in Figure 4.6.

Figure 4.6 Anchor cap with grout inlet

Figure 4.7 Concrete capping to anchorage

This cap normally has a small vent hole with the grout pumped in via an inlet tube attached to the anchor trumpet, although some post-tensioning systems have a large hole in the bearing plate to allow the grout to be pumped in via the anchorage cap. The second method is to provide a capping of concrete to encase the anchor head and protruding strand, as seen in Figure 4.7. This provides a watertight seal for the grouting operation with an inlet tube cast into the capping concrete to allow the grout to be pumped into the duct.

Inlet tubes are needed to allow the grout to be pumped into the duct and should have a minimum internal diameter of 25 mm. The inlet is provided at the lowest anchorage point allowing grout to be pumped 'uphill' into the duct, as this helps to expel any air along the length. Ideally, at the inlet anchorage the inlet tube should be located at the bottom of the anchorage arrangement with an outlet vent at the top of the anchorage to ensure that the anchorage is fully filled with grout. Outlet tubes are required along the tendon to vent off the air and to ensure that the ducts are full of grout. These can be 12 mm diameter, but it is normally recommended to make these 25 mm diameter so that they can also be used to pump grout into the duct should the need arise due to blockages elsewhere.

Vents are provided at each low point in the duct, both to drain out any standing water and to draw off any trapped air. Vents should also be provided at, and just after, each high point to allow air and water to be removed from the crest during the grouting operation, and after the grouting operation is finished these vents should be re-opened to again release any air or water build-up. A vent or outlet tube is provided at the 'far' end of the duct or far anchorage to ensure that the duct is filled along its full

Figure 4.8 Grout vents

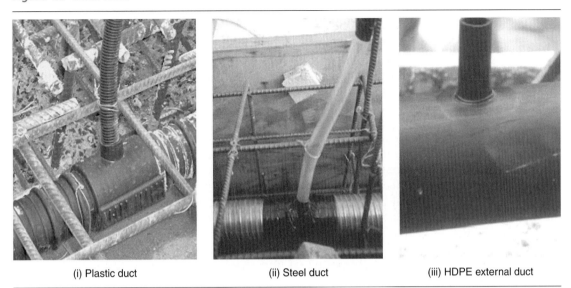

| (i) Plastic duct | (ii) Steel duct | (iii) HDPE external duct |

length. The outlet tube at the far end should be located at the top of the anchorage to ensure that the anchorage fully fills with grout. Typical arrangements for vent connections with plastic, steel and HDPE ducts are shown in Figure 4.8.

Vents should be provided at regular intervals along ducts with long 'flat' profiles to remove trapped air that can occur as the grout runs along the bottom of the duct. The maximum vent spacing required is typically less than 15 m.

Difficulties can occur with external tendons where the duct has a crest through a wide deviator or diaphragm such that the high point is hidden within the concrete and therefore inaccessible. To prevent air being trapped and forming a void in the duct, vents are positioned as close to the crest as possible on either side of the diaphragm. Although this does not always prevent a void, it should be small and the grout should still provide cover to the strand. Alternatively, a small diameter tube can be inserted through the top of the duct, with the connection positioned at the crest and the free end of the tube leading outside the concrete. This method is described by Smith and Wood (2001).

Grouting trials

Grouting trials are used to demonstrate the acceptability of the proposed grouting method, grout mix, equipment, operatives and vent arrangement before application in the permanent works. The tendon profiles chosen for the grouting trial should replicate the tendon profiles proposed in the permanent works. Several different profiles may need to be assembled to represent the different tendon profiles and tendon lengths found in the permanent structure. The longest tendon known by the author to have been part of a trial was 217 m between anchorages to match the tendon length to be used in the deck. In this case, the grouting trial was successful, demonstrating that the long tendon could be fully grouted in one operation with the equipment and material used.

After installation within the duct, the tendon should be sufficiently stressed to cause it to follow the required profile, typically in the region of 5% of its full stressing force. The anchor should then be capped and the tendon duct pressure-tested using air or water as appropriate before carrying out the grouting. The same operatives and supervisors who are assigned to the grouting of the permanent works should also undertake the grouting trial.

The grouting of the trial length should proceed without any delays in the process and without any blockages or leakages in the equipment. The trial can indicate where procedural refinements are required as well as providing useful information about the performance of different types of duct, vent fittings and grout material.

After the grout has set, the tendon and duct are cut or cored to allow inspection at critical sections. These should include the crests in the profile and at locations near the anchorages and vent positions. The grouted duct should not contain excessive voids likely to put the tendon at risk. Voids with dimensions of less than 5% of the duct diameter are often considered acceptable, especially in areas away from the strands. If the initial trials are unsuccessful, further trials should be carried out until all the problems are resolved. Lessons learned from the trial should be incorporated in the design and grouting procedures.

Pre-grouting test (pressure testing) of the ducting system

There is a need to prove the integrity of the duct system before grouting starts, to ensure that leaks do not occur. For internal tendons with plastic ducts, this is carried out before concreting and, for external tendons or internal tendons with corrugated steel ducts, before grouting. Traditionally, the integrity is proved by blowing air through the duct or by water testing, which involves the filling of the ducts with pressurised water to aid visual determination of any leaks. In many circumstances, the more rigorous investigation with air pressure testing gives better results, and is now a mandatory requirement for most work in the UK.

Pressure testing is both labour intensive and time-consuming, even for fully accessible external post-tensioning systems. Benefit would be gained from further development in the design and manufacturing of the ducting, to ensure that all joints become more positive and reliable. Much work has been done recently to develop a closed system of ducting to provide a fully watertight enclosure. These duct systems are manufactured using PVC or plastic for internal tendons and HDPE for external tendons. They are the only forms of duct currently permitted in the UK. Elsewhere in the world, it is still common to use corrugated steel ducting for internal tendons. Manufactured from steel strips and crimped together, corrugated steel ducts result in a non-watertight arrangement.

Air pressure testing

Air testing under pressure is used primarily on plastic and HDPE ducts to check the integrity and potential watertightness of the duct and fittings. The testing process can also indicate, by opening all the inlets and vents and allowing the air to flow through, that the vents, ducts, anchorages and other areas are all free from blockages. However, the passage of air is no guarantee that grout will also be able to flow freely.

Testing in two phases can be useful. First, the unvented system is tested to locate major leaks in the duct itself; and, second, testing is carried out with all vents and associated fittings added to determine the soundness of the vent work. The fittings for some plastic duct systems may not be able to sustain high pressures, even though the duct may still give an overall satisfactory performance, and this needs to be considered when choosing the system to be used.

Leak repair techniques normally involve patching, by placing a section of duct over a discontinuity and 'welding' it in place, or welding and sealing a discontinuity by applying a length of hot bead from a polyethylene strip. Either of these techniques, when undertaken carefully, produces repairs that should pass air testing; however, there may still be a risk of leaks during grouting. Alternatively, easily replaceable fittings, including prefabricated saddles and rubber collars, can be used. Duct discontinuities can then be resolved quickly and easily before concreting or grouting starts, as appropriate.

Leakages in an area of duct that lies within a deviator or anchorage location can prove difficult to resolve for external tendons. In these locations the duct is effectively hidden behind concrete and therefore excluded from any standard repair technique. If the duct is air tested before the tendon is stressed, it is possible to remove it from the deviator or anchorage and make a repair before re-installing. If the tendon has already been stressed then it would need to be either de-stressed to allow the duct to be removed and repaired, or repaired in situ if this is feasible.

The following issues are important during air testing.

(a) Construction sites are noisy. The air pressure selected should be high enough to allow rapid audible location of leaks.
(b) With regard to the issue of reasonable percentage loss in pressure, TR47 (Concrete Society, 1996, 2002) suggests that 10% of 0.01 N/mm^2 over five minutes is acceptable. Air loss through couplings in airlines may easily exceed this limit and should be taken into account when assessing the acceptability of any test. On large-scale grouting operations there is the opportunity to investigate acceptable loss levels and develop suitable criteria for the specific project parameters.
(c) Application of soap solutions to all couplers, vents and caps clearly indicates any leaks by producing bubbles where air emerges through small gaps.
(d) Micro-cracking of concrete in the anchorage zone is a feature of post-tensioning. Leakage through such cracks cannot be identified by post-concrete pressure testing of anchorage components before post-tensioning. However, tendons thus affected, which could not hold air pressure, could be grouted successfully. Making this judgement is a matter of experience.
(e) Air-testing pressure losses are not necessarily an indication of behaviour during grouting.

In summary, air testing does not necessarily indicate that water egress will occur at connections during grouting or that there is any risk of water ingress at connections later; however, the test is useful for determining the integrity of the ducting and in identifying potential problem areas.

Water pressure testing

This test is often used to identify leaks in internal corrugated steel ducts and is carried out before or after installing the tendons. It has been a common practice to water-test internal tendons in precast segmental construction, where the 'O' ring seals around the ducts at the segment joints are a potential weakness. It is also used in non-segmental construction and in cast in situ segmental construction. Water is pumped through the ducts and out of all the vents and outlets to ensure that an even flow can be achieved. The vents and outlets are then closed and the water pressurised to a level similar to that required during the grouting process. The pressure drop over a period of time is monitored as an indication of the integrity of the ducts and the seriousness of any leaks.

During water testing, locations where water emerges on the inside or outside faces of the concrete are easily identified and repaired. Cross-linking between tendons is also identified by the emergence of water from the vents of an adjacent duct.

After completion of the water testing, all residual water is removed from the duct by blowing air through and by providing drainpipes or vents at all low points where appropriate.

Leakages

Grout leakages can occur with any form of deck construction, although the pre-grouting tests described above should minimise their occurrence. For simple structures and straightforward prestress tendon layouts the risk of leaks is small, but for structures with more complex tendon arrangements the risks are significant.

In balanced cantilever construction, top tendons are commonly anchored on the end faces of the segments. Face anchors located at the top of the webs or in the top flange are usually contained within a thin-walled 'box' formed during concreting of the segment. This type of box is very prone to suffering grout leaks and cross-linking with other tendons close by.

It is desirable to grout tendons including their anchors in one go, feeding the grout through one anchor and letting it out at the other anchor. This allows the grout to be forced through all the interstices of the strands, filling all small holes to give a product that has minimum voids. However, it is common that the weakest area for holding grouting pressure is the face anchor box. This can be identified during pressure testing. Remedies include temporary patching where the leak is visible on the concrete and breaking out the wall of the box on the inside of the segment to affect a repair. The latter is not recommended, as breaking out concrete adjacent to live tendons can be hazardous.

Another solution is to fill the anchor box with grout to seal the problem area first and then grout the rest of the duct in a separate operation. This is best achieved by using a special thixotropic grout, which can be poured into the anchor box through the box vent, emerging through the top flange or the top of the web during the first stage of the grouting operation. The thixotropic grout, if not agitated, will fill the anchor box without passing through into the duct and jeopardising the second-stage grouting operation. Monitoring the quantity of grout poured in gives warning of any leaks. The disadvantage of this method is that the interstices of the strand will not be completely filled with grout in the anchorage zone.

Precast segmental concrete box girder bridges typically utilise relatively thin concrete members for the top and bottom slabs and walls. Ducts pass through the concrete emerging at the ends of segments at standard positions to minimise the disruption to the segment mould bulkhead during casting. In the top flange at pier locations and the bottom flange at mid-span, tendons are usually closely spaced. In addition to the risk associated with the tendons in close proximity, stripping and handling damage of the concrete can easily produce a leak or cross-linking with another tendon.

Good detailing and construction practice is vital to prevent cross-linking and grout leaking in segmental decks. Ducts should be as widely spaced as other design considerations permit. For pre-cast match-cast segmental construction, effective 'O' ring seals must be used around each duct at the segment joint, and the epoxy filler between the segment ends must be evenly spread over the faces to fill up all gaps.

If cross-linking between ducts occurs, this can be overcome by simultaneously grouting the affected ducts. Identification of the potential cross-linking by air or water testing and planning of the grouting process to control the flow of grout between the ducts is essential to minimise the risk of air becoming trapped.

The problems with leakages and cross-linking described above largely explain why there is still a moratorium on the use of internal tendons in pre-cast match-cast concrete segmental deck construction in the UK.

Grouting procedure

Once pressure testing has been carried out successfully, grouting of the duct can take place. Where possible, groups of tendons should be grouted together so as to minimise any risks of leakage between adjacent ducts and to simplify the overall grouting process.

If there is any water in the ducts this should be removed as much as possible before the start of grouting. It is not always practical to remove all the water from the ducts and a small quantity left behind does not usually present a problem, as it is expelled during the grouting process.

Grout is introduced via the inlet at a slow, even rate, maintaining a pressure of 0–1 N/mm^2, and grout is allowed to emerge sequentially from intermediate vents before they are sealed. For external tendons the rate of progress of the grout along the tendon should be established audibly by tapping of the duct.

At intermediate vents and at the anchorage outlet at the end of the tendon, a quantity of grout is allowed to vent, as shown in Figure 4.9, before the tap is slowly closed. All vents should be closed in sequence in the direction of grout flow, with the outlet being closed last. It is useful to hold the end of the vent pipe below the surface of the collected grout in a bucket. Any air being pushed out of the duct is seen as a series of bubbles in the bucket of grout. Grouting should continue until the

Figure 4.9 Grout being collected at an intermediate vent point

air bubbles are no longer visible. The outlet grout should be collected and subjected to a fluidity test using the flow cone discussed earlier, to ensure that it still complies with the specified requirements.

The grout in the duct system should be pressurised and held for a short period, allowing any air voids to rise up to the crests in the duct profile. The vents at the crests are then reopened one at a time to release any air present. Finally, with all vents closed again, the system is re-pressurised and held for another short period before finally being sealed.

Experienced operatives should undertake the grouting, and all procedures regularly checked as part of the quality-control requirements. Good communications are needed between the operatives, particularly where clear lines of sight between all involved are not possible.

Following grouting, vibration on the structure should be prevented for 24 hours to allow the grout to set and harden.

Post-grouting checks should include inspection of vent pipes to ensure that they remain full of grout. For external tendons, tapping of the duct can indicate if any air voids are present when a hollow sound is recorded. Investigations involving drilling into ducts should be carried out with care to avoid damage to the tendons.

Specifications usually require provision to be made for adequate flushing out of a duct with water if there is a problem during the grouting operation, such as equipment failure or grout blockage. The specification usually requires 'complete removal of the grout'; however, this can sometimes be very difficult to achieve. The washing-out water tends to flow over the top of the grout without fully removing it. Even the addition of compressed air to the water only affects a localised area. It is debatable whether flushing out should be attempted in favour of a later re-grouting exercise.

Grease and wax grout

Grease and wax grout is used with external prestressing to enable easy de-stressing or replacement of the tendons. The grease or petroleum wax is injected into the duct at between 80°C and 90°C, before cooling to become a soft, flexible filler. Additives are often included in the wax to help penetration through the wires and to inhibit corrosion.

With pre-assembled tendons or where tendons are individually sheathed, the wax is injected at the factory in controlled conditions. For larger and longer tendons it is often necessary to inject the wax on site, requiring specialist equipment for storage and heating of the wax. Wax grouting follows a similar procedure to cement grouting, except that injection pressures are typically $0.2 \, \text{N/mm}^2$.

Where external tendons are required to be re-stressable, sufficient length of strand is left protruding to allow a stressing jack to be re-attached. An elongated cap covers the protruding strand so that the length of tendon immediately behind the anchorage can be protected by grease or wax infill.

REFERENCES
AASHTO (1996–2002) Standard Specification for Highway Bridges, 16th edition. American Association of State Highway and Transportation Officials, Washington, DC.
BSI (1996) BS12:1991: Specification for Portland Cement. BSI, London.
BSI (2000) BS EN 197–1: Part 1 with amendments (2007), Composition, Specification and Conformity Criteria for Common Cements. BSI, London.

BSI (2007a) BS EN 445:2007: Grout for Prestressing Tendons – Test methods. BSI, London.

BSI (2007b) BS EN 446:2007: Grout for Prestressing Tendons – Grouting procedures. BSI, London.

BSI (2007c) BS EN 447:2007: Grout for Prestressing Tendons – Basic requirements. BSI, London.

Concrete Society (1996) *Technical Report no. 47. Durable bonded post-tensioned concrete bridges.* Concrete Society, Slough.

Concrete Society (2002) *Technical Report no. 47. Durable bonded post-tensioned concrete bridges.* 2nd Edition. Concrete Society, Slough.

FIP (1990) *Guide to Good Practice, Grouting of Tendons in Prestressed Concrete.* Thomas Telford, London.

Highways Agency (2009) *Manual of Contract Documents for Highway Works, Volume 1, Specification for Highway Works.* Norwich: HMSO, Clause 1711.

Smith LJ and Wood R (2001) Grouting of External Tendons – A Practical Perspective. *Proceedings of the Institution of Civil Engineers* **1(146)**: 93–100.

Prestressed Concrete Bridges, 2nd edition
ISBN: 978-0-7277-4113-4

ICE Publishing: All rights reserved
doi: 10.1680/pcb.41134.067

Chapter 5
Prestress design and application

Introduction

The design of the prestress for a bridge deck is integral to the overall design process, and it must take into account the behaviour of the structure as a whole under the applied loads and during the construction stages. Application of the prestress affects the structural behaviour, and the design process requires a clear understanding of the interaction between them in order to achieve a safe and efficient structure.

The general principle with prestressed concrete design is to ensure that the concrete is either in compression or within acceptable tensile stress limits under service load conditions. Prestress also enhances the shear and ultimate moment capacity of the concrete section.

This chapter describes the different aspects that are considered during the design of a prestressed concrete bridge. The initial sections deal with general design procedures, while the particular requirements of Eurocode (BSI, 2004a, 2005a), the UK design standard BS 5400 (BSI, 1990) and the USA standard specifications by AASHTO (1998–2011) are given towards the end of this chapter. Codes and Standards are continually being developed and updated, and reference should always be made to the latest requirements and the bridge owners' particular specifications before commencing a design.

The design of individual details and the analysis of prestressed concrete decks are discussed in Chapters 6 and 8 respectively, while the design of particular deck types is covered in Chapters 9 to 17.

Where formulae are used in this chapter, the units are in N and mm.

General approach

The structural form of the deck and the construction methods to be employed will influence the choice of prestress type and layout. Precast beams use pre-tensioned strands, wires or post-tensioned tendons, with typical arrangements as illustrated in Figure 5.1. Pre-tensioned strands or wires are usually straight, but can be deflected to give a draped profile. Post-tensioned tendons are draped to reflect the bending moments generated in the beam.

Prestressed bars are considered easier to handle and install, and are often used where they simplify construction. Such applications include the temporary prestress during the erection of precast segmental decks, as shown in Figure 5.2, or during the launching of box girder decks. Bars placed in ducts within the concrete are usually used straight between anchor points; however, large radius curved alignments can also be achieved.

Post-tensioned tendons are installed on site and have the advantage of being able to achieve complex tendon layouts and very high stressing forces. A typical layout for a continuous deck is illustrated in

Figure 5.1 Precast beam prestressing

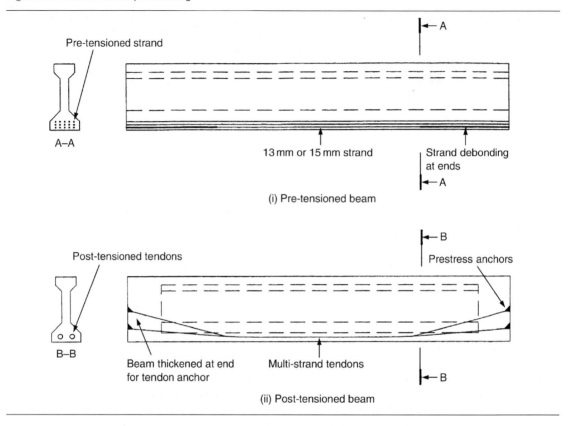

A–A

Pre-tensioned strand

A

13 mm or 15 mm strand

Strand debonding at ends

A

(i) Pre-tensioned beam

B

Post-tensioned tendons

Prestress anchors

B–B

Beam thickened at end for tendon anchor

Multi-strand tendons

B

(ii) Post-tensioned beam

Figure 5.3. Typical jacking systems normally require a tendon length of more than 5 m to distribute losses during lock-off, while special systems can be utilised to reduce the usual lock-off losses in a tendon if required. The friction losses become excessive for internal post-tensioned tendons longer than 120 m, even when double-end stressed. External post-tensioned tendons have much lower friction losses and are installed with lengths of over 300 m. In stage-by-stage construction, couplers are sometimes used to extend post-tensioned tendons across a construction joint.

External tendons, as seen in Figure 5.4, are used extensively for bridge strengthening as well as for new construction, giving designs where the tendons are readily inspectable and replaceable. External tendons are anchored at the diaphragms or on special anchor blocks along the span. The tendons are deflected at deviators to give a draped profile.

Prestress tendon layouts for most types of bridge structures are now well established. New designs normally start by considering a tendon arrangement similar to that used on previous structures of the same type: for example, straight strand in the bottom of precast beams; cantilever tendons and continuity tendons for balanced cantilever construction; or simple draped tendons for in situ, continuous decks.

For most types of deck, the prestress requirement is governed by the serviceability stress check. The number of tendons required at critical sections, such as mid-span and over supports, is derived by

Figure 5.2 Prestressing bars used for segment erection

calculating the stresses on the section due to the applied loads and construction effects, and then estimating the prestress required to keep these stresses within the allowable limits based on an initial estimate of the secondary moment. From this initial estimate of the number of tendons needed, a preliminary prestress layout can be drawn up. The difficulty at this stage is that the prestress secondary moments are not known, and therefore, after estimating the prestress layout required, it is necessary to calculate the actual secondary moments generated and to compare these with the values used in the initial estimate. It may take several iterations of estimating the secondary moment and adjusting the prestress arrangements before the required prestress stresses are obtained and the actual secondary moment matches the assumed secondary moment, to give the final design. After this, the ultimate moment is checked at the critical sections and, if necessary, additional tendons are added to ensure that adequate resistance can be mobilised. Where additional tendons are required a recheck of the serviceability stresses must be undertaken to ensure that they are still within the acceptable limits.

Figure 5.3 Typical post-tensioning layout

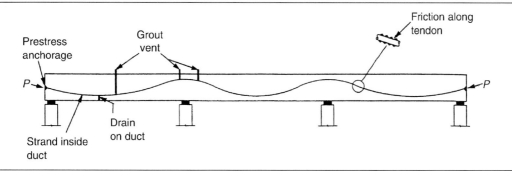

Figure 5.4 External tendons inside box girder

For structures with external, unbonded tendons the quantity of prestress required can be governed by the ultimate limit state, especially where the predicted increase in stress in the tendon at ultimate bending failure is small and there is no non-prestressed reinforcement available. This behaviour is described in the section 'Ultimate moment design', later in this chapter. In this case, the number of tendons required at critical sections is determined from the ultimate moment check, and the serviceability check then carried out to ensure that the stresses are within acceptable limits.

The magnitude of the secondary moments in statically indeterminate structures is dependent on the prestress layout. They can vary greatly with any type of structural form and are influenced by the construction techniques used. For bridge decks the secondary moment is usually sagging which reduces the magnitude of the design hogging moments over the piers and increases the magnitude of the design sagging moments in mid-span. For spans in the range of 35 m to 40 m, the prestress secondary moments for a typical box girder arrangement can be at least 5000 kNm at internal piers, while for heavily prestressed structures and for longer spans the secondary moments can be significantly higher.

Several countries now recommend that allowance should be made for the future installation of an additional 10% or more of prestress tendons when designing new post-tensioned decks. This can easily be incorporated using external tendons. The need for this extra provision has come about due to excessive loss of prestress or large deflections occurring on some existing structures, and because of the requirement to be able to upgrade or strengthen the structure in the future to take heavier loads. It would seem prudent to make provisions for future additional tendons on all new post-tensioned bridges.

Primary and secondary prestress effects
When the prestressing tendons apply load to the structure the resultant forces and moments generated can be considered as a combination of primary and secondary, or parasitic, effects, as illustrated in Figure 5.5.

Figure 5.5 Prestress in continuous decks

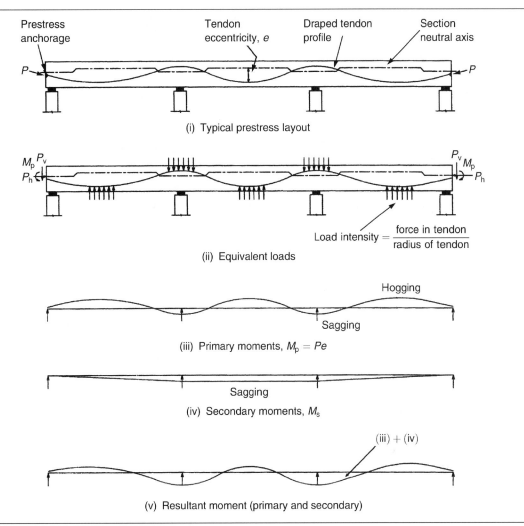

(i) Typical prestress layout

$$\text{Load intensity} = \frac{\text{force in tendon}}{\text{radius of tendon}}$$

(ii) Equivalent loads

(iii) Primary moments, $M_p = Pe$

(iv) Secondary moments, M_s

(v) Resultant moment (primary and secondary)

The primary effects are the moment, shear and axial forces generated at each section by the direct application of the force in the tendon at that section. The axial force, P, is the force in the tendons at the section being considered after all applicable losses have been taken into account. The primary moment is the tendon force multiplied by the distance of the force from the neutral axis, that is, $P \times e$, and the shear is the vertical force due to the inclination of the tendon.

Secondary effects occur when the structure is statically indeterminate and restraints to the structure prevent the prestressed element from freely deflecting when the prestress is applied. For statically determinate and simply supported decks the structure is free to deflect when prestressed and hence no secondary effects are set up. With continuous decks the intermediate supports restrain the deck from vertical movement, resulting in secondary moments and shears occuring when the prestress is applied.

Where decks are built integrally with abutments or piers the resulting restraints generate secondary effects from the prestressing, which must be taken into account in the design. An example of this is shown in Figure 5.6. The prestress compresses the portal beam causing it to shorten. This shortening is resisted by the stiffness of the columns, generating secondary moments and 'tensile' forces in the

Figure 5.6 Secondary effects due to built-in supports

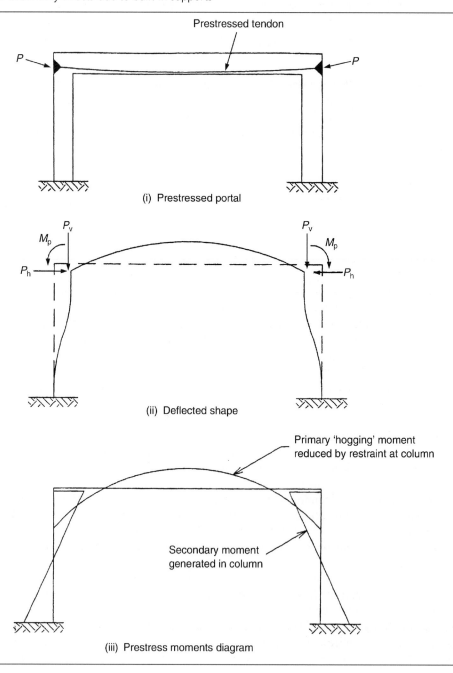

Prestressed tendon

P P

(i) Prestressed portal

M_p P_v P_v M_p

P_h P_h

(ii) Deflected shape

Primary 'hogging' moment
reduced by restraint at column

Secondary moment
generated in column

(iii) Prestress moments diagram

beam. Similarly, the prestress tends to bend the portal beam under its primary effects causing the beam to deflect, and the columns again resist this deflection, generating secondary effects.

The secondary moment set-up with post-tensioned tendons is dependent on the structural arrangement at the time of stressing. For bridges constructed in stages the structural arrangement changes as the deck is built out, and each stage of prestressing will act on a different structural system. The final secondary moment in the deck is based on the combination of all the secondary moments set up at each stage. This is complicated further by creep of the concrete and long-term losses in the prestressing. These cause a redistribution of the staged secondary effects with the changes acting on the modified structure.

Pre-tensioned strands installed in precast beams are stressed when the beam is 'simply supported' and no secondary moments are generated. However, if the beam is made 'continuous' or built into the supports after installation, creep of the concrete under the prestress load and redistribution of forces due to long-term stress losses in the prestressing steel will result in the development of secondary moments and shears.

Prestress secondary moments can be derived by a number of methods, as described in Chapter 8. One popular way is by applying the equivalent loads to a frame model. Other techniques include numerical methods such as using influence coefficient, or specialist analysis programs such as MIDAS that combine the prestress with the structural analysis.

When the equivalent load method is used, a combination of both the primary moments and secondary moments are produced; however, this approach becomes considerably more complex when the variations in prestress force along the tendon are included. The Influence Coefficient method is suited to the use of spreadsheets which simplify the calculation, and have the advantage that the prestress force profile is easily incorporated and the calculation can be quickly adjusted during the design process. Proprietary software programs are the easiest and quickest way to derive the combined primary and secondary effects on a structure, although many of those currently available do not take into account all the different aspects that must be considered during the design. Software that is able to include the tendons with the concrete elements and model long-term effects and stage-by-stage build-up of the structure should be used.

Primary and secondary moments both generate longitudinal bending stresses on the section, as indicated in Figure 5.7. The resultant stresses are taken into the serviceability stress check. The secondary moments and shears are taken into the ultimate limit state checks.

Post-tensioned tendons in horizontally curved decks generate torsions due to the secondary effects of the prestress. The behaviour of curved box structures is described in a paper by Garrett and Cochrane (1970). Intermediate supports along a curved structure resist the decks' tendency to twist, and, where the prestress generates secondary moments along the deck and vertical reactions on the supports, a complementary torsion is present.

For curved box girder decks with a total horizontal angle change of less than 20° between the torsional restraints, an estimate of the torsion is derived by computing the secondary moments, assuming that the bridge is straight, and dividing these by the horizontal radius of the deck. This 'M/R' diagram is then applied as a load to the structure, and the shears generated along the deck are the torsions that will be present in the curved structure. Where the angle change is over 20°

Figure 5.7 Prestress stresses on section

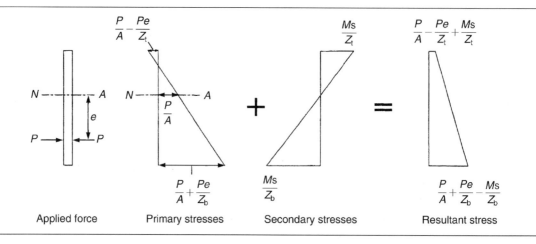

| Applied force | Primary stresses | Secondary stresses | Resultant stress |

the approximations made in this approach may not be valid. This method is presented in a paper by Witecki (1969).

Provided the tendons are symmetrical about the vertical axis of the section, the primary effects of the tendons are not affected by deck curvature; however, where this is not the case, such as with external tendons running straight between anchor and deviator points, transverse bending and further torsions can be generated.

In tightly curved box girders and continuous multiple-beam decks on a curved alignment the behaviour of the structure becomes complex as the beams twist and the deck slab distorts transversely. With these arrangements a three-dimensional finite element analysis should be carried out to determine the overall structural behaviour and the effects of the prestress.

Prestress force and losses

During stressing a tendon experiences losses in the force along its length due to friction with the duct. Further losses occur after lock-off, over the long term, due to elastic shortening, creep and shrinkage within the adjacent concrete and by relaxation of the tendon steel. The long-term stress in the pre-stressing steel is given by:

$$f_{pc} = f_p - l_E - l_r - l_c - l_s$$

Tendons may be stressed up to 80% UTS (ultimate tensile strength), although some codes limit the jacking force to 75% UTS, and the maximum force in the tendon after release by the jack to a maximum of 70% or 75% UTS for post-tensioning and pre-tensioning respectively.

External tendons are not bonded to the concrete, and the losses for elastic shortening, creep and shrinkage are estimated by considering the total concrete deformation between the tendon anchorages or other fixed points, and this is taken as being distributed uniformly along the tendon length.

Friction losses and tendon extension

Prestressing bars and pre-tensioned wires or strands are usually straight and encounter few friction losses. Multi-strand tendons that are placed inside ducting experience friction losses as the tendon is

Table 5.1. Typical friction coefficients

	k	μ
Bare strand in metal ducts	0.001 to 0.002	0.2 to 0.3
Bare strand in UPVC ducts	0.001	0.14
Bare strand in HDPE ducts	0.001	0.15
Greased and plastic sheathed strand in polyethylene ducts	0.001	0.05 to 0.07

stressed. These losses can be significant when the tendons follow a curved profile and cause the strands to press against the sides of the duct.

When the jack pulls a profiled multi-strand tendon, the movement of the strand is resisted by friction against the duct. The associated loss in the force transmitted along the tendon is expressed in terms of θ, the friction coefficient acting on the angle change in the tendon profile, and k, the wobble coefficient acting over the length of the tendon. The wobble coefficient is a measure of the unintentional local misalignment of the duct, which results in the tendon rubbing against the duct surface, causing additional friction losses. The force F at any point along the tendon, x metres from the stressing anchor, is given by:

$$F = F_{o}\, e^{-(\mu\theta + kx)}$$

The friction coefficients are obtained from the manufacturers' literature for the system being used. Typical values are given in Table 5.1. The actual value achieved on any particular project depends on a number of factors, including the surface condition of the steel, type, diameter, condition and support of the duct, and the installation method adopted.

For an external tendon, bar or pre-tensioned strand, $k = 0$ over its 'free' length. Where external tendons pass through deviators or diaphragms, friction losses occur although wobble is often ignored over the short lengths concerned.

The tendon force F is sometimes expressed as $F = F_{o}\, e^{-\mu(\theta + kx)}$, with the value of k being adjusted accordingly.

When the jack releases a multi-strand tendon a short length of reverse friction is set up as the strand pulls in and the wedges grip to lock the strand in place. The lock-off pull-in can be up to 7 mm, depending on the system being used. With a prestressing bar, anchorage is achieved by tightening a nut against the anchor plate after stressing the bar to the required force. Bar systems with fine threaded anchorages have minimal lock-off pull-in, while those with coarse threads can pull in a small amount.

Figure 5.8 shows a typical force profile along an internal multi-strand tendon stressed from one end. The force loss along the tendon is dependent on the tendon profile, and for tendons longer than 40 m and with large angle changes excessive losses can occur. This can be compensated for by double-end stressing, where the tendon is jacked from both ends. The maximum economic length for internal multi-strand tendons is usually about 120 m, above which the friction losses become too high, reducing

Figure 5.8 Tendon-force profile

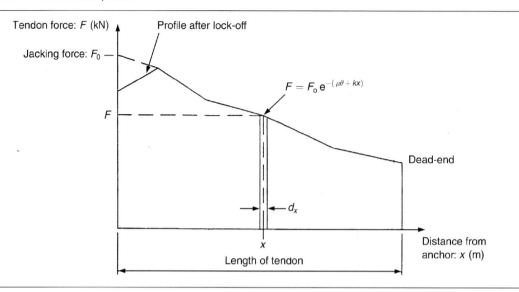

the effectiveness of the tendon. With external tendons the friction loss occurs only where the tendon passes through a deviator, and it is possible to use very long tendons without significant losses.

The extension of the tendon during stressing can be calculated from the force profile:

$$\text{Extension} = \sum F d_x / A_t E_t = \text{area under force profile} / A_t E_t$$

This gives the extension between the faces of the anchors, which is compared with the values measured on site during the stressing to confirm the satisfactory tensioning of the tendon. The measured extension at the jacking end must be adjusted to take into account any pull-in of the wedges at the dead-end of the tendon, which typically is between 2 mm and 4 mm, and also the extension in the strand between the measuring point behind the stressing jack and the face of the anchor. For cast-in dead-end anchors, shown in Figure 2.4, the predicted tendon extension is usually calculated up to a point midway between the end of the duct and the bulb at the end of the strands.

The E_t value used is determined from tests carried out on samples taken from the batch of strands actually used on the project being considered. With 7-wire strand, short lengths are tested by being held between fixed jaws in a testing rig. These tests give higher E_t values than those achieved with long lengths of strand, where the effect of the 'spiral' outer wires reduces the effective stiffness. The effective E_t value is also reduced in multi-strand tendons where the strands are often twisted around each other during installation and straighten out during stressing. For long multi-strand tendons, these effects reduce the E_t value by up to 5% compared to the values obtained by testing a short length of strand.

Elastic shortening and strains
Elastic shortening of the concrete during the stressing of the tendons causes stress losses in any tendons installed previously. The loss in stress is calculated by the following formula:

$$l_E = \sigma_c (E_t / E_c) \, \text{N/mm}^2$$

where σ_c is taken as the increase in stress in the concrete adjacent to the tendon, occurring after the tendon has been stressed.

In pre-tensioning where all the strands are anchored simultaneously the full prestress force will generate elastic shortening of the concrete and give a similar loss in each tendon at the time of force transfer. In post-tensioning each tendon at a section will suffer a different loss depending on when it is stressed in relation to the other tendons. It is usually sufficient in the design to calculate an average loss and apply this to all the tendons, with σ_c taken as half the total final stress in the concrete, averaged along the tendon length.

As the deck is loaded it deflects, causing changes to the strain in the concrete with any bonded tendons being subjected to a corresponding change in strain and stress. The tendons are usually near the 'tension' face of the concrete and the stress in the tendons increases under further loading; however, this effect is small and often neglected when considering the overall prestress design.

Relaxation of tendon steel

After the force is applied to the prestressing steel and the tendon anchored, relaxation occurs in the steel which results over time in a reduction of the force in the tendon. Typical values of relaxation after 1000 hours for tendons stressed to 70% UTS at 20°C are given in Table 2.1. The amount of relaxation of the stress in the prestressing steel depends on the initial stress present and steel characteristics, as indicated in Figure 5.9.

Magura *et al.* (1964) proposed the following equation for calculating the tendon stress relaxation under constant strain:

$$f_{pt}/f_p = 1.0 - (\log(t)/c)(f_p/f_y^* - 0.55)$$

$c = 10$ for stress relieved strand with expected relaxation of 7% after 1000 hours when stressed to 70% UTS

Figure 5.9 Relaxation losses in strands plotted against initial stress

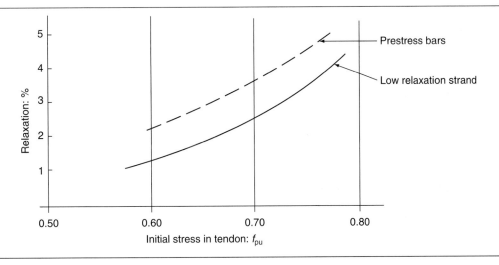

$c = 45$ for low relaxation strand with expected relaxation of 1.5% after 1000 hours when stressed to 70% UTS

Up to 50% of the 1000-hour relaxation percentage can occur within the first 24 hours. For low relaxation steel the total relaxation during a structure's life can be greater than three times the 1000-hour value. When the initial stress in the strand is less than $0.55f_y^*$ the relaxation losses become small. Near the stressed end of a tendon the initial stress after lock-off can be greater than 70%, while further along the tendon the stress may well be less than 50% UTS if significant friction losses occur.

Eurocodes 2 (BSI, 2004a, 2005a) gives the relaxation in prestress tendons with low relaxation steel to be:

$$\Delta\sigma_{pr} = \sigma_{pi}[0.66\rho_{1000}\, e^{9.1\mu}(t/1000)^{0.75(1-\mu)}]10^{-5}$$

where μ is the ratio of the initial stress in the tendon to the characteristic strength of the tendon.

BS 5400 part 4 (BSI, 1990) sets the loss of force to be taken in the design as the maximum relaxation after 1000 hours for a jacking load equal to that imposed at transfer. This may be conservative in some areas, but may underestimate the long-term losses in other situations. The normal practice in the UK is to estimate the average relaxation that is likely to occur in all the tendons and apply this value throughout.

AASHTO (1998–2011) gives guidance on the relaxation losses to be taken in the design with a simplified method taking $17\,\text{N/mm}^2$ for low relaxation strand; or a more refined method as follows for low relaxation strand:

Losses up to deck placement, $\Delta f_{pR1} = f_{pt}\,[f_{pt}/f_{py} - 0.55]/30$

Losses after deck placement, $\Delta f_{pR2} = \Delta f_{pR1}$

Both Δf_{pR1} and Δf_{pR2} can be assumed as $8\,\text{N/mm}^2$

Again, this is conservative in some cases, but may underestimate the long-term losses in other situations.

Relaxation losses can increase with high temperatures and steam curing of precast elements can affect the losses in any pre-tensioned steel. For the normal range of temperatures the effects on the relaxation can be ignored. In the hot regions of the world, such as the Middle East where temperatures can exceed 50°C, the increase in relaxation is usually not significant enough to affect the overall prestress design.

Creep losses
As concrete shortens due to creep under the compressive stresses, the change in concrete strain causes a corresponding change of strain and stress in the tendons. The loss of stress in the tendon due to concrete creep is given by:

$$l_c = \phi\sigma_c E_t/E_c\ \text{N/mm}^2$$

The creep factor of the concrete is ϕ and is primarily dependent on the concrete mix, the environmental conditions, the member thickness and the maturity of the concrete at the time of loading. In normal bridge construction, ϕ is in the range 1.6 to 2.5.

Eurocode 2 (BSI, 2004a) provides a chart for deriving the ϕ factor based on the concrete age, the thickness of the concrete, the strength of the concrete and different cement classes, and a range of humidity, which gives the ϕ factor typically between 1.0 and 2.0 for typical bridgeworks. For cases where the concrete adjacent to the tendon is subjected to stresses greater than 45% of the cylinder compressive strength of the concrete at that time, the ϕ factor is increased by $e^{1.5(k\sigma - 0.45)}$, where k_σ is the ratio between the applied stress in the concrete and the cylinder compressive strength.

BS 5400 part 4 (BSI, 1990), Appendix C, provides suitable charts for deriving ϕ taking into account the different parameters, while section 6 of the same document gives the following estimates for the creep per unit length in the concrete, expressed in terms of the stress in the concrete adjacent to the tendons.

For pre-tensioned strand in humid or dry exposure:

with $f'_{ci} \geqslant 40 \text{ N/mm}^2$ creep $= 48 \times 10^{-6}$ per N/mm^2

with $f'_{ci} < 40 \text{ N/mm}^2$ creep $= (40/f_{ci}) \times 48 \times 10^{-6}$ per N/mm^2

For post-tensioned strand in humid or dry exposure and tensioned between seven and 14 days:

with $f'_{ci} \geqslant 40 \text{ N/mm}^2$ creep $= 36 \times 10^{-6}$ per N/mm^2

with $f'_{ci} < 40 \text{ N/mm}^2$ creep $= (40/f_{ci}) \times 36 \times 10^{-6}$ per N/mm^2

where the stress in the concrete at transfer is $>0.33f_{ci}$ the above-estimated creep should be increased pro-rata up to 1.25 times when the stress is $0.5f_{ci}$.

AASHTO (1998–2011) provides formulae for the creep loss based on the creep factor in a similar way to Eurocode 2 (BSI, 2004a) and BS 5400 (BSI, 1990).

The presence of reinforcement can reduce the creep of the concrete, but with prestressed members the percentage of reinforcement in the concrete is usually low, and its effect on the creep losses usually ignored.

The actual creep that occurs depends on many factors and, as the design and construction of prestressed concrete decks become more sophisticated, the creep effects need to be assessed as accurately as possible. The estimated losses given above should be used only for simple designs that are not sensitive to small changes in prestress forces. For more accurate consideration of the creep effects and associated losses in the prestress, specialist software such as MIDAS is able to incorporate the creep into the analysis and combine both the stress in the concrete and the stress in the tendons at each stage of the structure's life.

Shrinkage losses

Shortening of the concrete due to shrinkage causes corresponding shortening and stress reduction in the tendons. Shrinkage losses are given by:

$$l_s = \Delta_{cs} E_t \text{ N/mm}^2$$

Δ_{cs} is the shrinkage strain within the concrete that has occurred since the tendons were installed, and it is dependent on similar parameters to those for the creep factor described above.

Eurocode 2 (BSI, 2004a) gives the shrinkage strain in two parts, the drying shrinkage and the autogenous shrinkage as the concrete is hardening. The total drying shrinkage for concrete in a humidity of 80% and with a cube strength of $50 \, N/mm^2$ is given as 240×10^{-6} ($\pm 30\%$), modified by a factor of 0.7 for concrete thickness greater than 500 mm and between 1.0 and 0.7 for concrete thickness between 100 mm and 500 mm. The drying shrinkage that has occurred after any given time, t, is derived by applying the factor $(t - t_s)/[(t - t_s) + 0.4\sqrt{h_o^3}]$, where t_s is the age at the end of curing of the concrete and h_o is the nominal size of the concrete. The autogenous shrinkage is given by $[1 - e^{(-0.2t^{0.5})}] [2.5(f_{ck} - 10)]10^{-6}$.

BS 5400 part 4 (BSI, 1990), Appendix C, gives guidance on deriving Δ_{cs}, while section 6 provides an estimate of the shrinkage strain, based on the relative humidity (RH) and age of concrete at transfer, as follows:

	Humid exposure (90% RH)	Normal exposure (70% RH)
Pre-tensioned strand with transfer between three and five days	100×10^{-6}	300×10^{-6}
Post-tensioned tendon with transfer between seven and 14 days	70×10^{-6}	200×10^{-6}

With Aashto (1998–2011), shrinkage losses are estimated in a similar way to Eurocode 2 (BSI, 2004a) and BS 5400 part 4 (BSI, 1990), with the shrinkage in the concrete adjacent to the tendons used to derive the losses in the prestress.

The presence of reinforcement can reduce the shrinkage of the concrete, but with prestressed members the percentage of reinforcement in the concrete is usually low, and its effect on the shrinkage losses usually ignored.

As with creep losses, the above estimates should be used only for simple designs that are not sensitive to small changes in prestress forces. For most cases the shrinkage losses should be estimated based on the actual concrete mix and environmental conditions expected using the guidance given in Eurocode 2 (BSI, 2004a), BS 5400 part 4 (BSI, 1990), Appendix C, or similar. Alternatively, the modern software packages such as MIDAS will derive the shrinkage losses based on a number of codes that can be selected, and then take these losses into account in the analysis during every step of the construction and throughout the bridge's life.

Tendon eccentricity in ducts

Strand tendons with a profiled layout will move to the inside face of any curved duct. This usually results in the tendon sitting at the bottom of the duct over the piers and at the top of the duct along the span. This offset between the duct centreline (C/L) and the tendons can be significant and depends on tendon and duct sizes. Typical offset values to be taken into account when calculating the tendon eccentricity in the design are given in Table 5.2.

Table 5.2. Tendon offset inside duct

No. of 13 mm strands	Internal diameter of duct: mm	Offset from duct C/L to strand centre: mm	No. of 15 mm strands	Internal diameter of duct: mm	Offset from duct C/L to strand centre: mm
1	20	4	1	25	5
7	55	8	7	60	10
12	65	10	12	80	15
19	80	12	19	100	17
27	95	14	27	110	20
31	100	15	31	130	22
37	120	25	37	140	25
42	130	25			
55	140	25			

Note: C/L = centreline

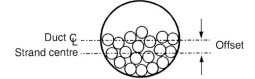

Serviceability limit state stress check

Longitudinal stresses in a prestressed concrete deck change at each stage of construction and throughout the structure's design life. The stress in the concrete must be kept within allowable limits for both compression and tension. It is usual to keep the concrete in compression across the full section under permanent loads and normal live loading with tensions of between 2 and 3 N/mm^2, permitted with unusual loading conditions or under the influence of temperature differentials. Typical allowable stresses in the concrete under service load when designing to Eurocode (BSI, 2004a, 2005a), BS 5400 (BSI, 1990) or AASHTO (1998–2011) are given later in this chapter.

In precast segmental construction with match-cast joints the longitudinal reinforcement is not continuous, and it is normal to ensure that the joints remain in compression under all loading conditions both during construction and in the permanent structure. For precast construction, with elements connected using unreinforced concrete or mortar infill, it is normal that the complete area of the connection is kept in compression with a minimum stress of 1.5 N/mm^2 to ensure that the joint is held together.

The stress levels along the length of a deck are checked at all the critical stages in the structure's life. This would normally include

(*a*) at transfer of the prestress to the concrete
(*b*) during each stage of construction, with temporary loads applied
(*c*) at bridge opening, with full, live load
(*d*) after long-term losses in the prestress and full creep redistribution of moments have occurred.

Figure 5.10 Stress distribution due to shear lag

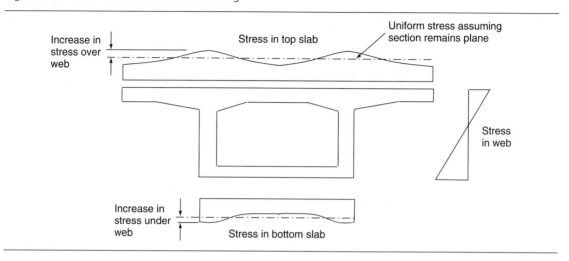

The unfactored nominal force in the tendons, after all associated losses are considered, is used when carrying out the serviceability limit state checks.

Where external tendons are used, the design of the deck should be checked for the situation whereby any one pair of the tendons are removed, lest it be necessary to replace them in the future. In this case, it is normal to consider a reduced live loading to reflect traffic management systems that may be implemented.

For a beam or box subjected to bending, the cross-section is assumed to remain plane with a linear variation in stress over the section depth and a constant stress at any one level. When flanged beams are subjected to high shear, such as adjacent to supports, the bending in the section can result in a non-uniform distribution of longitudinal stresses across the flanges, as illustrated in Figure 5.10. This effect, where the stresses in the flanges are higher over the webs than elsewhere, is known as shear lag. With prestressed concrete decks, the moments due to prestress, dead load and superimposed dead load tend to balance each other out and hence reduce the magnitude of any shear lag that occurs. An indication of this is the deflections that occur in a deck. With just the deck self-weight and the prestress applied then it is usual that the deck deflects upwards, and even with the superimposed dead load applied the resultant deflection in the deck is often still upwards. The moments in a deck are related to the deck curvature which creates the deck deflections, and the deflections are a general indication of the moments present. For this reason, shear lag is generally not significant in prestressed concrete bridge decks of standard dimensions.

When shear lag becomes significant the change in stress distribution around the section is quantified using a three-dimensional finite element analysis, or as described in the Cement and Concrete Association (C&CA – no longer in existence) Technical Report for box girders by Maisel and Roll (1974).

When analysing the structure to derive forces and deflections the effects of shear lag are usually neglected, although, with unusual arrangements such as very wide decks with widely spaced webs or long-edge cantilevers, shear lag may become more significant and should be considered in the analysis.

Figure 5.11 Effective flange width

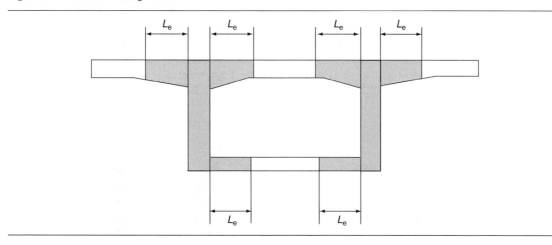

Eurocode 2 (BSI, 2004a), BS 5400 (BSI, 1990) and AASHTO (1998–2011) all include the concept of using reduced flange widths in the design of a section to simulate the effects of shear lag. By applying the moments and forces to a section with a reduced flange width the derived stresses are increased, in a manner similar to shear lag behaviour. The 'effective flange width' L_e, as illustrated in Figure 5.11, is given in BS 5400 (BSI, 1990) as the lesser of $0.1 \times$ distance between points of zero moment or the actual flange width either side of the webs. Eurocode 2 (BSI, 2004a) is similar but with an added width of $0.1 \times$ the clear distance between the webs. With continuous beams this results in a much smaller effective flange width near the support compared to mid-span, where the effects of shear lag are much less.

In AASHTO (1998–2011), the effective flange width is given as:

(i)   For precast beam and in situ slab, total effective width of an interior beam taken as the lesser of:
Total flange width $<0.25 \times$ span length of beam; or
$<12 \times$ flange thickness plus the greater of the web width or $1/2$ of the top flange width; or
$<$ average spacing of adjacent beams
(ii) For precast beam and in situ slab, with flange on one side only, width taken as the lesser of:
Total flange width $<0.125 \times$ span length of beam; or
$<6 \times$ flange thickness plus the greater of $1/2$ the web width or $1/4$ of the top flange width; or
$<$ width of overhang
(iii) For box girders,
Treated as a pair of beams; or
Full flange widths are taken, with shear lag determined by analysis

One of the problems with using effective flange widths is the difficulty in accurately accounting for the prestress effects. With post-tensioned construction, in the region beyond the anchor zone, the axial force will act on the complete deck section; while the primary and secondary moments will act to counter the dead load and may be applied to the reduced area. For this reason, the full analysis of a prestressed concrete bridge deck usually uses the full deck section rather than the effective flange width, with the effects of shear lag catered for by adjusting the stresses across the deck at the supports based on a 3D FE analysis of the structure.

Warping of thin-walled box girder decks under torsional and distortional behaviour causes redistribution of the longitudinal bending stresses at a cross-section. This behaviour is described in the C&CA Technical Report by Maisel and Roll (1974), which presents several methods to quantify the effect. Under asymmetrical and torsional loading the box section distorts and the cross-section undergoes out-of-plane displacements, which cause additional longitudinal stresses. When a full-length three-dimensional finite element model is used this behaviour is automatically included in the analysis.

Deflections and pre-camber

Deflections of the concrete deck occur due to self-weight and prestress, and from the weight of the permanently applied loads. Additional movements occur due to long-term creep of the concrete and losses in prestress. The deflections due to the prestress are often greater than, and in the opposite direction to, the dead-load deflections. This results in the deck deflecting upwards on application of the prestress and removal of the formwork. Subsequent application of the superimposed dead-load and long-term losses in the prestress force reduce this upwards deflection or can result in a net sag deflection.

The deck should be cast and erected so that the required profile is achieved on completion. Concrete creep will cause the deck to change its profile over its design life and it is usual to aim to achieve the desired alignment at the end of construction or at the time of bridge opening, although the long-term changes in deflections should be checked to ensure that they are not excessive.

The adjustment made to the profile during casting to achieve the desired shape is called the pre-camber, and it is governed by the overall design, the construction sequence and the concrete properties.

The balance between the dead loads and prestress usually results in small net deflections. Deflections are generally not significant for normally proportioned prestressed concrete decks designed with stresses within the allowable limits, and they do not need to be checked other than to confirm the pre-camber values to be catered for.

From a visual aspect, it is desirable for a span to appear to be 'hogging' rather than 'sagging'. To ensure this on straight decks a small upwards profile is often built into the deck along each span by adjusting the pre-camber. The upward profile imposed is usually only a few centimetres at mid-span.

Vibrations and fatigue in tendons

Vibrations from traffic or wind seldom create a problem for most prestressed concrete decks and it is not normally necessary to consider this in detail. The fluctuation in the direct stress in the tendons due to live and other loading is also very small, and fatigue is generally not critical. However, where external tendons are used, in certain circumstances vibrations of the individual tendons can occur due to traffic or deck movements.

The details and arrangements of external tendons should be developed so that they are not subjected to vibrations that are likely to give rise to fretting of the strand or bending stresses that could cause fatigue problems. This is achieved by ensuring that the frequency of the free length of tendons, between anchors or deviator points, is not close to the natural frequency of the deck or of the traffic using the bridge. Approximate values for frequencies can be taken as follows:

Frequency of the external tendon $= (\pi/L_T)\sqrt{(F/m_t)}H_z$

Frequency of the deck $= (k_f^2/(2\pi L_{sp}^2))\sqrt{(EI/m_d)}H_z$

k_f depends on span continuity and is π for simply supported spans and between 3 and 4.5 for continuous deck.

Vibration frequency of highway traffic is often taken as being between 1 Hz and 3 Hz. For rail traffic the frequency depends on train speeds and axle spacing, along with the bogey design and deck behaviour, and requires a complex rolling stock analysis to derive the vibrations that could be applied to the tendons.

To prevent vibrations it is usually sufficient to limit the free length of the tendon to 12 m or less, which should result in the tendon frequency being different from that of the deck and traffic.

Ultimate moment design

As the bending moment on a prestressed beam increases, the compression on one side goes up, while on the other side the concrete goes into tension. When the tensile strength of the concrete is exceeded, cracking will occur and the load is transferred either to the prestressing tendons or to any non-prestressed reinforcement present. As the moment increases further, the cracks in the concrete open and propagate across the section with a pure couple set up between the compression in the concrete, and tension in the tendon and the non-prestressed reinforcement. The maximum moment is reached when either the concrete or tendon and reinforcement fail.

When considering the ultimate moment of resistance the full width of the flanges are taken into account to resist the compressive forces generated. Shear lag effects are neglected.

The moment of resistance, or capacity, at a section is derived by comparing the balance of the tensile force in the tendons and reinforcement with the compressive force in the concrete. The strain distribution is assumed to be linear across the section, with the point of zero strain being the effective neutral axis.

On the compressive side, away from the neutral axis the stress in the concrete builds up rapidly until it reaches a peak of the compressive strength in bending, shown as $0.67f_{cu}$ in Figure 5.12(iii), before reducing slightly with increasing strain. A maximum strain in the concrete of 0.0035 is taken at failure, with this occurring at the extreme compression fibre.

On the tensile side, the stress in the tendons and passive reinforcement is a function of its distance from the neutral axis, while the concrete is assumed to have no tensile strength. Where several different layers of tendons and reinforcement are present, the strain, and hence the stress, in each is different. For the design of the completed structure the stress in the tendon, f_p, is considered after all long-term losses have occurred to give a 'worst case' situation.

The stress–strain relationships of concrete, prestress tendons and reinforcement are all well established. Typical curves for prestress tendons and reinforcement can be obtained from the material suppliers, while the behaviour of concrete is described in various publications, such as the book by Neville

(1995). Idealisation stress–strain curves for design purposes are given in Eurocode 2 (BSI, 2004a), Figures 3.3, 3.7 and 3.9, and in BS 5400 part 4 (BSI, 1990), Figures 1, 2 and 3.

The typical design stress and strain distributions across a section with internal, bonded tendons when subjected to bending are illustrated in Figure 5.12. The section is assumed to remain plane under bending. The initial stress and strain in the tendons, reinforcement and concrete is that generated by

Figure 5.12 Ultimate moment of resistance

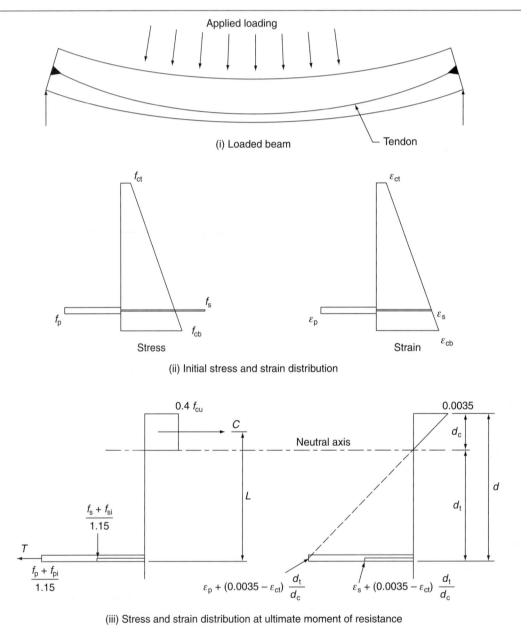

the prestress and permanent loading. As the applied loading increases, the stresses and strains change until they reach the ultimate design state indicated in Figure 5.12(iii).

To determine the ultimate moment of resistance or strength capacity of the section an iterative approach is used, with a first estimate of d_c taken, from which C and T are calculated. 'd' is then adjusted and C and T recalculated until $C = T$, giving the ultimate moment of resistance, $M_r = T \times L$. In practice, the stress/strain profile shown in Figure 5.12(iii) is usually simplified to a rectangular stress block. In BS 5400 part 4 (BSI, 1990) a uniform stress of $0.4f_{cu}$ is usually taken over the full depth to the neutral axis, while in Eurocode 2 (BSI, 2004a) the stress of $0.67f_{ck}$ is taken over a reduced depth of block equal to $0.8 \times d_c$ with a concrete strength of $50\,\text{N/mm}^2$, with the stress and depth of block decreasing as the strength of concrete increases.

The moment of resistance of any section along the deck must exceed the bending moment generated by the applied loads to give a sufficient factor of safety against failure. This is achieved by applying appropriate factors to reduce the calculated capacity of the section and to increase the effects of the applied loads.

With external tendons, unbonded from the concrete over long lengths, the change in the steel stress under ultimate loads is directly related to the increase in the tendon length between its fixed points as the deck deflects. The increase in strain in the unbonded tendon may be small depending on the position of the fixed points and the tendon layout. This limits the increase in stress in the tendon and ultimate failure is usually governed by the compression in the concrete which can give rise to a brittle failure mechanism. Estimate of the ultimate moment of resistance follows the same procedure for internal tendons as discussed above, although for long tendons installed over several spans $f_{pi} = 0\,\text{N/mm}^2$ is taken when the average strain increase is small. For short tendons anchored within one span, f_{pi} depends on the tendon layout and deck arrangement. For very short tendons, crossing the section being considered, the increase in stress can be up to $450\,\text{N/mm}^2$, which is similar to the increase that can be expected for internal, bonded tendons.

When a deck with external tendons deflects, the tendons will remain straight between their restraints and, at a section away from the anchors or deviators, the tendons move towards the neutral axis. This has the effect of reducing the strain in the tendon while also reducing the lever arm, L, decreasing the moment of resistance. This can become significant if the free length of external tendons is greater than a quarter of the span length and the tendons are unrestrained near the critical sections.

As with the serviceability stress check, a deck containing external tendons should have adequate capacity to carry a reduced live loading with any two of the tendons removed.

It is usually sufficient to check the ultimate moment only at the critical points along the deck, such as over each of the intermediate piers and at each mid-span.

A more detailed description on checking the ultimate moment and flexural strength to Eurocode 2 (BSI, 2004a), BS 5400 (BSI, 1990) and AASHTO (1998–2011) is given later in this chapter.

Shear design

The vertical loads on the deck impart a shear onto the section that must be catered for in the design. The resistance of any section to the imposed shears is complex and comprises of components from the uncracked concrete, the aggregate interlock across any cracked interface, the vertical or inclined shear reinforcement, and dowel action of any longitudinal reinforcement.

Prestressing tendons assist in resisting the shear in two ways: first by enhancing the shear-carrying capacity of the concrete; and, second, by providing a vertical component of force that counteracts the shear when the tendon is inclined in the appropriate direction. The prestress force considered in the design of the completed structure is that which remains after all long-term losses have occurred to give a worst-case scenario.

There are two types of shear failure associated with prestressed concrete beams: with the section either uncracked or cracked in flexure. Where the section is uncracked in flexure, shear failure is caused by web cracks forming where the principal stresses exceed the tensile capacity of the concrete. Where sections are cracked in flexure it is possible for the shear to combine with the flexural crack to result in failure. The capacity of the section under both types of failure is usually considered with the lowest value taken for the design.

Shear is only considered at the ultimate limit state: when the shear resistance of the section must exceed the shear forces applied to give an adequate factor of safety. The entire shear on a section is assumed to act on the webs, with only the web resistance taken into account in the design. The maximum allowable shear stress on a section is limited to keep the principal stress in the concrete to acceptable values, and this limit often governs the minimum width required for the webs.

Shear capacity at any section is the sum of the ultimate shear resistance of the concrete V_c, and the ultimate shear resistance of the reinforcement present, V_s. This can include both the non-prestressed reinforcement and any bonded prestressing tendon not being utilised at the ultimate limit state for other purposes.

Shear in the deck is checked adjacent to each pier and at regular intervals along each span, with reinforcement provided to give adequate resistance. Shear reinforcement is normally provided in the form of links or stirrups.

In long-span box girders, where the deck is haunched, the longitudinal bending in the deck gives rise to compression in the bottom slab which acts parallel to the soffit, as illustrated in Figure 5.13. This compression has a vertical component that acts against the shear forces and can be deducted from the applied shear force, V.

Allowance is made for the reduction in effective web width where internal post-tensioned tendons are placed within ducts in the webs. Traditionally, when using BS 5400 part 4 (BSI, 1990), before grouting the tendons the full duct diameter is deducted from the web width 'b', while after grouting two-thirds of the duct diameter is deducted. Eurocode 2 (BSI, 2004a), with grouted ducts a width of half the duct diameter, is deducted, while with ungrouted or grouted plastic ducts a width of $1.2 \times$ duct diameter is deducted.

When checking the ultimate shear capacity at the ends of pre-tensioned beams, allowance is made for the loss of prestress over its anchorage length, l_t, as defined in Chapter 6. It is normally sufficient to consider the prestress as varying linearly over this length when deriving V_c. If the shear resistance based on the beam being simply reinforced, and ignoring the prestress, is greater than that calculated, assuming a prestressed beam, then the higher value can be taken.

Figure 5.13 Contribution of haunched slab to shear resistance

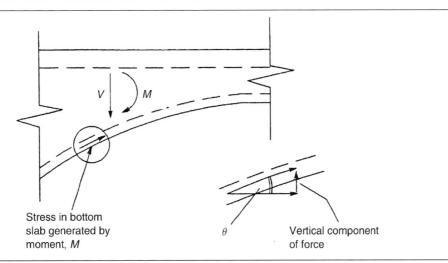

Stress in bottom
slab generated by
moment, *M*

θ

Vertical component
of force

With external tendons the shear capacity at any section should be adequate to carry a reduced live loading with any two tendons removed, to allow for future tendon replacement if necessary.

Later in this chapter the design procedures are described for shear when using Eurocode 2 (BSI, 2004a), BS 5400 (BSI, 1990) and AASHTO (1998–2011).

Torsion design

If the deck is subjected to non-symmetrical loading or is on a horizontal curve the twisting that can occur sets up a torsion within the prestressed concrete members. Torsion is normally considered at the ultimate limit state and does not generally govern the dimensions of the concrete section, but may require additional reinforcement to be provided.

In the webs the torsional stresses in the transverse plane can be added to the vertical shear stresses. The resultant stresses and associated force is designed for the ultimate limit state following the procedure for shear, but with additional longitudinal reinforcement provided, to cater for the longitudinal component of the torsion. In the flanges the torsional stresses are calculated and both transverse and longitudinal reinforcement provided as appropriate.

For rectangular sections, the torsional shear stress, v_t, is given by:

$$v_t = 2T_u/(h_{min}^2(h_{max} - h_{min}/3))$$

For box sections the stress is different where the width of webs and slabs varies, and at any particular point around the box is given by:

$$v_t = T_u/(2h_{wo}A_o)$$

With precast segmental decks, where no reinforcement passes through the joints, the concrete should be in compression at all times and the longitudinal tensile stresses from torsion need to be overcome by

Figure 5.14 Longitudinal shear

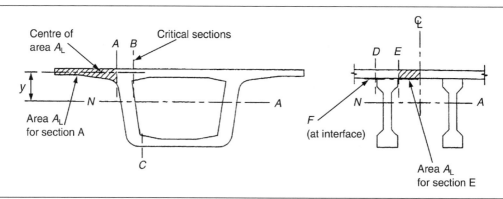

a residual longitudinal compression from the prestress. For a concrete box girder, the additional stress required is given by $T_u/(2h_{wo}A_o)$.

The procedures for checking the torsional stresses and calculating the reinforcement requirements to Eurocode 2 (BSI, 2004a), BS 5400 (BSI, 1990) and AASHTO (1998–2011) are given in the section 'Ultimate torsion', later in this chapter.

Longitudinal shear

As the bending moments change, the flow of stress through the section gives rise to longitudinal shear forces within the concrete which are checked at the slab-web interfaces, as indicated in Figure 5.14.

The longitudinal shear acting on section A is equal to the difference in the forces acting on the two ends of area A_L over the length considered. The difference in force is equal to the area multiplied by the change in longitudinal stress acting on the section, which is itself a function of the change in bending moment. This can be expressed in terms of the vertical shear force, with the ultimate longitudinal shear force per unit length given by:

$$V_L = V(A_L\, y/I)$$

where A_L and y refer to the area of concrete outside the section being considered.

Where the concrete is monolithic, or is able to transmit shear across the critical section, it generates some resistance to the longitudinal shear with reinforcement provided to give the remaining resistance.

The requirements for reinforcement to cater for the longitudinal shear when designing to Eurocode 2 (BSI, 2004a), BS 5400 (BSI, 1990) or AASHTO (1998–2011) are given in the relevant sections later in this chapter.

Partial prestressing

The design of fully prestressed structures is based on the concrete being uncracked under service loading: either with the stresses in compression under all loading conditions or with small tensions allowed under transient loading which are kept below the tensile strength of the concrete. For partially prestressed structures the principle is to allow the concrete to crack under service loading and to limit the crack widths to the normal allowable values for reinforced concrete. This is often related to a hypothetical allowable tension in the concrete of up to 6 or $7\,\text{N/mm}^2$ when carrying out the longitudinal stress check. Several countries, including Denmark and Australia, have successfully incorporated this approach in their bridge design standards.

The ULS or capacity checks of the moment and torsion are carried out in the normal manner, while for shear the concrete resistance is calculated as cracked in flexure.

A fatigue check of the prestress is carried out to ensure that the service stress fluctuation in the strand is not critical for the level of stress present. A check to ensure that the stress range is below $120\,\text{N/mm}^2$ is normally sufficient.

The advantages of partial prestressing include reduction in the quantity of prestress, full utilisation of non-prestressed reinforcement, smaller deflections due to prestress and reduced creep in the concrete. The disadvantages include the need for more non-prestressed reinforcement and a possibility that the durability of the structure will be reduced in harsh environments, although this has not currently been highlighted where partially prestressed structures have been used.

Construction sequence and creep analysis

The way a bridge is built affects the moments and shears generated in the structure, and this is taken into account during the design process. The structure is checked for strength, stability and the serviceability requirements at each stage of construction. The final moments and shears derived should reflect the construction sequence followed. Figure 5.15 illustrates the dead-load bending moments in

Figure 5.15 Creep redistribution of moments

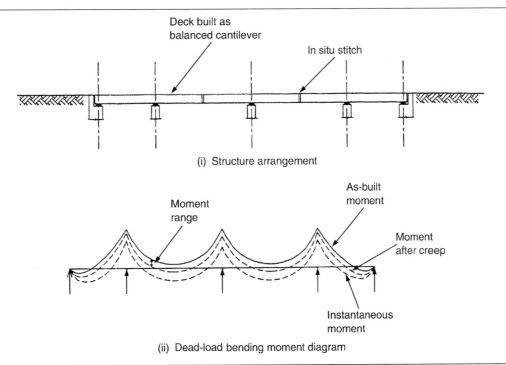

(i) Structure arrangement

(ii) Dead-load bending moment diagram

a four-span deck constructed in stages as a series of balanced cantilevers. At the end of the deck construction the bending moment diagram follows the cantilever moments with large hogging at the piers and little moment at mid-span. If the deck is built 'instantaneously', the balance between the hogging at the piers and the sagging at mid-span is more even. After completion of the construction the creep of the concrete adjusts the moments with the final moments being between the as-built moments and the moment if the deck was built instantaneously.

When the statical system of a concrete structure changes during construction, creep of the concrete will modify the as-built bending moments and shear forces towards the 'instantaneous' moment and shear distribution; the amount of the change being dependent on the creep factor, ϕ, of the concrete which is defined as:

$\phi = $ creep strain/elastic strain

The value of ϕ depends on the concrete constituents, mix proportions, section details, environmental conditions and the age of concrete, and typically range from 1.3 to 3 or more. For precast construction ϕ is normally around 1.6, while for in situ construction ϕ would normally be between 2.0 and 2.5. Eurocode 2 (BSI, 2004a) clause 3.1.4 and BS 5400 part 4 (BSI, 1990), Appendix C.2, gives the derivation of ϕ in more detail.

Where the change to the statical system is sudden, such as when connecting balanced cantilevers with a mid-span stitch or jacking up the deck, the modification to the moments is:

$M_{final} = M_{as\text{-}built} + (1 - e^{-\phi})(M_{inst.} - M_{as\text{-}built})$

Where the change is gradual, such as in the differential shrinkage between precast beams and in situ top slab or differential settlement between adjacent piers, the modification to the moments becomes:

$M_{final} = M_a(1 - e^{-\phi})/\phi$

Shears are modified in a similar way.

The effect of creep can be significant, as indicated by the following comparison between the ϕ factors:

ϕ	$(1 - e^{-\phi})$ sudden change	$1 - e^{-\phi}/\phi$ gradual change
1.6	0.80	0.50
1.8	0.83	0.46
2.0	0.86	0.43
2.2	0.89	0.40
2.4	0.91	0.38

Considering in situ concrete with a ϕ equal to 2.2 and using balanced cantilever construction with a 'sudden change' when the cantilevers are stitched together, the final moments will have crept from the as-built condition 89% of the way towards the instantaneous condition. Even with a gradual change, such as in long-term settlement of the piers, the final moments and shear built up in the deck are only 40% of the moments and shears that would have been generated if the change had occurred instantaneously.

Both the dead load and prestress secondary effects are affected by the stage-by-stage build-up of the deck and by the creep that occurs during the construction period and afterwards.

Temperature effects

Changes in the effective temperature of the deck will cause it to expand or contract, while differential temperature gradients through the concrete cause stress variations that must be considered in the design.

Where a deck is free to expand or contract, the overall change in effective temperature will not give rise to any forces in the structure, although the movement must be catered for in the bearing and expansion joint design. When a restraint exists that restricts free movement, such as integral decks or multiple fixed piers, forces will be generated throughout the structure. These forces can be determined by modelling the change in effective temperature as a strain differential in a frame analysis program.

Under a uniform temperature change, the change in deck length, $\Delta_L = \Delta_t \propto L$.

The co-efficient of thermal expansions, $\propto$, is taken as $12 \times 10^{-6}/°C$ with most aggregates or $9 \times 10^{-6}/°C$ with limestone aggregates.

Differential temperature effects are generated through the concrete section as the outer surface heats or cools more quickly than the rest. This was extensively researched by the TRL as reported by Emerson (1977), and from this profiles of the temperature gradient through a deck in different situations were developed. Eurocode 1, part 1–5 (BSI, 2003) section 6, BS 5400 part 2 (BSI, 2006), Figure 9, and AASHTO (1998–2011) Chapter 3 section 3.12.3, all include a number of simple temperature gradients that can be used with different structural types.

When a rapid increase in temperature occurs, for example during sunrise, the outer skin of the concrete heats up more quickly than the concrete core. As the outer skin tries to expand it is restricted by the core, which results in compressive stresses building up at the edges of the section and tensile stresses in the centre. Where there is a rapid decrease in temperature, such as at sunset, the outer concrete cools more quickly than the inner core, resulting in tensile stresses at the edges of the concrete and compressive stresses in the centre.

A typical simplified temperature gradient through a section is indicated in Figure 5.16(i). The profile of the gradient through a section is similar with both a positive or negative temperature change. The different temperatures through the concrete section cause different strains and therefore different stresses to be generated.

The stresses can be converted into equivalent forces and give rise to an out-of-balance axial force and moment, as indicated in Figure 5.16(iv) (a) and (b).

Provided the ends of the deck are not restrained, the force and moment will relieve at the free ends of the structure, as indicated in Figure 5.16(v), and the stresses will be adjusted accordingly along the deck

Figure 5.16 Stresses generated from temperature gradient

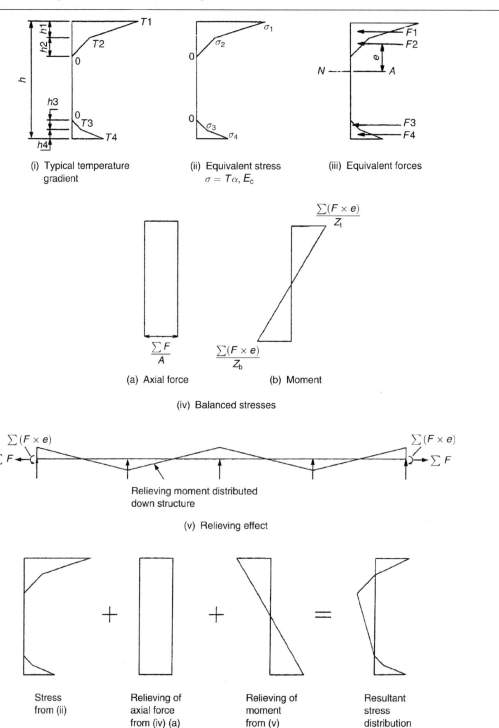

(i) Typical temperature gradient

(ii) Equivalent stress $\sigma = T\alpha, E_c$

(iii) Equivalent forces

$\dfrac{\sum F}{A}$

(a) Axial force

$\dfrac{\sum(F \times e)}{Z_t}$

$\dfrac{\sum(F \times e)}{Z_b}$

(b) Moment

(iv) Balanced stresses

$\sum(F \times e)$

$\sum F$

$\sum(F \times e)$

$\sum F$

Relieving moment distributed down structure

(v) Relieving effect

Stress from (ii)

+

Relieving of axial force from (iv) (a)

+

Relieving of moment from (v)

=

Resultant stress distribution

to give the resultant stress distribution across the section. If the ends of the deck are restrained, for example with integral structures, the moments are prevented from fully relieving.

The stress in the section is the equivalent stress from the temperature change, less the stress from the relieved equivalent axial force, plus the stress from the relieving of the out-of-balance moment. These stresses are included in the design when considering the serviceability stress check of the prestressed concrete.

The relieving of the out-of-balance moment, illustrated in Figure 5.16(v), generates shear forces in the spans, and these are considered during the ultimate shear design process.

Traditionally, only the differential temperature across the vertical plane – that is, between the top and bottom of the section, when considering the longitudinal direction – is taken into consideration in the design. Depending on the orientation of the deck and the position of the sun, differential temperatures can also occur between the two sides of a deck, giving rise to differential transverse effects. This has not been investigated in detail and no guidance is given in the BS 5400 part 2 (BSI, 2006) or AASHTO (1998–2011) design codes. Eurocode 1 (BSI, 2003) provides guidance applying a temperature gradient of 5°C between the two sides of the deck, although further research is required to quantify the true significance of this effect.

Eurocode 1 (BSI, 2003) also notes that significant differences in temperature can exist between outer and inner walls of a box girder and recommends a linear temperature difference of 15°C.

A further effect that can occur in a concrete deck is differential temperature through the slabs and webs, giving rise to stress, forces and moments in the transverse direction. There is no guidance on this given in any of the codes and it is frequently ignored in the design, although it can be significant in the transverse design, particularly when the deck is transversely prestressed. It has less effect in a totally reinforced concrete deck where the reinforcement in the transverse direction controls the local distribution of stresses caused by this differential temperature, while in a transversely prestressed deck the stresses set up need to be included in the stress checks on the sections. In this circumstance it would be reasonable to consider a linear distribution of temperature difference between the inside and outside faces of the slabs or webs of 15°C, although further research is needed to understand the precise distribution of temperature around and through concrete decks.

Concrete properties

The strength of concrete is specified in terms of its cube strength, f_{cu}, or its cylinder strength, f_c'. Cube strengths are determined from crushing $150 \times 150 \times 150$ mm samples, while with cylinders a 150 mm diameter by 300 mm-long sample is crushed.

The characteristic strength of the concrete is related to its crushing strength 28 days after casting; however, for prestressed concrete the strengths earlier than this and in the longer term are also important. Many factors affect the development of strength within the concrete, as described by Neville (1995), but for design purposes the strength at any time is often taken as $f_{ci} = e^{0.25[I - \sqrt{(28/t)}]} f_{cu}$, where t is the concrete age in days.

Some codes, such as the British Standard BS 5400 (BSI, 1990), base their design rules on the cube strength of the concrete, while others, such as the American AASHTO (1998–2011) standards and European Eurocode 2 (BSI, 2004a), use the cylinder strength of concrete. The relationship between

Figure 5.17 Cube versus cylinder strengths

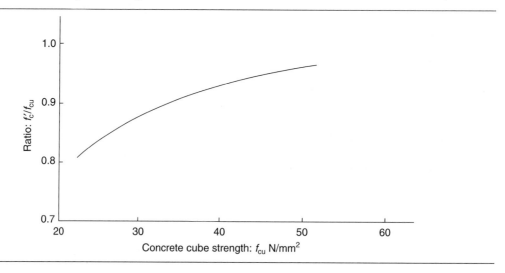

cube and cylinder strength varies depending on the concrete strength and its properties, but, as an approximate guide, the comparison given in Figure 5.17 can be taken.

It is common to consider the cylinder strength as being $0.8 \times$ the cube strength, while Eurocode 2 (BSI, 2004a) Table 3.1 provides a comparison between the two for design purposes.

The required strength of the concrete is determined by the compressive stresses generated in the concrete by the prestress and applied forces. The strength must be sufficient to ensure that the allowable stress limits exceed the stresses present. A minimum strength, f_{cu} of 45 N/mm^2, or f'_c of 35 N/mm^2, is typical for prestressed concrete; however, it is becoming more common to use higher strengths. Concrete with f_{cu} of 50 or 60 N/mm^2, or f'_c of 40 or 50 N/mm^2, are frequently used on projects. It is common on precast segmental viaducts, where high early strength of the segments is required for stripping purposes, for the concrete in the segments to achieve 28-day cubes strength of 70 N/mm^2; in fact, a recent project regularly achieved a cube strength of 100 N/mm^2. It is likely that concrete design strengths will continue to rise as quality controls improve and durability requirements lead to a denser concrete being achieved.

The rate of gain of strength of the concrete is important as this governs the time at which the prestress can be applied. At transfer of the prestress force to the concrete, it is normally required that the concrete strength f_{ci} be a minimum of 30 N/mm^2 (f'_{ci} of 25 N/mm^2). This requirement can vary depending on the tendon and anchor arrangement and the magnitude of force applied.

High early concrete strengths are required for precast segmental construction to be able to achieve a daily casting cycle for the segment production cell. Typical concrete strengths of 12 N/mm^2 are achieved after ten hours to allow stripping of the shutters, with 25 N/mm^2 achieved after 36 hours to allow the segments to be lifted and transported to the storage yards.

The modulus of elasticity of the concrete is related to the concrete strength, as well as being affected by the material properties of the aggregates used. The modulus also reduces as the stress on the concrete increases. For normal concrete, an approximate design value for the modulus can be taken as $4800\sqrt{f_{ci}}$. Typical values range from 28 000 to 36 000 N/mm^2 as the concrete strength, f_{ci}, increases from 30 to 60 N/mm^2, although these can vary by up to ±20%.

To minimise creep and shrinkage losses in the prestressing, the cement content and water/cement ratio of the concrete should be kept to a minimum, compatible with the high concrete strengths required. Admixtures are often used to improve the workability of the concrete and allow the water/cement ratio to be reduced.

Application of the prestress

With pre-tensioning the full prestress is applied to the concrete at an early age, while with post-tensioning the application of the prestress is often in stages and may occur over a period of some months.

Where possible, when using post-tensioning, it is advantageous to apply part of the prestress soon after concreting. The minimum concrete age when the prestress can be applied is restricted by the strength of concrete required to enable the anchorages to be loaded. However, stressing some of the tendons as soon as possible can help to limit the tendency for early thermal or shrinkage-induced cracking to occur. Early prestressing can be achieved by partially stressing some of the tendons to, say, 50% of their design load. By only applying part of the load to the anchor it is possible to apply the prestress with the concrete at a lower strength than that needed to apply the full load.

When a deck is prestressed it normally 'lifts up' along the span, reducing any load on the falsework supporting the concrete. In stage-by-stage construction, if the deck is cantilevering over a pier when it is prestressed the end of the cantilever is often deflected downwards and increases the load on the falsework beneath. To avoid overstressing the falsework the prestress can be applied in stages. The first-stage prestress should be just adequate to support the concrete dead load without causing any significant deflections. The falsework can then be removed after which the rest of the prestress can be installed.

Design procedures to Eurocode 2 (BSI: 2004a, 2005a, 2004b and 2005b)

The following section highlights specific design requirements when using Eurocode 2, supplementing the sections above. Eurocode 2 is based on a limit-state approach with the structure checked against criteria at both the serviceability limit state (SLS), beyond which the service criteria of the structure are no longer met, and the ultimate limit state (ULS) associated with collapse or other failure of the structure.

The Eurocode 2 documents consist of part 1.1 General Rules (BSI, 2004a) along with its national annex (BSI, 2004b), and part 2 Concrete Bridges (BSI, 2005a) along with its national annex (BSI, 2005b).

The imposed affects on a structure are referred to as actions within Eurocode, with both applied loads and imposed deformations being considered as permanent, variable or accidental actions to create persistent, transient or accidental design situations. The actions to be applied to the bridge are given in the various parts of Eurocode 1 (BSI, 1991).

At SLS the characteristic combinations of actions and the unfactored behaviour of the structure and material are considered, while at ULS partial factors on the actions, modelling and material are used to ensure that an adequate overall factor of safety is achieved. The partial factors, γ, applied to the actions for both the SLS and ULS conditions, are given in Eurocode 0 (BSI, 2002a, 2002b).

Design to Eurocode 2 (BSI, 2004a) is based on the 28-day concrete cylinder strength, f_{ck}.

Serviceability limit state stress limitations

The general approach within the Eurocodes is to limit the crack widths or to keep the concrete in compression adjacent to the prestressing tendon, depending on the loading and the environmental condition.

For bridges with bonded internal tendons in ordinary exposed locations, the area around the ducts is required to be kept in compression under frequent load conditions. The Eurocode (BSI, 2005a) part covering bridgeworks recommends that this compression is kept for a distance of 100 mm from all parts of the tendon or duct, while the UK National Annex (BSI, 2005b) sets the distance as the required cover based on the environmental conditions. Outside the decompression area, crack widths limits of 0.2 mm are acceptable. This is a change from the historic approach of designing prestressed concrete elements which limited the tensile stress in the concrete irrespective of where the tendons are, and is more rational in its approach to protecting the tendons, which are usually the critical structural element. The Eurocode (BSI, 2004a) can be compared with the partial prestressing approach adopted in some countries where hypothetical tensile stresses and allowable crack widths are used to maintain the durability of the structure.

For a bridge with bonded tendons in sheltered locations, or for inner surfaces such as in box girders, the decompression limit is required under quasi-permanent combination of loading only, with 0.2 mm crack widths allowed outside of that.

For bridges with unbounded external tendons, the concrete is treated as an ordinary reinforced member with crack widths of 0.3 mm allowed.

On the compression side of a member, the maximum compressive stress is limited to $0.6f_{ck(t)}$ unless confinement reinforcement or additional cover is provided.

The allowable stress in the prestressing steel for both pre-tensioned and post-tensioned tendons is given as:

Prior to seating (before lock-off)

Recommended values, least of: $0.8f_{pk}$ or $0.9f_{p0.1k}$

Overstressing permitted when stressing force measured to an accuracy of $\pm5\%$, with the maximum recommended value of $0.95f_{p0.1k}$

Stress in the tendon immediately after transfer (lock-off)

Recommended values least of: $0.75f_{pk}$ or $0.85f_{p0.1k}$

The allowable stress in the prestressing under Eurocodes is generally higher than that in other codes, which will give some saving in prestress quantities when designing bridges to the new codes. Caution should be taken if allowing the stress in the tendon to go above $0.8f_{pk}$ during the stressing operation, due to an increase in the risk of wire or strand breakage or slippage. Many prestressing specialist contractors are reluctant to stress tendons beyond $0.78f_{pk}$ during installation, bearing in mind that the stress in the parts of the tendon extending into the jack is 2% or 3% higher than the stress in the tendon at the anchorage.

When considering the serviceability limit state, Eurocode (BSI, 2004a) recommends that allowance be made for possible variations in the prestress force by adopting upper and lower bound factors for the characteristic values of 1.05 and 0.95 for pre-tensioned or unbonded tendons, and 1.1 and 0.9 for post-tensioned tendons. However, the UK National Annex (BSI, 2004b) recommends that the unfactored value of the prestress force derived from the analysis is used.

Ultimate moment resistance

The ultimate moment of resistance is derived by determining the failure condition of the section being considered, with the compression and tension chords being balanced by adjusting the height of the neutral axis, as indicated in Figure 5.12(iii).

When considering the ultimate moment of resistance under the normal persistent and transient situations, material partial factors γ_c and γ_s are applied with 1.5 for concrete and 1.15 for prestress tendons and reinforcement. Under accidental situations such as vehicle impact, the factors become 1.1 and 1.0 respectively.

For concrete, Eurocode (BSI, 2004a) recommends the maximum design compressive strength, f_{cd}, of $0.67f_{ck}$, while the UK National Annex (BSI, 2004b) adopts $0.57f_{ck}$ which is equivalent to $0.45f_{cu}$. This design compressive strength is then applied to the section as a uniform stress over a reduced depth of compression. The maximum strain in the concrete is limited to 0.0035. Referring to Figure 5.12(iii), the stress in the concrete is represented by a rectangular stress block, with the stress being f_{cd} for concrete of strength up to f_{ck} of 50 N/mm^2 and $[1 - (f_{ck} - 50)/200] \times f_{cd}$ for concrete strengths up to f_{ck} of 90 N/mm^2. The depth of the rectangular stress block is taken as $0.8 \times d_c$ for concrete of strength up to f_{ck} of 50 N/mm^2 and $[0.8 - (f_{ck} - 50)/400] \times d_c$ for concrete strengths up to f_{ck} of 90 N/mm^2.

The stress in the prestressing and reinforcement on the tension side of the element subjected to the moment is derived from the consideration of strain compatibility with the adjacent concrete, with the maximum stress limited to the characteristic strength divided by the material factors noted above. For prestressing steel, the stress–strain profile can be simplified with a constant horizontal profile of $0.87f_{p0.1k}$ after that stress is reached; similarly, with the reinforcement present when it reaches a stress of $0.87f_{yk}$.

With external, unbounded, tendons the increase in stress in the tendon is determined by considering the overall deformation of the structure rather than the strain compatibility at any particular section. The overall length increase of each tendon is determined and the stress increase averaged out over the length of the tendon. Eurocode (BSI, 2004a) allows the simplified approach of adopting an increase in stress in the unbounded tendons of 100 N/mm^2 without considering the overall structure deformations in detail, although this may not be conservative in some circumstances, such as long straight tendons passing through several spans.

Ultimate shear resistance

The presence of the prestress force on the section increases the shear resistance of the concrete and where the tendons are inclined the vertical component of the prestress usually counters the applied shear force.

The vertical component of force from inclined prestress tendons is treated as part of the load. Eurocode (BSI, 2004a) recommends a favourable partial factor of 1.0, while the UK National Annex (BSI, 2004b) recommends 0.9, reducing the prestress force to take into account possible variations in the losses in the tendon.

In deriving the shear resistance of the concrete without considering the reinforcement, the prestress increases the resistance by $k_1\sigma_{cp}b_wd$ to give the total resistance of the concrete as:

$$V_{Rd,c} = [C_{Rd,c}k(100\rho_1 f_{ck})^{1/3} + k_1\sigma_{cp}]b_wd$$

with a minimum of $V_{Rd,c} = (v_{min} + k_1\sigma_{cp})b_wd$.

The definitions and values for $C_{Rd,c}$, k, ρ_1 and v_{min}, which make up the plain concrete resistance, can all be found in Eurocode (BSI, 2004a):

k_1 is recommended as 0.15

σ_{cp} is limited to $0.2f_{cd}$

For sections uncracked in bending, defined as when the flexural tensile stress in the concrete is less than $f_{ctk,0,05}/\gamma_c$, the shear resistance is given by:

$$V_{Rd,c} = [Ib_w/S] \times \sqrt{[(f_{ctd})^2 + \alpha_1\sigma_{cp}f_{ctd}]}$$

$\alpha_1 = 1.0$ for post-tensioned tendons and pre-tensioned tendons outside of the transmission zone, and the ratio of the distance from the start of the transmission to the upper bound transmission length within the transmission zone.

This uncracked shear resistance limit is not applied to sections nearer to a support than where a 45° line drawn from the edge of the support intersects the section neutral axis.

The ultimate shear resistance at any section requiring shear reinforcement is derived using a truss analogy, with a similar approach adopted for prestressed concrete as for reinforced concrete and just the formulae applied adopted to suit. The shear is taken on a compression strut in the concrete which is assumed to be inclined between 45° and 22° to the horizontal, with tension and compression chords in the concrete. Where reinforcement is required to increase the shear resistance this usually takes the form of vertical bars placed between the tension and compression chords.

Where shear reinforcement is provided in bridge structures it is usually perpendicular to the member axis, and in the formulae given below the angle of the shear reinforcement has been taken as 90°.

The design resistance of the reinforcement to be added to the shear resistance of the concrete is given as follows:

$$V_{Rd,s} = (A_{sw}/s)zf_{ywd} \cot \theta$$

$$Z = 0.9d$$

θ = assumed angle of compression strut

From this it can be seen that the smaller the angle of the compression strut with the horizontal the greater the resistance of the reinforcement provided, as there are more legs of the links crossing the potential failure plane. This results in a reduction in the vertical reinforcement required, but is offset by an increase in the reinforcement to cater for the additional longitudinal force, ΔF_{td}, associated with the applied shear force which is given by:

$$\Delta F_{td} = 0.5V_{Ed} \cot \theta$$

The maximum tensile force in the longitudinal reinforcement is limited to that required to cater for the maximum bending moment along the deck. At the point of maximum moment the shear force is minimal.

Where the concrete resistance is adequate to cater for all the applied shear, a minimum percentage of shear reinforcement equal to $(0.08\sqrt{f_{ck}})/f_{yk}$ is required.

The maximum shear force on any section must be limited to prevent crushing of the concrete in the compression strut, with the maximum shear resistance of the section, $V_{Rd,max}$, given by:

$$V_{Rd,max} = \alpha_{cw}b_w z v_1 f_{cd}/(\cot \theta + \tan \theta)$$

α_{cw} and v_1 coefficients are given in Eurocode (BSI, 2005a), but are modified in the UK National Annex (BSI, 2005b).

It can be seen that if a shallow angle of the compression strut is chosen this will reduce the maximum shear resistance the section can generate. The choice over the angle of the compression strut, between 45° and 22°, will depend on the governing factors in the section. Using a higher angle will allow thinner webs to be adopted in the section, while lower angles will reduce the amount of shear reinforcement needed.

Eurocode (BSI, 2004a) does not differentiate between internal and external tendons and applies the same shear design approach to both types of prestressing.

Ultimate torsion resistance

The torsional resistance is considered on the basis of a thin-walled closed section. Solid sections are represented by equivalent thin-walled closed sections, while open beam sections are considered as a series of thin-walled sections, totalled to get the overall torsional resistance:

The shear stress in any wall is given by $T_{ed}/(2A_k t_{ef,1})$

The shear force, $V_{ed,l}$, in any wall is given by $\tau_{t,i} t_{ef,1} z_i$

The effects of both the shear and torsion are combined and the design carried out as for the ultimate shear resistance.

The additional longitudinal reinforcement per metre around the perimeter required for the torsion is given by:

$$A_{sl} = T_{ed} \cot \theta / (2 A_k f_{yd})$$

This additional longitudinal reinforcement is added to the other tension reinforcement required, but can be reduced in the compression areas of the section subject to the compressive force available.

When considering the maximum capacity of a section, the torsion and shear resistance are considered together such that:

$$T_{Ed}/T_{Rd,max} + V_{Ed}/V_{Rd,max} \leqslant 1.0$$

For a closed thin-walled box structure, $T_{Rd,max}$, the design torsional resistance moment is given by:

$$T_{Rd,max} = 2 v \alpha_{cw} f_{cd} A_k t_{ef,I} \sin \theta \cos \theta$$

v and α_{cw} are those values adopted in the shear design.

Longitudinal shear

Eurocode recognises the need to design for the shear along the interface between the webs and flanges of a section, as indicated in Figure 5.14. The shear along the interface is derived by considering the change in the normal force on the flange along the element. By calculating the compressive stress on the flange at the ends of a selected length of deck, the shear force can be determined and the shear stresses checked.

The shear reinforcement required across the interface is given by the shear force divided by $(\cot \theta_f \times f_{yd})$. The angle θ_f is the chosen angle of the compression strut in the flanges and is limited to between 45° and 26.5° for compression flanges, and between 45° and 38.6° for tension flanges.

Design procedures to BS 5400 (BSI, 1990)

The following section highlights specific design requirements when using BS 5400 (BSI, 1990), supplementing the sections above. BS 5400 (BSI, 1990) is based on a limit-state approach with the structure checked against criteria at both the serviceability limit state (SLS) and the ultimate limit state (ULS).

At SLS, the factored nominal forces and unfactored behaviour of the structure are considered, while at ULS load factors and material factors are used to ensure that an adequate overall factor of safety is achieved.

The load factors, γ_{fl} applied for both the SLS and ULS under different load combinations, are given in BS 5400 part 2 (1990) (BSI, 2003) and modified in the UK by the Highways Agency's standard BD 37/01 (Highways Agency, 2001). An additional analysis factor, γ_{f3}, is applied to increase the loading effects and is 1.0 for the SLS and 1.1 for the ULS when elastic methods are used in the analysis of the structure.

Design to BS 5400 (BSI, 1990) is based on the 28-day concrete cube strength, f_{cu}.

Serviceability limit state stress check
In the UK, prestressed concrete bridge decks are usually designed to Class 1 under permanent loads and normal live loading and to Class 2 for all other load combinations.

The allowable stresses in the concrete under service load are:

Compressive stresses: $0.5f_{ci} \leqslant 0.4f_{cu}$ with triangular stress distribution or
$0.4f_{ci} \leqslant 0.3f_{cu}$ with uniform stress distribution

Tensile stresses: $0 \, \text{N/mm}^2$ with permanent loads and normal live load
$0.45\sqrt{f_{cu}}$ with full loading and other effects on pre-tensioned members
$0.36\sqrt{f_{cu}}$ with full loading and other effects on post-tensioned members

Allowable tensile stresses during construction are:

$1 \, \text{N/mm}^2$ during erection with prestress and co-existent dead and temporary loads
$0.45\sqrt{f_{cu}}$ for pre-tensioned members under any loading
$0.36\sqrt{f_{cu}}$ for post-tensioned members under any loading

The allowable stress in the prestressing steel after anchoring the tendon is $0.7f_{pu}$ for post-tensioned tendons and $0.75f_{pu}$ for pre-tensioned tendons; however, the jacking force may be increased to give stresses up to $0.8f_{pu}$ during stressing.

Ultimate moment resistance
The stress in the concrete, prestress tendons and reinforcement is reduced by a partial factor for strength, or material factor, γ_m, which is 1.5 for concrete and 1.15 for prestress tendons and reinforcement. Referring to Figure 5.12(iii), the stress in the concrete becomes a maximum of $0.45f_{cu}$ and the stresses in the tendons and reinforcement are divided by 1.15.

When calculating the ultimate moment of resistance in the case of rectangular sections, or flanged sections with the neutral axis in the flange, the compressive stress in the concrete is taken as an average of $0.4f_{cu}$ over the full depth of compression, with the concrete strain limited to 0.0035 at the outermost fibre. The increase in stress in the tendon, f_{pi}, and non-prestressed reinforcement, f_{si}, is derived by considering the change in strain caused by the bending within the beam and is obtained by reference to the tendon stress–strain curves given in Figures 2 and 3 in BS 5400 part 4 (BSI, 1990), with $\gamma_m = 1.15$.

To determine the ultimate moment of resistance, the compressive force in the concrete, C, and the tensile force in the tendons and reinforcement, T, are calculated as follows:

$C = 0.4f_{cu}d_cb_s$

$T = A_t(f_p + f_{pi})/1.15 + A_p(f_s + f_{si})/1.15$

As described in an earlier section, the position of the neutral axis is determined to give the situation where the forces C and T balance, with the ultimate moment of resistance being the moment generated between them.

Where the neutral axis lies within the web of a flanged section the compressive force is calculated by considering the components of force generated in both the flange and the web above the neutral axis. The compressive stresses are based on the stress–strain curves with the material factor applied.

If the ultimate moment of resistance is less than 1.15 times the applied ultimate moment then the strain in the tendon, at the centroid of the outermost 25% of the total tendon area, should be at least $0.005 + f_{pu}/(E_s\gamma_m)$. This is to ensure that the section is 'under-reinforced', with the tendons yielding significantly before failure of the concrete in compression.

Ultimate shear resistance

With internal tendons, the ultimate shear resistance of the concrete V_c is the lesser of the ultimate shear resistance of a section uncracked in flexure, V_{co}, and the ultimate shear resistance of a section cracked in flexure, V_{cr}:

$$V_{co} = 0.67bh\sqrt{(f_t^2 + f_{cp}f_t)} + V_p$$

$$V_{cr} = 0.037bd\sqrt{f_{cu}} + (M_{cr}/M)V$$

where:

$$M_{cr} = (0.37\sqrt{f_{cu}} + f_{pt})I/y$$

The values for V_p, f_{cp} and f_{pt} are based on the prestressing forces after all losses have occurred and have been multiplied by the partial safety factor, $\gamma_{fL} = 0.87$. Where the vertical component of prestress, V_p, resists shear at the section being considered, it is ignored when calculating V_{cr}.

The maximum shear force V is limited to $0.75\sqrt{f_{cu}}bd$, but no greater than 5.8 N/mm^2, to prevent excessive principal stresses occurring in the concrete. With the use of concrete of strength significantly higher than 60 N/mm^2, the limit of 5.8 N/mm^2 appears to be conservative, although further research and testing are needed to justify increasing this limit.

BS 5400 part 4 requires that minimum shear reinforcement is provided such that:

$$(A_{sv}/S_v)(0.87f_{yv}/b) = 0.4 \text{ N/mm}^2$$

When V exceeds V_c shear reinforcement in the form of vertical links is required such that:

$$(A_{sv}/S_v) = (V + 0.4bd_t - V_c)/(0.87f_{yv}d_t)$$

Longitudinal reinforcement is provided in the tensile zone with:

$$A_{sl} \geqslant V/2(0.87f_{yl})$$

This quantity can include both the non-prestressed reinforcement and any bonded prestressing tendon not being utilised at the ultimate limit state for other purposes.

The shear reinforcement requirements are based on a truss analogy with the longitudinal reinforcement or prestress needed to form the tensile chord in the truss. Clark (1984) published a paper arguing that the above formula for the longitudinal reinforcement was in error and showing that it should be:

$$A_{sl} \geqslant \cot \theta V / 2(0.87f_{yl})$$

where θ is the angle of the assumed included strut in the concrete. As the point being considered gets closer to the support, θ increases and the required A_{sl} reduces. With prestressed concrete decks, where the prestress tendons are usually anchored on the ends of simply supported decks or continue fully over the intermediate supports for continuous decks, there is normally no need for additional longitudinal reinforcement for the shear action.

In the UK, the Highways Agency's standard BD 58/94 (Highways Agency, 1994) requires that sections with unbonded external tendons be designed as reinforced concrete columns subjected to an externally applied load. The shear stress on the section is given by:

$$v = (V - 0.87V_p)/(bd) \text{ where } v \text{ should not exceed the lesser of } 0.75\sqrt{f_{cu}} \text{ or } 5.8 \text{ N/mm}^2$$

The reinforcement required, A_{sv}, is given by:

When $v \leqslant \xi_s v_c$ $A_{sv} \geqslant 0.4bs_v/0.87f_{yv}$

When $v > \xi_s v_c$ $A_{sv} \geqslant bs_v(v + 0.4 - (1 + 0.05(0.87P)/A_c)\xi_s v_c)/0.87f_{yv}$

v_c and ξ_s are given in BS 5400 part 4, Tables 8 and 9 respectively.

The resulting reinforcement requirements using this method appear overly conservative when compared to other international design codes, and further testing and research is needed to justify the application of these formulae with external tendons.

Ultimate torsion

For torsion, it is necessary to calculate the torsional shear stresses generated by the ultimate loads and where these stresses exceed $v_{t\,min}$, as given in Table 10 of BS 5400 part 4 (BSI, 1990), reinforcement is provided by means of transverse links and longitudinal bars. The stresses generated by torsion and shear are added together and checked to ensure that the sum does not exceed $0.75f_{cu}$ and is less than 5.8 N/mm^2.

For rectangular sections, torsional reinforcement is provided such that:

$$A_{st}/S_v \geqslant T_u/(1.6x_i \, y_i(0.87f_{yv}))$$

$$A_{st}/S_L \geqslant (A_{st}/S_v)(f_{yv}/f_{yl})$$

where T and I sections are used the torsion is considered as acting on the individual rectangular elements, with the section divided up to maximise the sum of $(h_{max}h_{min}^3)$ of each rectangle. Each rectangle is then designed to carry a proportion of the torsion based on its value of $(h_{max}h_{min}^3)$ in relation to the sum of the values for all the rectangles. The reinforcement is determined as for normal rectangular sections and detailed to tie the individual rectangles together.

For box sections, reinforcement is provided such that:

$$A_{st}/S_v \geqslant T_u/(2A_o(0.87f_{yv}))$$

$$A_{st}/S_L \geqslant (A_{st}/S_v)(f_{yv}/f_{yl})$$

Where a part of the section is in compression the compressive force may be used to reduce the A_{sl} required.

This reduction is given by:

$$A_{sl}/S_L = (f_c h_{wo})/(0.87f_{yl})$$

Longitudinal shear

To cater for the longitudinal shear stresses at the critical sections, BS 5400 requires that sufficient reinforcement cross the plane to ensure that V_L is not greater than the lesser of either:

(a) $k_1 f_{cu} L_S$

or

(b) $v_L L_S + 0.7 A_r f_y$

Values for v_L and k_1 are given in BS 5400 part 4 (BSI, 1990), Table 31. For precast beam-and-slab construction, a minimum area of reinforcement of 0.15% of the contact area is required across the interface between the beam and slab. BS 5400 part 4 allows reinforcement that is provided for other purposes to be utilised to resist the longitudinal shear, as long as it is fully anchored beyond the section being considered.

Partial prestressing

Classified in BS 5400 part 4 (BSI, 1990) as Class 3, partial prestressing of concrete is not usually adopted on bridgeworks in the UK.

When using BS 5400 to design a partially prestressed deck, the hypothetical SLS stresses are given in part 4, Table 25. The ultimate moment and torsion design is carried out in the normal manner and for ultimate shear the concrete resistance is calculated as cracked in flexure, with:

$$V_{cr} = (1 - 0.55(f_{pe}/f_{pu}))v_c bd + M_o V/M$$

where $M_o = f_{pt} I/y$.

Any vertical component of prestress is ignored.

f_{pe} and f_{pt} are based on the prestressing force after all losses have occurred, and then multiplied by the partial safety factor $\Upsilon_{fl} = 0.87$. f_{pe} should be $\leqslant 0.6f_{pu}$.

v_c is given in BS 5400 part 4 (BSI, 1990), Table 8, in which the value of A_s is the total area of prestressed and non-prestressed reinforcement in the tensile zone.

Design procedures to AASHTO standard specifications (AASHTO, 1998–2011)

The following section highlights specific requirements for the design when using AASHTO (1998–2011), and it supplements the earlier general sections on the different design aspects of prestressed concrete bridges. The AASHTO (1998–2011) standard specification provides the basic requirements, but may be supplemented by other documents such as the Guide Specification for the Design and Construction of Segmental Concrete Bridges (AASHTO, 1999).

AASHTO (1998–2011) is based on a load and resistance factor design (LRFD) method, where the loads are increased by suitable factors and the material resistance reduced by appropriate factors to reflect the statistical nature of loads and structural performance.

With the LRFD approach, individual load factors and strength capacity reduction factors, ϕ, are used to ensure that an adequate overall factor of safety is maintained. The loading adopted in AASHTO (1998–2011) and the load factors for the different load combinations are given in section 3, while the strength reduction factors are given in the individual design sections, prestressed concrete being under section 5.

In the analysis and design, the concrete section is assumed to behave with a linear elastic strain distribution over its depth and with stresses linearly proportional to the strains. After the concrete has cracked the tension in the concrete is ignored.

Design to the AASHTO (1998–2011) specifications is based on the 28-day cylinder strength of concrete, f_c.

AASHTO (1998–2011) no longer recognises the use of dry jointed precast segmental construction, with clause 5.5.4.2.2 stating that joints between precast units shall be either cast-in-place closures or match-cast and epoxied joints.

Service limit state

Allowable stresses in the concrete under service load are:

Temporary stresses before losses due to creep and shrinkage

Compression:
Pre-tensioned concrete $\leqslant 0.60 f'_{ci}$
Post-tensioned concrete $\leqslant 0.60 f'_{ci}$

Tension for bridges other than segmental construction:
Pre-compressed tensile zone   none specified
Tension areas with no bonded reinforcement   1.38 N/mm^2 or $\leqslant 0.25\sqrt{f'_{ci}}$
Where tensile stress exceeds this, bonded reinforcement to be provided to cater for the tensile force in the concrete assuming an uncracked section, maximum tensile stress $\leqslant 0.63\sqrt{f'_{ci}}$

Tension for segmental bridges:
At joints with minimum bonded reinforcement through
the joint, sufficient to carry the tensile force, maximum tensile stress $\leqslant 0.25\sqrt{f'_{ci}}$
At joints and other areas without minimum bonded reinforcement
through the joint   No tension

At other areas with bonded reinforcement to be provided to cater for the tensile force in the concrete assuming an uncracked section, maximum tensile stress $\leqslant 0.50\sqrt{f'_{ci}}$

Principal tensile stress at neutral axis in web $\leqslant 0.289\sqrt{f'_{ci}}$

Stress at service load after losses have occurred:

Compression
All load combinations $\leqslant 0.60 f'_c$ (reduced if section slender) except
Prestress + Dead load $\leqslant 0.45 f'_c$
Live load + 50% (Prestress + Dead Load) $\leqslant 0.40 f'_c$

Tension for bridges other than segmental construction:

Pre-compressed tensile zone:
    members with bonded reinforcement* $0.50\sqrt{f'_c}$
    severe exposure conditions $0.25\sqrt{f'_c}$
    members with unbonded tendons 0
    * includes bonded prestressing strands

Tension for segmental bridges:
At joints with minimum bonded reinforcement through the joint,
sufficient to carry the tensile force, maximum tensile stress $\leqslant 0.25\sqrt{f'_{ci}}$
At joints and other areas without minimum bonded reinforcement
through the joint   No tension
At other areas with bonded reinforcement to be provided to cater
for the tensile force in the concrete assuming an uncracked section,
maximum tensile stress $\leqslant 0.50\sqrt{f'_{ci}}$

Principal tensile stress at neutral axis in web $\leqslant 0.289\sqrt{f'_{ci}}$

The allowable stress in the prestressing steel is as follows:
   Pre-tensioned members, prior to transfer
      Low-relaxation strands $0.75 f'_s$
      Stress-relieved strand $0.70 f'_s$
   Post-tensioned members
      At anchorage and couplers after lock-off $0.70 f'_s$
      Elsewhere after lock-off $0.74 f'_s$
      For short periods prior to lock-off $0.90 f^*_y$
      At service load after losses $0.80 f^*_y$

Flexural strength

AASHTO (1998–2011) requires that the factored moment at a section, M, be greater than the nominal moment strength, M_r, reduced by the appropriate strength capacity reduction factor, that is, $M \leqslant \phi M_r$, where ϕ is as follows:

Conventional construction

   Tensioned controlled prestressed concrete members $\phi = 1.00$

Segmental construction

> Fully bonded tendons $\phi = 0.95$
> Unbonded or partially bonded tendons $\phi = 0.90$

By varying the different ϕ values, AASHTO (1998–2011) recognises the comparative performances between internal, fully bonded and external tendons, and with conventional and segmental construction.

The nominal moment strength of a section can be calculated following the procedure outlined earlier in the chapter. AASHTO (1998–2011) provides guidance on estimating the stress in the prestressing tendons at the flexural resistance, along with formulae for deriving the flexural strength with flanged and rectangular sections.

Shear capacity

The factored shear force at a section V should be less than $\phi(V_c + V_s)$, with V_c and V_s the nominal shear strengths of the concrete and of the reinforcement respectively.

AASHTO (1998–2011) allows the use of a simplified approach where V_c is the lesser of the capacities of the concrete cracked or uncracked in flexure, with the shear strengths given as:

Uncracked in flexure $V_{cw} = bd(0.16\sqrt{f_c'} + 0.3f_{pe}) + V_p$

Cracked in flexure $V_{ci} = 0.05bd = \sqrt{f_c'} + V_d + (M_{cr}/M)V$

where:

$$V_{ci} \geqslant 0.16bd\sqrt{f_c'}$$

$$M_{cr} = (I/y_t)(0.5\sqrt{f_c'} + f_{pe} - f_d)$$

M and V are the factored moment and co-existent shear from the load combination causing maximum moment at the section:

$V_s = A_{sv}f_{yv}d/S_v$ when the shear reinforcement is vertical

Alternatively, AASHTO (1998–2011) provides a more sophisticated approach based on the truss analogy to derive the shear resistance, which generally results in more efficient sections and reduced shear reinforcement requirements.

The strength reduction factor, ϕ, is given below:

> Conventional construction
> tensioned controlled prestressed concrete members $\phi = 0.90$

> Segmental construction
> Fully bonded tendons $\phi = 0.90$
> Unbonded or partially bonded tendons $\phi = 0.85$

AASHTO treats shear design in the same way for both internal and external tendons when calculating V_c and the shear reinforcement to be provided. With external tendons the vertical component of the prestress is considered as an applied load and reduces the applied shear where the deviators are in the outer third of the span.

REFERENCES

AASHTO (1998–2011), Load and resistance factor design (LRFD) bridge design specifications, 4th edition. AASHTO, Washington, DC.

AASHTO (1999) Guide specification for design and construction of segmental concrete bridges, 2nd edition. AASHTO, Washington, DC.

BSI (1990) 5400: Steel, concrete and composite bridges. Part 4. Code of practice for the design of concrete bridges. BSI, London.

BSI (1991) BS EN 1991: Eurocode 1: Actions on structures. BSI, London.

BSI (2002a) EN 1990:2002: Eurocode 0: Basis of structural design, incorporating amendment no. 1. BSI, London.

BSI (2002b) NA to BS EN 1990:2002: UK national annex for Eurocode 0: Basis of structural design. BSI, London.

BSI (2003) BS EN 1991–1-5:2003: Eurocode 1 incorporating corrigendum no. 1: Actions on structure. Part 1–5. General actions – thermal actions. BSI, London.

BSI (2004a) BS EN 1992–1-1:2004: Eurocode 2: Design of concrete structures. Part 1–1. General rules and rules for buildings. BSI, London.

BSI (2004b) NA to BS EN 1992–1-1:2004: UK national annex to Eurocode 2: Design of concrete structures. Part 1–1. General rules and rules for buildings. BSI, London.

BSI (2005a) BS EN 1992–2:2005: Eurocode 2: Design of concrete structures. Part 2. Concrete bridges – design and detailing rules. BSI, London.

BSI (2005b) NA to BS EN 1992–2:2005: UK national annex to Eurocode 2: Design of concrete structure. Part 2. Concrete bridges – Design and detailing rules. BSI, London.

BSI (2006) BS 5400: Steel, concrete and composite bridges. Part 2. Specification for loads. BSI, London.

Clark LA (1984) Longitudinal shear reinforcement in beams. *Concrete* **18(2)**: 22–23.

Emerson M (1977) TRRL Laboratory report 765: Temperature difference in bridges: Basis of design requirements. Transport and Road Research Laboratory, Crowthorne.

Garrett RJ and Cochrane RA (1970) The analysis of prestressed concrete beams curved in plan with torsional restraints at the supports. *The Structural Engineer* **3**: 128–132.

Highways Agency (1994) (MDRB) Departmental standard BD 58/94: The design of concrete highway bridges and structures with external and unbonded prestressing. Norwich, HMSO.

Highways Agency (2001) (MDRB) Departmental standard BD 37/01: Loads for highway bridges. HMSO, Norwich.

Magura DD, Sozen MA and Siess CP (1964) A Study of Stress Relaxation in Prestressing Reinforcement. *PCI Journal* **9(2)**.

Maisel BI and Roll F (1974) Methods of analysis and design of concrete box beams with side cantilevers. C&CA, London.

Neville AM (1995) *Properties of Concrete*, 4th edition. Longman, London.

Witecki AA (1969) Simplified method for the analysis of torsional moment as an effect of a horizontally curved multispan continuous bridge. *American Concrete Institute, First International Symposium on Concrete Bridge Design*: 193–204.

Prestressed Concrete Bridges, 2nd edition
ISBN: 978-0-7277-4113-4

ICE Publishing: All rights reserved
doi: 10.1680/pcb.41134.111

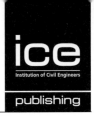

publishing

Chapter 6
Design of details

Introduction

Prestressed concrete has a number of general features, such as prestressing components, anchorages, couplers and tendon ducts, while other details such as deviators, blisters and diaphragms are common elements. This chapter looks at the design and detailing of these different features.

Where formulae are used in this chapter the units are in N and mm.

Anchorages

All prestressing tendons are anchored against or within the concrete. These anchorages exert large concentrated forces on the concrete and prestressing components, and give rise to high local stresses that must be catered for in the design. Anchorage of pre-tensioned strand is achieved by a bond between the individual wires and the concrete, while for post-tensioning the anchorage is with anchor blocks bearing against the concrete or by embedment of cast-in dead-end anchors within the concrete.

Pre-tensioned strand

When pre-tensioned strand is released from the jacking frame the force transfer is achieved through a bond between the wires and the concrete. At the end of the beam, as seen in Figure 6.1, the strand tries to slip into the concrete with the force transferred as the bond gradually builds up away from the end face until the concrete takes up the total force in the strand. The distance over which the force transfer takes place is called the transmission length. The slip-in at the end is typically between 2 mm and 3 mm.

Eurocode 2 (BSI, 2004) gives the transmission length, $l_{pt} = \alpha_1 \alpha_2 \emptyset \sigma_{pm0}/f_{bpt}$, where:

$\alpha_1 = 1.0$ or 1.25 for gradual or sudden release respectively

$\alpha_2 = 0.25$ or 0.19 for tendons with circular section, or 3- and 7-wire strands respectively
$f_{bpt} = \eta_{p1}\eta_1 f_{ctd(t)}$ with:

$\eta_{p1} = 2.7$ or 3.2 for indented wires, or 3- and 7-wire strands respectively

$\eta_1 = 1.0$ or 0.7 for good bond conditions or 0.7 otherwise

For good bond conditions and a gradual release, a typical 15.2 mm diameter 7-wire strand with a stress of 70% of UTS and a concrete strength of 35 N/mm², the transmission length based on the above is 789 mm. Eurocode 2 (BSI, 2004, 2005) requires a consideration of ±20% in the transmission length to give a worst case for the condition being considered.

Figure 6.1 Pre-tensioned strand at the end of a precast beam

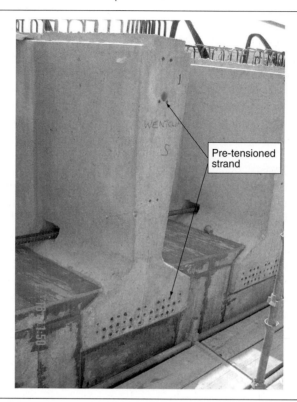

BS 5400 part 4 (BSI, 1990) defines the transmission length as $l_t = k_t D_t / f_{ci}$, when $F_o \leqslant 75\%$ UTS and $f_{ci} \geqslant 30\,\text{N/mm}^2$. k_t depends on the type of wire or strand as follows:

    for plain, indented or crimped wire with a wave height of less than $0.15D_t$, $k_t = 600$
    for crimped wire with a total wave height greater than $0.15D_t$, $k_t = 400$
    for 7-wire standard and super-strand, $k_t = 240$
    for 7-wire drawn or compacted strand, $k_t = 360$

With a typical 15.2 mm diameter 7-wire strand with a stress of 70% of UTS and a concrete strength of $35\,\text{N/mm}^2$, the transmission length based on BS 5400 (BSI, 1990) is 617 mm.

AASHTO LRFD specifications (AASHTO, 1998–2011) are dependent on the stress in the strand under the condition being considered and give the development length as:

$$l_d = k\,(0.15f_{ps} - 0.097f_{pe})d_b$$

where $k = 1.0$ or 1.6 for members of depth less than or greater than 1.6 m respectively.

The build-up of stress in the strands can conservatively be assumed to vary linearly from zero at the ends of the beam to its full stress after the transmission length.

No reinforcement is needed around the strand to counter the local stresses; however, depending on the overall distribution of the strand around the section, it may be necessary to provide spalling and equilibrium reinforcement in a way similar to the anchors of post-tensioned tendons. The AASHTO LRFD specifications (AASHTO, 1998–2011) require that vertical reinforcement be provided as close to the end of the prestressed beam as practicable to resist 4% of the total prestress force at transfer, with the stress in the reinforcement being limited to $140 \, \text{N/mm}^2$. This requirement is similar to the end face spalling reinforcement detailed in the Ciria guide (Ciria, 1976). The AASHTO LRFD specifications (AASHTO, 1998–2011) also requires 32 mm diameter reinforcement bars at 150 mm centres to enclose the strand in the bottom flange for a distance of $1.5d$ from the ends of the beams, although this requirement does not exist in the other design standards.

Post-tensioned tendons

When post-tensioned tendons, either strands or bars, are anchored they apply a large concentrated force to the concrete which must be contained. The guide published by Ciria (1976) describes the behaviour and design of post-tensioned concrete anchor blocks for simple arrangements. The force from the tendon is first transferred directly to the anchor plate and then into the adjacent concrete. The region of concrete immediately behind the anchor block is highly stressed and reinforcement is needed to prevent the concrete splitting or spalling. The force flows from the anchor out into the surrounding concrete, as indicated in Figure 6.2(ii), until, at a distance away from the end face, the stress distribution matches that derived from simple bending theory. The distance taken to reach this 'normal' stress distribution is usually taken as equal to the depth of the anchor block, although the distance can be shorter than this with multi-anchor arrangements spread out on the end face.

Behind each anchor, splitting forces and tensile stresses occur, which are resisted by reinforcement as indicated in Figure 6.2(iii) and quantified in large concentrated force to the concrete which must be contained. The guide published by Ciria (1976) bases the design on the jacking force and suggests reinforcement be provided as follows.

Bursting reinforcement

Provided as a spiral or series of links around each individual anchor to counter the bursting forces set up perpendicular to the line of the tendon, with:

$$A = F_{bs}/(0.87f_y)$$

The bursting force, F_{bs}, depends on the end block arrangement and is given by:

y_{po}/y_o	0.3	0.4	0.5	0.6	0.7
y_{po}/y_o	0.23	0.20	0.17	0.14	0.11

The force on the anchor, F_o, varies from the jacking force, reducing as the tendon is locked off and suffers long-term losses. $0.87f_y$ is often replaced by a stress of $200 \, \text{N/mm}^2$ or less to control cracking in the concrete, although higher stresses are usually acceptable for the short-term loading before tendon lock-off.

Bursting reinforcement is provided as a spiral or links uniformly spread out over a distance between $0.2y_o$ and $2y_o$ from the anchor face, and is detailed to enclose a cylinder or prism with dimensions 50 mm larger than the face of the anchor plate. The bursting forces, in two directions perpendicular to the tendon, are determined and reinforcement provided to suit. Where spirals or links are used,

Figure 6.2 End block design

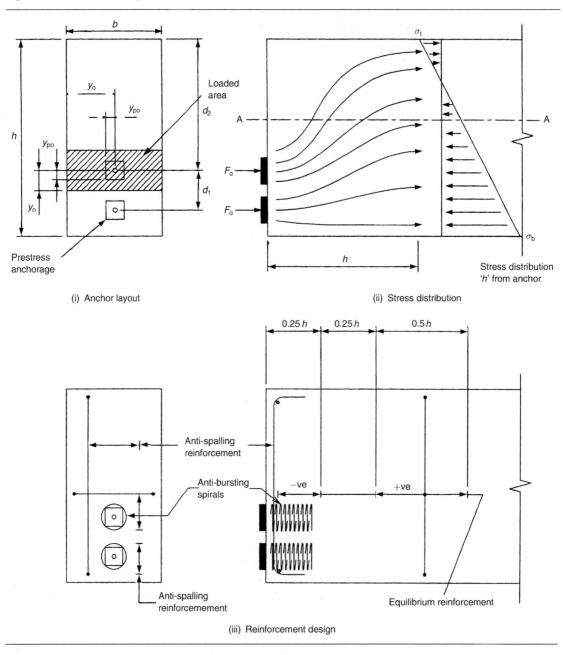

(i) Anchor layout

(ii) Stress distribution

(iii) Reinforcement design

the area of reinforcement resisting the bursting is taken as the total area crossing the plane being considered.

The loaded area used for determining y_o is taken as symmetrical about the anchor extending to the nearest edge of the concrete or the midpoint between adjacent anchors, as illustrated in Figure 6.2(i).

Figure 6.3 End block reinforcement with internal ducts

Figure 6.3 shows the bursting reinforcement provided on each of the three tendons by spirals placed just behind the anchor-block face, on the left side of the picture.

Spalling reinforcement

Provided to prevent spalling of the end-face concrete around the anchor plate, with:

$$A = (0.04F_o)/(0.87f_y)$$

To control cracking in the concrete, the stress in the reinforcement should not exceed $200\,\mathrm{N/mm^2}$, while the bars should be placed as near to the end face as possible and anchored around the concrete edges. Corners and edges of the concrete anchor block are particularly vulnerable and the spalling reinforcement should be placed with its minimum cover in these areas.

Where the anchor is positioned non-symmetrically on an end face, additional spalling stresses are set up and additional reinforcement is provided in the unsymmetrical face such that:

$$A^1 = 0.2((d_1 - d_2)/(d_1 + d_2))^3(F_o/0.87f_y)$$

Equilibrium reinforcement

Provided to maintain the overall equilibrium of the end block.

The force from the anchor block is assumed to have fully spread out at a distance h from the anchor face, as indicated in Figure 6.2(ii). By consideration of the concrete block with the anchor

Figure 6.4 Spread of stress into flanges

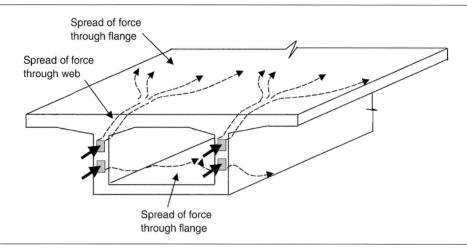

force on one side and the stress distribution on the other, the equilibrium of any horizontal plane, such as A–A, is checked and vertical reinforcement provided to resist the out-of-balance moment and shear. Based on the method presented in the Ciria guide (1976), reinforcement is provided to resist a force equal to the moment divided by a lever arm of $h/2$. The reinforcement is then placed over the distance $0.25h$ or $0.5h$, as indicated in Figure 6.3(iii), depending on the direction of the out-of-balance moment.

The shear on each horizontal plane is checked to ensure that the shear stress does not exceed $(2.25 + 0.65\rho f_y)$ N/mm^2, where ρ is the ratio of the reinforcement crossing the plane and is equal to A_s/bh.

Where flanged box or beam sections are used, the stress behind the anchor has to spread out over the width of the flange. The situation where the tendons are anchored in the webs is indicated in Figure 6.4. The flow of stresses into the flanges is checked for 'equilibrium' effects and reinforcement provided, as described above. For complex concrete shapes and anchor block arrangements it is necessary to carry out 3D finite element analysis of the end block to determine the 'equilibrium' requirements, with reinforcement being provided to counter any tensile stresses set up.

Eurocode 2 (BSI, 2004, 2005) provides a strut and tie approach to designing the end blocks of post-tensioned tendons and provides rules for establishing the tensile reinforcement to be provided for bursting effects around the anchorages. It has little guidance on spalling and equilibrium reinforcement requirements. Based on a design force of 1.2 × the maximum force in the tendon and the yield strength of the reinforcement, it suggests limiting the stress in the bars to 250 N/mm^2 to avoid the need to check crack widths in the concrete.

AASHTO (1998–2011) bases the design of the anchorage zone on a Strength Limit State with a force of 1.2 × the maximum tendon jacking force. It gives guidance resulting in similar reinforcement being provided, allowing either elastic stress analysis, strut and tie models, or the results from experimental tests to be used to derive the forces and reinforcement required.

When several tendons are anchored at the same section the bursting, spalling and equilibrium requirements change as each tendon is stressed. The order of stressing of the tendons should be specified by the designer and the design checked for the situation after each tendon is installed.

The reinforcement in the anchorage zone has to be carefully detailed to ensure that all the necessary reinforcement is included without causing congestion, which would hinder placing of the wet concrete and possibly result in poorly compacted concrete. Bursting reinforcement is provided by a spiral or links placed around the tendon close to the anchor face and inside the other spalling and equilibrium reinforcement near the concrete surfaces. The spalling reinforcement normally consists of small diameter bars placed as close to the end face as possible and around all vulnerable corners. Equilibrium reinforcement is placed around the perimeter of the section over the appropriate region in the end block. Figure 6.3 shows a typical reinforcement arrangement for the end block of a precast concrete beam. As well as the anti-bursting spirals behind the anchors, the spalling reinforcement can be seen with horizontal and vertical bars near the end face. The closer spacing of the vertical links indicates the equilibrium reinforcement.

To ensure that the tendon force is applied squarely to the anchor block the tendons must be accurately positioned within the anchorage area and be straight for a minimum distance behind the anchor plate. The minimum distance required is a function of the tendon size or force in the tendon, and is typically 800 mm for a 7×15 mm strand tendon, 1000 mm for a 19×15 mm strand tendon and 1500 mm for larger tendons.

Cast-in dead-end anchors for post-tensioned tendons

Where dead-end anchors are encased in concrete as 'blind anchors', illustrated in Figure 2.4, the force is transferred to the concrete by the bond between the strand and concrete and by the wire 'bulbs' at the end. The tension ring around the tendon as it emerges from the duct resists the outward forces generated by the angle change in the individual strands as they fan out. Figure 6.5 shows an alternative dead-end anchor arrangement for a multi-strand tendon with the strands looped at the end to assist in the anchorage. In this case, spiral reinforcement is provided to counter any potential bursting forces. The overall reinforcement requirements are governed by the arrangement of the strand and are usually specified by the suppliers of the system being used.

Depending on the concrete shape and arrangement of the dead-end anchors, it may be necessary to provide equilibrium reinforcement similar to that described above and to tie the anchor into the concrete behind it to prevent tension cracks forming.

Anchor blisters or blocks

Anchorage blisters or blocks are used mainly in concrete box girder decks, where they are cast on the side of the concrete member as seen in Figures 2.13 and 2.14. As well as the usual bursting and spalling reinforcement, additional bars are needed to tie the blister or block into the main body of concrete, illustrated in Figure 6.6(i).

Bursting and spalling reinforcement quantities are calculated as above and are shown for a top anchor blister in Figure 6.7. Additional tie-back reinforcement is required to prevent cracks occurring behind the anchor due to the tensile forces generated to achieve strain compatibility in the concrete. It has been traditional to provide tie-back reinforcement to cater for a force equal to 50% of F_o, although finite element analysis can show that significantly less reinforcement than this is needed in some cases. AASHTO (1998–2011) specifies a requirement for a minimum of 25% of F_o for the tie-back force

Figure 6.5 Dead-end anchorage reinforcement

along with a reduction based on the compression in the concrete behind the anchor, although the compression can have limited effect locally and can be conservatively ignored. Basing the design on the tendon jacking force it is prudent to limit the stress in any tie-back reinforcement to 200 N/mm^2 to limit cracking in the surrounding concrete.

The blister is normally kept as small as possible to reduce the interference with the shutters used to form the inside of the box or sides of the web. The anchors are positioned at the interface between the webs and slabs close to the concrete faces and with the minimum clearance that is needed to position the stressing jack. The blister is sized to give the minimum edge distance to the anchor plate.

With blisters the force from the anchor follows the tendon, flowing directly through the blister and into the concrete section. The tendon is normally anchored at between 10° and 20° to the web or slab surface

Figure 6.6 Blister and anchor block design

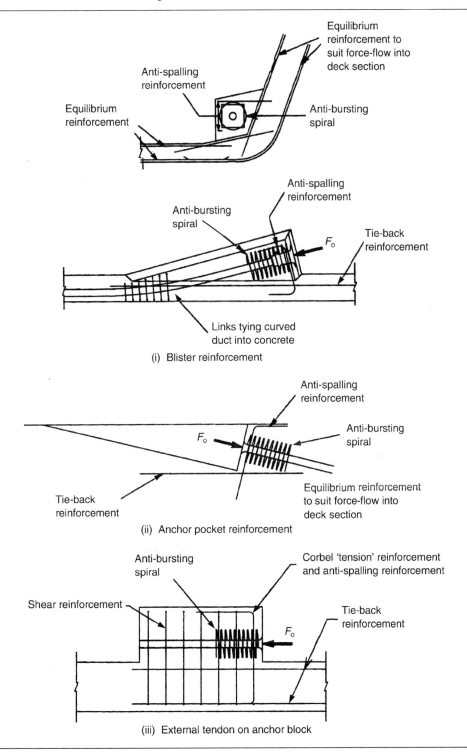

Equilibrium reinforcement to suit force-flow into deck section

Anti-spalling reinforcement

Equilibrium reinforcement

Anti-bursting spiral

Anti-spalling reinforcement

Anti-bursting spiral

F_o

Tie-back reinforcement

Links tying curved duct into concrete

(i) Blister reinforcement

Anti-spalling reinforcement

Anti-bursting spiral

F_o

Tie-back reinforcement

Equilibrium reinforcement to suit force-flow into deck section

(ii) Anchor pocket reinforcement

Anti-bursting spiral

Corbel 'tension' reinforcement and anti-spalling reinforcement

Shear reinforcement

Tie-back reinforcement

F_o

(iii) External tendon on anchor block

Figure 6.7 Blister reinforcement

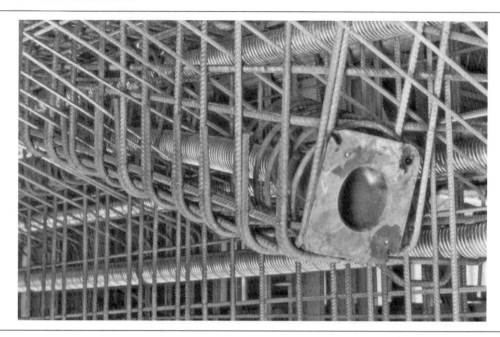

and additional reinforcement is provided to tie the inclined force into the concrete alongside. As it passes from the blister into the concrete web or slab the tendon usually sweeps through a tight curve. Over the length of this curve the tendon is tied into the concrete with links as described in the section on ducts, later in this chapter.

Anchor blocks for external tendons carry the load from the anchorage into the web or slab concrete. Depending on their size and layout, they act either as a corbel or as a cantilever and are designed accordingly, as indicated in Figure 6.6(iii).

Equilibrium effects occur for both blister and anchor block arrangements as the force distributes across the section and into the webs and slabs, requiring reinforcement that can be quantified as described in 'Post-tensioned tendons', earlier in this chapter.

The spread of stresses throughout the concrete section can be determined using 3D finite element modelling with solid elements, as illustrated in Figure 6.8, which shows a short slice of a longer model set up to analyse a blister in a box girder deck. From this analysis, the tensile stresses and forces at each section are determined and reinforcement provided where needed.

Anchor pockets

Figure 6.6(ii) illustrates an arrangement where a recess or pocket is provided in the concrete to anchor the tendons. The large anchor forces tend to cause cracking of the concrete behind the recesses and similar tie-back reinforcement should be provided as for the anchor blisters, as well as the normal bursting and spalling reinforcement. Equilibrium reinforcement is needed in the webs and slabs to counter the tensile stresses as the force spreads out.

Figure 6.8 Three-dimensional modelling of anchor blister

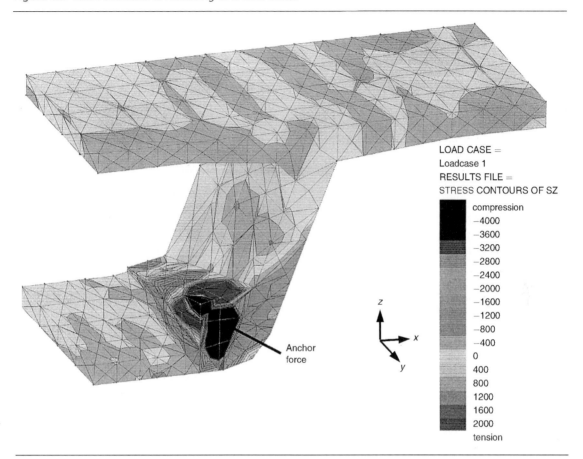

LOAD CASE =
Loadcase 1
RESULTS FILE =
STRESS CONTOURS OF SZ

compression
-4000
-3600
-3200
-2800
-2400
-2000
-1600
-1200
-800
-400
0
400
800
1200
1600
2000
tension

Anchor force

Couplers

Where the tendons are joined together using couplers, normally between one section of deck already concreted and the next section to be built, the local design in front of the coupler is similar to a post-tensioned anchor block. With the arrangement illustrated in Figure 2.5, the normal anchor plate, as seen on the right side, is cast into the first stage of concrete and, once the tendon is installed and stressed, the reinforcement requirements around the anchor block for bursting, spalling and equilibrium are as defined in 'Post-tensioned tendons' above.

When the tendon in the next section of deck is installed it is coupled to the back of the first tendon and stressed. The force on the anchor plate is reduced as the forces in the tendons either side become balanced. The governing criterion for the design of the reinforcement is during the first stage with only the first section of tendon anchored. There is normally no requirement for any additional reinforcement in the second-stage concrete behind the coupler as there are no forces imparted locally to the surrounding concrete.

Couplers require the concrete section to be large enough to house the arrangement and still maintain sufficient concrete cover to the surrounding reinforcement. This is seldom a problem with bar couplers, as illustrated in Figure 2.6, due to their compact size; however, with multi-strand tendons the concrete section often needs thickening around the coupler.

Eurocode 2 (BSI, 2004, 2005) and AASHTO (1998–2011) limit the amount of prestress tendons anchored at any one section to 50% of the total prestress at that section. AASHTO (1998–2011) requires staggered couplers to be located a minimum distance of twice the section depth apart, while Eurocode 2 (BSI, 2004, 2005) specifies a spacing of between 1.5 m for sections of depth less than 1.5 m and 3.0 m for section depths of 3.0 m or greater. BS 5400 (BSI, 1990) contains no restriction on the staggering of couplers and allows all couplers to be located at a single location. Many bridges have been built with all the tendons coupled at a single location, and when adequately detailed and designed have performed well in the past.

Ducts

Duct sizes are governed by the practicalities of having sufficient space to thread the tendon through and to allow the grout to flow freely around the strands or bars. The internal area of a duct is normally at least twice the tendon area. Typical duct sizes are indicated in Table 2.2, which shows the different standard systems available.

Internal tendon ducts, as shown in Figure 6.3, must be supported and held in place during the concreting operation. This is to prevent the ducts from being dislodged or 'floating' up in the fluid concrete. Reinforcing bars are normally placed beneath and around the duct and fixed to the rest of the reinforcement cage to provide a rigid support. The required spacing of these supports depends on the duct size and stiffness, and is normally between 0.5 m and 1.0 m.

To ease concrete placing, to prevent local crushing of the concrete and to ensure that a dense concrete surround is achieved, ducts cast into the concrete should have sufficient space around them. Eurocode 2 (BSI, 2004, 2005) requires a minimum clear spacing between the ducts of at least

(a) duct internal diameter
(b) aggregate size for vertical spacing and aggregate size + 5 mm for horizontal spacing
(c) 40 mm for vertical spacing and 50 mm for horizontal spacing.

The concrete cover to the duct should be sufficient to protect the ducts and tendon inside from the environment as well as prevent any spalling or bursting out of the tendons when stressed. Eurocode 2 (BSI, 2004, 2005) specifies a minimum cover to the ducts equal to the duct diameter, while BS 5400 part 4 (BSI, 1990) states that the cover should not be less than 50 mm, although this depth may need to be increased when the ducts are curved either perpendicularly or in the plane parallel to the concrete surface. Table 36 of BS 5400 part 4 (BSI, 1990) gives an indication of the minimum cover required for different duct radii. Below these values, reinforcement may be needed to prevent the tendon from bursting out of the duct.

With curved tendons, the strand bears against the edge of the duct and exerts a pressure on the concrete, causing splitting forces that must be considered in the design. The tendency to split the concrete depends on the force in the tendon and the radius of the curve. The force per metre length on the concrete is equal to the force in the tendon divided by the duct radius, in metres. When the radius is large enough, no additional reinforcement is needed, as the tensile strength of the concrete is sufficient. With tighter curves, reinforcement should be provided to restrain the tendon and transfer the force back into the surrounding concrete, as indicated in Figure 6.9.

Figure 6.9 Restraining curved ducts

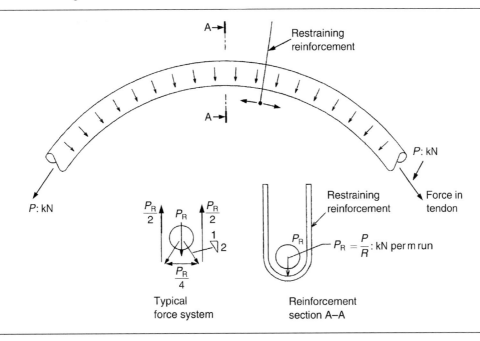

$$P_R = \frac{P}{R} : \text{kN per m run}$$

Typical
force system

Reinforcement
section A–A

Curved tendons running parallel to each other in the plane of curvature are at risk of bursting into the adjacent duct. If the radius of the ducts is large and there is sufficient spacing between the ducts, as noted above, the concrete tensile strength should be adequate to prevent any bursting. Table 37 of BS 5400 part 4 (BSI, 1990) gives an indication of the minimum radii and spacing of the ducts below which reinforcement may be needed.

Where a curved tendon exerts a force on the concrete section it is sometimes necessary to provide additional reinforcement. With a tendon placed inside the bottom slab of a haunched box girder a resultant downward force is generated, as indicated in Figure 6.10. This will produce bending and shears in the bottom slab and tension in the webs.

Multi-strand tendons can follow a profile with tight curves, the minimum radii being governed by the ability of the duct to bend during fixing. Individual strands can be bent to radii less than 0.5 m. With tendons the minimum radius in standard corrugated metal or plastic ducts varies from 3 m for a 4×15 mm strand system, 5 m for 12×15 mm strands, and up to 8 m for a 31×15 mm strand system. Tighter radii can be used if metal pipe ducts are used, pre-bent to the required radius.

External tendon ducts are easily accessible between the deviators and anchorage points, but they should be checked during the design to ensure that they do not clash with each other or push against the concrete surface as they follow their intended alignment. Figure 6.11 shows a series of external ducts being laid out before fixing in position and threading the tendons. Each duct is well clear of its neighbour, allowing easy inspection, and wide spacing as they pass through the deviator gives plenty of room to fix the reinforcement around each hole.

Figure 6.10 Tendons in curved bottom slab

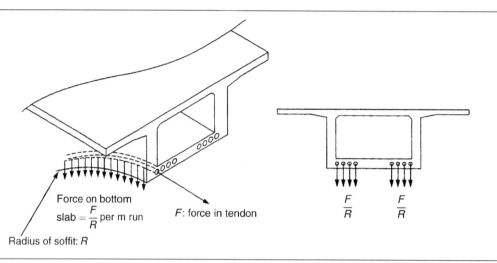

Force on bottom
slab $= \dfrac{F}{R}$ per m run

Radius of soffit: R

F: force in tendon

$\dfrac{F}{R}$ $\dfrac{F}{R}$

Figure 6.11 Ducts for external tendons

Diaphragms

Diaphragms are generally used in the deck at points of support to transfer the vertical load from the webs into the bearings below and to provide transverse stiffness to the deck section. Typical diaphragm-wall arrangements for box girders and I beam decks are illustrated in Figure 6.12. For the arrangement shown in Figure 6.12(i), a truss analogy is normally used to model the force transfer

Figure 6.12 Diaphragm arrangements

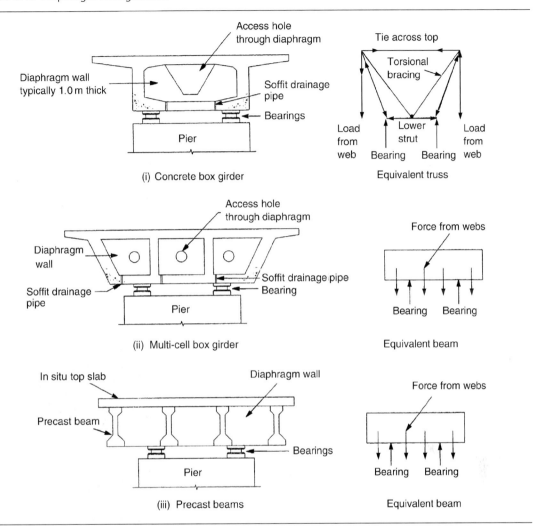

(i) Concrete box girder — Equivalent truss

(ii) Multi-cell box girder — Equivalent beam

(iii) Precast beams — Equivalent beam

from the webs into the bearings. The force is resisted by a 'tie' across the top of the diaphragm and a strut towards the bearing. In Figures 6.12(ii) and 6.12(iii) the diaphragms behave as a beam and bending moments and shears can be determined, and reinforcement provided in the normal way.

Where a web is positioned over a bearing the vertical load goes directly into the support. Elsewhere, 'hanging reinforcement' may be needed to take the force from the web up to the top of the diaphragm 'beam' to ensure that the flow of force accurately reflects the assumed structural behaviour.

When a bearing is offset from the web by less than half the deck depth the hanging reinforcement can be proportioned to carry just the shear force in the webs below the line A–A, indicated in Figure 6.13(i). The remainder of the force is transferred through the concrete using shear friction or corbel action. Where the bearing offset is greater than half the deck depth, hanging reinforcement is provided to carry the full shear force on the diaphragm.

Figure 6.13 Diaphragm hanging reinforcement

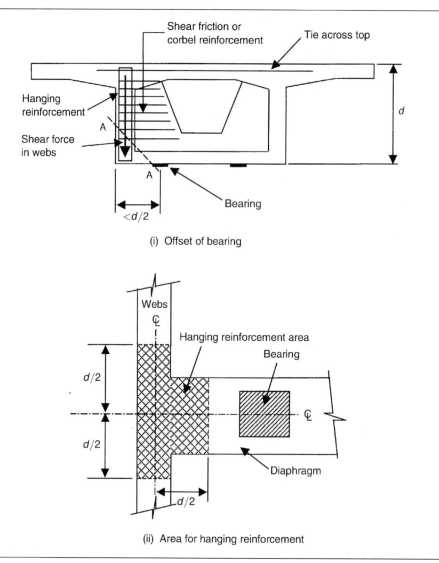

Shear friction or corbel reinforcement

Tie across top

Hanging reinforcement

d

Shear force in webs

A

A

Bearing

$<d/2$

(i) Offset of bearing

Webs
Ç

Hanging reinforcement area

Bearing

$d/2$

$d/2$

Ç

Diaphragm

$d/2$

(ii) Area for hanging reinforcement

Hanging reinforcement is provided in the form of vertical bars distributed over part of the diaphragm and webs, as indicated in plan in Figure 6.13(ii). The reinforcement provided in the webs for vertical shear can be utilised as part of the hanging requirement, while draped tendons running along the deck can be used to carry some of the vertical load up to the top of the diaphragm.

Where significant torsion exists in a box girder deck, it is necessary to consider the horizontal forces present in the top and bottom slabs adjacent to the diaphragm and to transfer these forces into the bearings. Reinforcement should be provided across the diaphragm to resist these forces and to tie the slabs together. The reinforcement can be derived based on a strut-tie approach to transfer the forces through the diaphragm.

Figure 6.14 Vertical prestress in diaphragm

As an alternative to providing reinforcement for the 'hanging' and tie forces, vertical and transverse prestressing tendons can be used. Using prestressing to resist the forces in the diaphragm reduces congestion of the reinforcement, making concreting easier, as well as reducing potential cracking by keeping the concrete in compression. Figure 6.14 shows an arrangement where vertical bar tendons were used for a diaphragm in a precast segmental box girder deck.

Where diaphragms are incorporated into the deck over bearings they should be designed with provisions for temporary jacks to be installed to allow the replacement of the permanent bearings in the future. This may require additional reinforcement above the future jacking points, as well as extra strut-tie and hanging reinforcement if the temporary jacking point is more onerous than the bearing location. If traffic loading can be controlled during the bearing replacement the load in the temporary jacks may be less than that in the bearings.

The concrete above both the permanent and temporary bearings is designed in a manner similar to that for an anchor end block, with spalling and bursting reinforcement being provided accordingly. With the temporary bearings it is usually assumed that a reduced live load would occur when the bearing replacement takes place and if a reduced live load is assumed it should be stated on the drawings or in the maintenance manual.

Figure 6.15 shows a half-depth diaphragm arrangement for a precast beam deck. In this case each of the beams is supported directly by the bearings on the pier crosshead and the small half-depth

Figure 6.15 'U' beam diaphragm

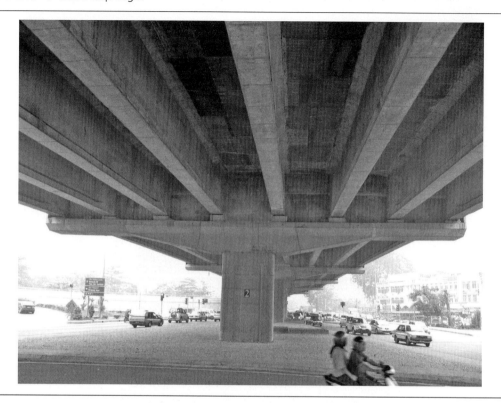

diaphragm provides transverse restraint and rigidity, but is also used for jacking up the deck should the bearings need replacing. This simple arrangement provides an efficient structural system. The governing criterion for the design is the situation with temporary jacks placed under the diaphragm during replacement of the bearings.

Transverse diaphragms are sometimes used for precast beam decks with longer spans at 'quarter' or 'third' points in the span to connect the beams transversely and improve the distribution of the applied loads across the deck. These intermediate diaphragms are usually of minimal thickness and lightly reinforced, and the design forces, moments and torsions are derived from a grillage or 3D analysis of the deck.

In box girder decks, diaphragms are usually provided at each end support and over each intermediate pier, with a typical arrangement shown in Figure 6.16. The diaphragm walls are typically 1.0 m to 1.5 m wide to extend over the width of the bearings, plus an allowance for any relative movement between the deck and the support. The position of the temporary jacks during bearing replacement is usually just inside the permanent bearings, beneath the diaphragm wall and on top of the pier.

Transverse diaphragms are sometimes used within the span for multi-cell box girders to provide transverse rigidity to the box arrangement and to distribute the applied loading across the complete box section.

Figure 6.16 Typical box girder intermediate diaphragm (reproduced courtesy of Hyder Consulting (UK) Ltd, copyright reserved)

Diaphragms are used to anchor external tendons. The thick diaphragm walls provide an ideal location for the anchorages and with suitable reinforcement can easily accommodate the high forces imposed. The anchors should be placed as near to the deck webs or slabs as possible, to minimise the bending generated in the diaphragm, which acts as a beam or wall spanning between the webs and slabs. Access holes through the diaphragms result in uneven load distributions from the external tendons to the deck section. As a simplified approach to the design, the anchor load can be considered to spread out across the diaphragm at 45° towards the 'supports' consisting of the webs and slabs, with the load in each direction proportioned depending on the concrete arrangement. Reinforcement is then provided in each direction to cater for the associated bending moments and shears.

If the diaphragm thickness is more than 50% of its minimum span between slabs or webs it behaves as a deep beam and reinforcement is provided as described in the Ciria Guide No. 2 (1977).

More refined analysis and design are undertaken using a detailed 3D finite element model which is used to derive the distribution of forces and stresses from the anchors to the deck section.

Deviators

Deviators are used to deflect external tendons to give the desired prestressing profile and as such are subjected to large applied forces. These forces from the tendons must be tied into the deck section. Figure 6.17 illustrates several different types of deviators that are commonly used. The concrete

Figure 6.17 Deviator arrangements

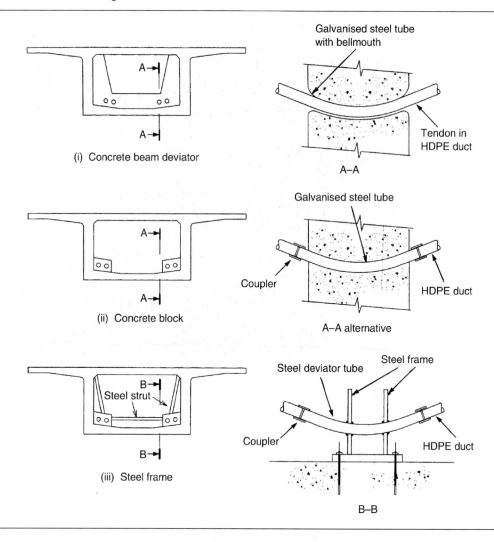

(i) Concrete beam deviator

Galvanised steel tube with bellmouth

Tendon in HDPE duct

A–A

(ii) Concrete block

Galvanised steel tube

Coupler

HDPE duct

A–A alternative

(iii) Steel frame

Steel strut

Steel deviator tube

Steel frame

Coupler

HDPE duct

B–B

beam is the most common arrangement and can cater for very high loads from the deviated tendons with the force being transferred into the deck section either by a 'strut and tie' action or by bending and shear through the beam. The capacity of the concrete-block arrangement is less than with the beam arrangement as the force from the tendon is carried locally into the adjacent concrete with bending in the slabs and webs. Steel deviator arrangements are often used on existing structures where external tendons are being retrospectively fitted to provide strengthening. For new construction, it is usually easier to design and construct concrete deviators, while for existing structures it is often easier to install steel deviators.

Concrete beam and block deviators are shown in Figures 6.18 and 6.19 respectively. The external tendons pass through holes or tubes left in the concrete that guide the tendons through the required alignment. The radii of the tendons through the deviator typically range from 2.5 m for a 7×15 mm

Figure 6.18 Concrete beam deviator

Figure 6.19 Concrete block deviator

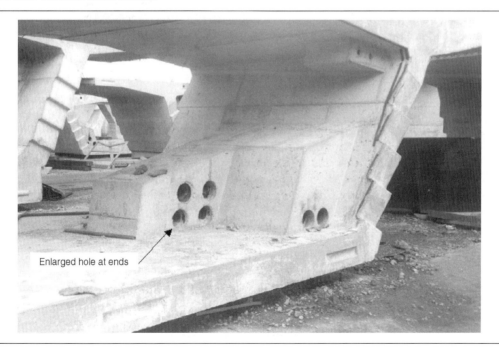

Enlarged hole at ends

Figure 6.20 Steel deviator (reproduced courtesy of Hyder Consulting (UK) Ltd, copyright reserved)

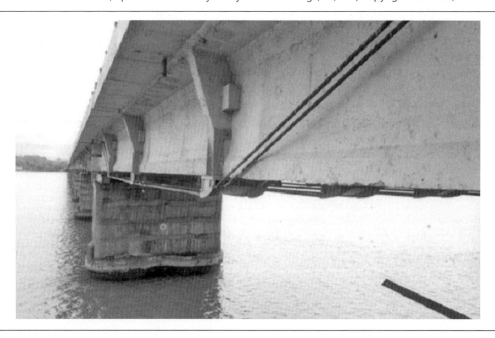

strand tendon, 3.0 m for a 19×15 mm strand tendon and up to 5.0 m for 27×15 mm strand tendons. In Figure 6.18 the beam deviator has also been used to anchor some of the external tendons, while the hole in the centre of the beam is for a future additional tendon should this be required.

Figure 6.20 shows a steel deviator clamped to the underside of a prestressed concrete beam, used during the strengthening of an existing bridge. In this example, the steel deviator transfers the load from the deflected tendon into the concrete webs via a concrete nib; alternatively, the steel could be fixed to the concrete using prestressing bars.

The holes in the deviators and the continuity of the tendon ducting can be arranged in several different ways. The most common method is to concrete in oversized steel or stiff plastic sleeves through which the external tendons are threaded while still inside their own protective duct system, as illustrated in Figure 6.17(i). This has the advantage of allowing the external tendon duct to be continuous over its full length. The difficulty with this arrangement is in lining up the hole correctly when casting the concrete, and 'misalignments' are common occurrences. Any misalignments result in the tendon bearing against the concrete edge where it emerges from the deviator, causing kinks in the tendons and spalling of the concrete. This effect can be reduced by enlarging the hole over the end sections, as seen in Figure 6.19, or by including a small radius on the ends of the sleeves so as to move the point of contact back from the end face and reduce the risk of kinks in the tendon.

An alternative arrangement is to place a steel tube into the deviator and to join the external tendon ducting to this on either side, as illustrated in Figure 6.17(ii) and (iii). The disadvantage of the tube method is again the difficulty in achieving the required tendon alignment. When the deviators are being constructed the external tendons are not in place, and setting out the sleeve to the correct angle and position at both ends is difficult.

On some projects the holes through deviators have been formed directly in the concrete without sleeves or tubes, as was the case in Figure 6.18. A shaped void former is used to create the hole in the concrete during casting and is removed soon after the concrete begins to harden. A bell-mouth arrangement at the ends of the deviator hole allows the tendons to deviate through the required angle and provides some tolerance should small misalignments occur.

The design force on the deviator is equal to the force in the tendon multiplied by the 'angle change' (in radians). The tendons generate a 'radial' force that is tied into the concrete by links positioned around the hole and extending back into the deck section. The force on the deviator generates bending and shears in the adjacent deck section, which must be designed for and reinforcement provided as necessary. With Eurocode 2 (BSI, 2004, 2005), the design forces are based on the maximum tendon force increased by 20% to provide a worst-case condition, with the reinforcement designed limiting the stress to $250\,\text{N/mm}^2$ to limit cracking. Where BS 5400 part 4 (BSI, 1990) and the Highways Agency design standard BD 58/94 (Highways Agency, 1994) are used, the design is carried out at the ultimate limit state with the applied ultimate force based on the characteristic strength of the tendon. Alternatively, the design of the deviators is based on the maximum load carried by the tendon, with the stresses in the reinforcement limited to permissible stress levels and low enough to minimise the risk of cracks forming in the concrete.

Figure 6.21 Three-dimensional finite element analysis of deviator

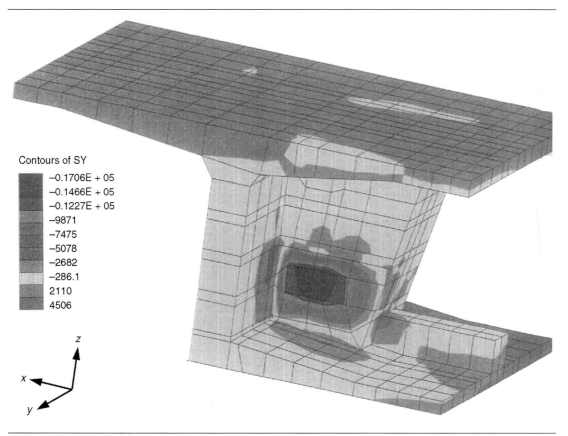

Where concrete deviators are cast against the sides of webs and flanges, the forces from the deflected tendons flow into the adjacent concrete both longitudinally and transversely. For more complex arrangements it may be necessary to set up a 3D finite element model, as illustrated in Figure 6.21, to determine the distribution of force into the concrete. By using solid elements and modelling a sufficient length of the structure the principal stresses can be derived and the reinforcement provided to cater for any tensile forces that occur.

REFERENCES

AASHTO (1998–2011) Load and resistance factor design (LRFD) bridge design specifications, 4th edition. AASHTO, Washington, DC.

BSI (1990) BS 5400: Steel, concrete and composite bridges. Part 4. Code of practice for the design of concrete bridges. BSI, London.

BSI (2004) BS EN 1992–1-1:2004: Eurocode 2: Design of concrete structures. Part 1–1. General rules and rules for buildings. BSI, London.

BSI (2005) BS EN 1992–2:2005: Eurocode 2: Design of concrete structures. Part 2. Concrete bridges – design and detailing rules. BSI, London.

Ciria (1976) *Guide No. 1. A guide to the design of anchor blocks for post-tensioned prestressed concrete members*. Ciria, London.

Ciria (1977) *Guide No. 2. The design of deep beams in reinforced concrete*. Ciria, London.

Highways Agency (1994) (MDRB) Departmental standard BD 58/94: The design of concrete highway bridges and structures with external and unbonded prestressing. HMSO, Norwich.

Prestressed Concrete Bridges, 2nd edition
ISBN: 978-0-7277-4113-4

ICE Publishing: All rights reserved
doi: 10.1680/pcb.41134.135

Chapter 7
Concept design of prestressed concrete bridges

Introduction

There are many factors affecting the choice of bridge type, span arrangement and general layout. Prestressed concrete decks include a wide range of construction forms with span lengths ranging from less than 25 m for single spans to over 400 m with cable-stayed bridges. For spans below 25 m, reinforced concrete is likely to be more economical, while for spans above 400 m, steel or composite cable-stayed decks are more likely to be used to reduce the deck weight. Between these two extremes prestressed concrete often gives an economical, aesthetic and simple solution. Figure 7.1 indicates the typical span range for different types of prestressed concrete deck construction, defined in terms of the type of construction and deck arrangement.

The choice of deck form is influenced by the particular site constraints, and the advantages and disadvantages of each type need to be carefully considered to determine the optimum solution.

This chapter briefly reviews the different types of prestressed concrete bridges and considers the factors affecting the choice of concept design for an individual project.

Deck types

Prestressed concrete is used in a wide variety of ways in the construction of bridge decks. It is formed as beams or boxes, cast in situ or precast, lifted or launched into place, and it utilises either pre-tensioned or post-tensioned tendons. As well as simply supported or continuous span arrangements, concrete decks are also used on cable-stayed structures.

For short spans, in situ voided slabs are sometimes used, with the void formers reducing the deck weight and the section area to improve the efficiency of the prestress. The tendons are placed in the concrete 'webs' between the voids or in the slabs. For short spans, precast beams are usually more economical than voided slabs. This has resulted in few voided slab types being constructed in recent years.

Precast beams with an in situ concrete deck slab are popular for short- and medium-span bridges, especially where existing precasting facilities result in cost savings being achieved. For spans up to 30 m precast beams compete favourably on cost terms with in situ reinforced concrete or steel beam construction. Beams utilising either pre-tensioning or post-tensioning have been developed with a range of different shapes, including 'I', 'T' and 'U' beams. For long multi-span viaducts, as illustrated in Figure 7.2, precast beams offer cost savings due to the repetition and speed that is achieved during their construction.

135

Figure 7.1 Span ranges for different deck types

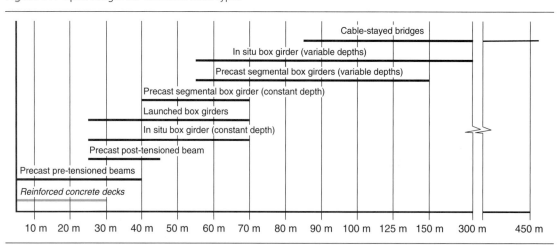

Figure 7.2 Precast beams (reproduced courtesy of Hyder Consulting (UK) Ltd, copyright reserved)

Single- or multi-cell concrete box girders are well suited for medium- to long-span bridges. They are ideal for curved decks due to their torsional rigidity and are able to cope with complex geometric requirements. They provide efficient structural solutions and combine with a range of different construction techniques to suit most site conditions.

Construction depths greater than 1200 mm are needed to give the height necessary to construct the boxes, although the need for reasonable access into the box for inspection and maintenance requires a depth of 1800 mm or more. This tends to limit the use of box girders to spans greater than 35 m.

Figure 7.3 In situ box girder

Spans less than 35 m are more likely to be constructed using precast beams, in situ reinforced concrete or steel beam construction.

In situ box girder decks with spans up to 50 m are usually built on falsework or launched into position. For longer spans the concrete is cast in situ as a balanced cantilever, using form travellers as shown in Figure 7.3.

Precast single-cell or twin-cell box girders are cast in short segments or in full-span lengths. The cost of precasting, transporting and erecting the segments requires a significant length of deck to be constructed for the technique to be economic. Full-length precast units are very heavy and require large equipment to transport and position them; however, they offer an efficient form of construction for long viaducts where savings are achieved through repetition in the construction process and a reduction in construction time.

Short precast segments with match-cast joints, as shown in Figure 7.4, have become popular where rapid construction and minimum site disruption is a requirement. The financial investment in setting up a casting yard and in the purchase of transportation and erection equipment generally requires deck lengths greater than 2 km before this form of construction becomes economical. In some countries where this form of construction has become popular, the setting up of permanent precasting yards with specialist casting companies has resulted in this type of construction becoming popular for smaller projects.

Figure 7.4 Precast segmental box girder (reproduced courtesy of H&T Associates, copyright reserved)

Where access is difficult beneath a bridge deck, or environmental concerns restrict the construction options, box girders are launched into position, as seen in Figure 7.5. The deck is cast in sections behind one of the abutments and incrementally launched into position. Design requirements of the deck during the launch phase usually result in more concrete, reinforcement and prestress than for normal in situ construction. This is offset by savings in the temporary works and equipment when compared with other forms of construction.

For longer spans the depth of a concrete box girder can become excessive. Box girders built by the in situ balanced cantilever technique have been used for spans up to 300 m. At this span deck depths at the piers exceed 14 m and it is difficult to achieve an aesthetically pleasing solution.

Only relatively shallow decks are required with a cable-stayed arrangement, as seen in Figure 7.6. Prestressed concrete spans over 500 m long have been built using this form of construction. For spans shorter than 200 m a cable-stayed deck is usually more expensive than alternative deck types, although where a shallow deck is needed for clearance or aesthetic reasons cable-stayed arrangements produce elegant solutions. For twin pylon cable-stayed bridges with spans over 500 m, steel or composite construction usually offers more economic solutions in the current market.

Prestressed concrete is also used in other types of construction, such as stressed ribbon decks, trusses, extra-dosed and arch bridges. These less common types of deck solutions are usually adopted due to particular site constraints, project requirements or aesthetic considerations.

Figure 7.5 Incrementally launched box girder

Figure 7.6 Cable-stayed bridge

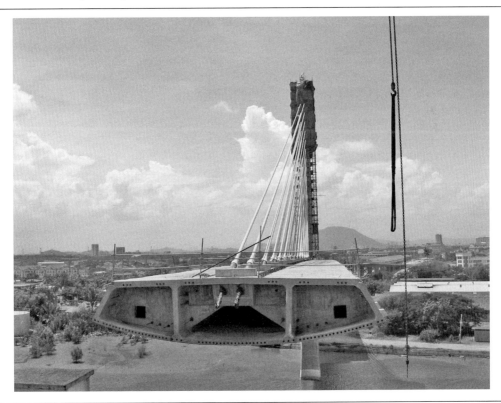

The advantages and disadvantages of the main deck types are summarised as follows:

Deck type	Advantages	Disadvantages
Slab	Simple construction	Inefficient section, limited to short spans
Voided slab	Simple construction	Inefficient section, limited to short spans
Precast beams	Minimum site disruption Rapid construction Economical for spans 20–35 m Repetitive Reduced on-site work Beams cast in factory conditions	Good access to site needed Heavy lifting equipment Complex with continuous construction Less aesthetically pleasing Limited use with curved decks
In situ multi-cell box girder	Efficient deck section Aesthetically pleasing Simple construction Economical for short lengths and medium spans Can cope with complex geometry	Labour-intensive Significant temporary works Slow construction Good access needed
In situ single-cell box girder	Efficient deck section Suitable for tight curves and complex geometry Aesthetically pleasing Simple construction Economical for medium to long spans	Labour intensive Significant temporary works Slow construction
Precast segmental box girder	Efficient deck section Suitable for tight curves and complex geometry Rapid construction Aesthetically pleasing Minimal disruption to site Segments cast in factory conditions Economical for long viaducts	Expensive casting yard and equipment High level of technology required
Precast full length span box girder	Efficient deck section Rapid construction Aesthetically pleasing Minimal disruption to site Segments cast in factory conditions Economical for long viaducts	Expensive casting yard and equipment High level of technology required Not suitable with difficult access
Incrementally launched box girder	Minimum site disruption Minimal environmental impact Repetition Simple construction Minimum temporary works	Higher concrete, reinforcement and prestress quantities Deeper box section
Cable-stayed deck	Good aesthetic appearance only option for very long spans	More expensive for short spans Design more complex Sophisticated construction

The typical arrangements and construction techniques for these different deck types are discussed in more detail in Chapters 9 to 17.

Selecting the deck arrangement

It may appear that the designer is presented with a wide choice of bridge arrangements to select from when he considers a particular bridge site, but in practice the choice is usually more restricted once all the restraints and local factors are taken into account.

As an indication of the construction choices for different span lengths, Figure 7.7 shows the typical deck options for a single-span prestressed concrete bridge. Two- or three-span 'overbridges' would follow a similar choice. Where a deck crosses a railway a common solution is to use precast beams in order to minimise any disruption and reduce the construction risk to the railway system. Similarly, for short river spans precast beams minimise the impact on the river and do not require falsework. For longer spans, box girder construction is usually the most efficient arrangement, while for short lengths of deck these are most likely to be built in situ. For longer spans over a railway or river the box girder can be incrementally launched into position or built with precast segments; however, these are relatively expensive forms of construction for short lengths of deck.

For long multi-span bridges and viaducts, the span lengths are usually greater than 30 m for aesthetic and economic reasons. These can be built using precast beams or with box girders and Figure 7.8 indicates the most common options used. Precast beams are generally only used for spans less than

Figure 7.7 Choice for single-span decks

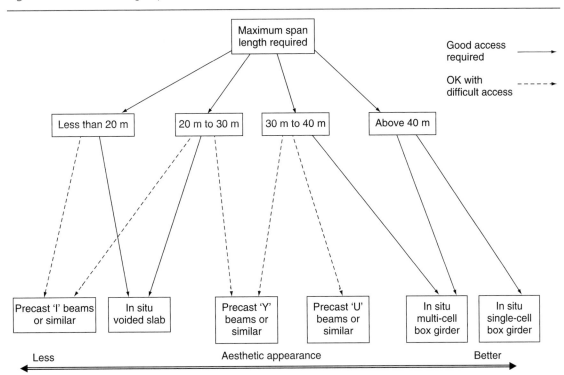

Figure 7.8 Choice for multi-span viaducts

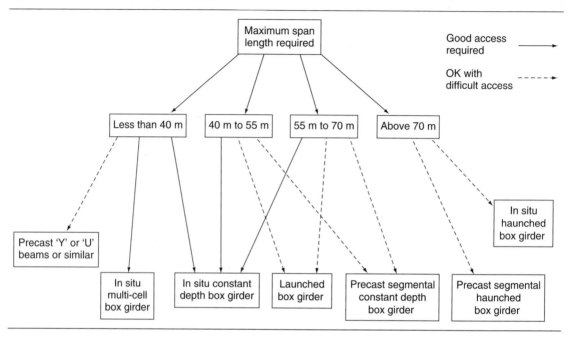

40 m as above this their heavy weight requires specialist lifting equipment. Launched box girders offer good solutions where a deep valley or other obstruction is to be crossed, while precast segments provide a versatile form of construction but usually need at least 2 km of deck to be economical.

When initially planning a structure and sizing up the concrete section, a good starting point is to select a suitable deck layout, cross-section and concrete thickness, based on previous similar projects. These can then be developed to take into account the different loading, design codes and ground conditions of the particular site being considered. The examples in Chapters 9 to 17 may provide a good basis for selecting an initial arrangement.

For precast beams, the manufacturers usually provide guidance on the span range and beam spacing for the different types of beams available. For other types of deck construction there are many papers and articles in the professional journals and in the other publications available. The internet is also building up a growing collection of papers and project reports which can provide useful data. Appendix C contains references to several useful websites and other sources of information on prestressed concrete bridges.

Aesthetics, durability and buildability all influence the deck arrangement and details adopted. Bridges should fit into their surroundings and present a pleasant appearance; good practice is presented in a publication by the Highways Agency (1996). The issues of durability are discussed in Chapter 3 and are often associated with particular details of a deck rather than affecting the overall choice of arrange-ment. Bridge decks often present challenges in their construction due to their locations and complexity, and buildability is a governing factor in the choice of structure. Many of the issues of buildability are covered in the other chapters, while guidance is given in a Ciria report (1996).

Articulation and span arrangements

Where bridges have an overall length less than 60 m it is common practice to build the deck continuous and integral with the piers and abutments. This reduces the number of expansion joints and bearings, and eliminates the maintenance problems that can occur with these elements. For longer structures the deck should be made continuous for as long as practically possible, which depends on the structural form. Concrete box girders have been constructed with continuous lengths up to 1.7 km and concrete cable-stayed bridges with decks up to 800 m in length between expansion joints.

Expansion joints in the deck should be positioned above a pier or at an abutment where the ends of the deck can be fully supported. The arrangement must be detailed to give good access to all parts of the expansion joint. Half joints in the deck should be avoided, if possible, unless there is sufficient space for access to the bearings and expansion joints for inspection, maintenance and replacement. Half joints, hinges or other discontinuities in the span should also be avoided where they might allow long-term deflections from creep and prestress losses to adversely affect the profile of the deck.

The need for bearings between the deck and supports depends on the structural form and stiffness of the substructure. Launched box girder decks require bearings to facilitate the launch process, while precast segmental construction usually incorporates bearings to simplify erection. In situ concrete decks can be constructed on bearings or built integrally with the substructure, provided the deck movements do not generate excessive forces in the piers and foundations.

Bridge decks with horizontal curvature give rise to particular behaviour that need to be addressed in the design. A bridge deck will always try to expand or contract towards its effective 'fixed' point, and with curved decks this can set up unusual horizontal load distributions in the guided bearings if they are orientated parallel to the deck above. For this reason some designers prefer to align guided bearings towards the point of fixity; however, for long decks or tight curves this can cause problems due to the relative transverse movements that might occur across the expansion joints.

Curved alignments can generate significant torsion and overturning effects in the deck structure. Where this is critical this may require the bearings to be offset transversely from the structure centreline or the deck tied down to counteract the effect.

Span arrangements are usually governed by the nature of the obstruction being bridged, but where possible the spans should be arranged to suit the type of deck being proposed. With precast beams it is preferable to keep a constant beam length throughout to standardise the construction equipment. For continuous box girder decks the piers should be arranged to give an end span that is approximately 70% of the length of the adjacent internal span in order to create a balance between the hogging and sagging moments. Where a deck is built by the balanced cantilever method the length of end spans should be reduced to approximately 60% of the internal spans to minimise out-of-balance moments on the end pier cantilever. If the end span of a continuous deck is less than the above recommendations there is a possibility that uplift will occur at the end bearings, in which case a tie-down arrangement will be required.

The minimum span length and height of deck above the ground is usually dictated by the clearance requirements of the road, rail or river passing underneath. A ratio of span length to deck height above the ground of approximately three is considered optimum to achieve a good visual balance, although this is not always practical with very high or very short piers. For shorter deck lengths where only a few spans are required a balanced arrangement of the crossing with either one, three or five spans is considered to give a better visual appearance than an even number of spans.

Guidance on the best arrangements from aesthetic considerations is given in a publication by the Highways Agency (1996).

Post-tensioning with internal or external tendons

In the past, both internal and external post-tensioned tendons have suffered problems with their design, construction and durability. These problems have been well documented and current practices in detailing and construction have established a basis on which both systems can be used successfully. As part of the overall design concept a designer must choose between internal or external tendons, or a combination of both.

The advantages of external tendons over internal tendons are as follows.

- The casting of segments is simplified by eliminating the presence of ducts in the webs and flanges.
- The formwork, fixing of reinforcement and concreting are all made easier.
- Reduced deck web thickness due to the absence of ducts results in a saving on deck weight and substructure costs. They allow the use of dry joints with precast segmental construction, where appropriate. This makes deck erection simpler and faster as well as reducing the cost of construction.
- Installation is made easier by having access to all parts of the duct.
- Placing of the strands is more straightforward and not prone to the blockages of the ducts that can be associated with internal tendons.
- Tendon layouts are simplified.
- Grouting of tendon is easier due to the better access available and because individual tendons can be grouted without the leakage or blockage problems that can occur with internal tendons in some forms of construction.
- Reduced friction losses within the tendon results in a higher effective force in the strand.
- It is easier to inspect tendons during construction and for long-term maintenance.
- The tendons are replaceable in the future.
- Provisions to add tendons later to upgrade or strengthen the deck are easily incorporated.

The disadvantages of external tendons over internal tendons are as follows.

- They are protected by HDPE ducts which are fully cement grouted or wax filled, which result in a higher initial material cost for the prestress system.
- Reduced tendon eccentricity at critical points increases the prestress quantity required.
- Under ultimate bending conditions external tendons require more prestress to generate the same moment of resistance.
- They rely on anchorage integrity to maintain the prestressing force. If the anchor fails the tendon 'fails' along its complete length.
- Anchorage points and deviators are subjected to high concentrations of forces which need to be tied into the deck structure.
- Fatigue or fretting can be a problem if the free length of the tendon is too long, making the tendon susceptible to vibrations.

- The exposed tendons are more vulnerable to accidental damage.
- Better access is required to the tendons, deviators and anchorages for installation, future inspection and replacement of the tendons if necessary.

Internal and external tendons both produce durable and efficient structures and the choice is often influenced by local requirements and practices. Many recent bridges have used a combination of both internal and external tendons to combine the advantages of both systems.

Bridge costs

In general, the shorter the span length the less cost per m^2 of deck. Prestressed concrete bridges are unlikely to be economical for spans less than 20 m, where plain reinforced concrete structures would be preferred. Precast prestressed beams offer suitable economical solutions for shorter spans if access is limited, such as when crossing a river or railway. As the deck span gets longer, the quantities of concrete, reinforcement and prestressing steel used per m^2 of deck increase, causing a rise in the cost of construction. If significant temporary works are required, this also increases the overall cost.

Where complex or expensive foundations are needed, such as over poor ground or to avoid obstructions below, the substructure costs become a higher proportion of the overall cost and adjust the balance in favour of longer spans. Other factors influencing the cost of a bridge include the location, ground conditions, design, labour costs, material costs and site restrictions.

In the UK, a typical single-span highway overbridge or underpass costs approximately £850 per m^2 of deck (in 2010) with the abutments making up a significant portion of the overall cost. For a typical three-span overbridge, with spans up to 20 m, the cost is approximately £700 per m^2, as the substructure costs are spread out over a greater deck area. As the span lengths increase, the cost per m^2 of deck increases.

A typical multi-span viaduct with 40 m spans could cost approximately £850 per m^2 of deck, while with 70 m spans the cost could increase to approximately £1200 per m^2 of deck. For cable-stayed bridges the costs can be over £2500 per m^2 or significantly greater with unusual or architecturally enhanced structures. Footbridges tend to cost more per m^2 due to the smaller deck area and relatively high abutment and support costs (all prices are based on the time of writing).

When developing the concept design for a bridge, it is usual to carry out a cost comparison for different deck forms and span arrangements. Preliminary estimates of deck and substructure quantities and construction costs for the different options give an indication of the most cost-effective solution, although other considerations such as aesthetics and environmental issues may influence the final choice.

With long multi-span viaducts, the choice for the optimum span length to give the most cost-effective solution is a balance between the deck and substructure costs. With longer spans the deck quantities and cost increase, but, by reducing the number of piers, savings can be made in the substructure. Deck costs are usually directly proportional to the span length. The cost of the substructure, expressed as a cost per m^2 of deck, generally reduces as the span gets longer, but, where piles are used, steps in the cost occur as another pile is added to take the increase in load. The cost of the deck and the substructure can be plotted on to a graph, as illustrated in Figure 7.9, thus enabling the optimum span length to be selected.

Figure 7.9 Optimising bridge costs

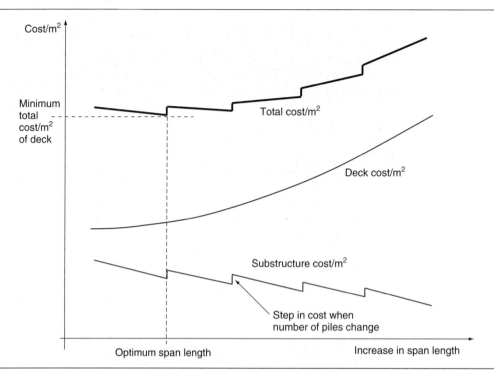

External tendons cost more per tonne than the equivalent internal tendons, but this is offset against savings elsewhere. With the tendons outside the concrete section the concrete section sizes, particularly the webs, are reduced in thickness. In addition, reinforcement fixing and concrete placing are made easier. The use of external tendons with segmental construction in conjunction with span-by-span erection has resulted in rapid assembly of long viaducts, giving overall cost savings by reducing the construction period.

Material quantities

The quantity of concrete, reinforcement and prestressing used in a deck is normally in a reasonably well-defined band for given span lengths and deck types, although every bridge is different and the deck width, live loading and other aspects all influence the final quantities achieved. During the design process it is useful to compare the material quantities for the deck with those experienced on similar projects as an indication of the efficiency of the arrangement proposed. Figure 7.10 indicates the typical range of material quantities that can be expected in a deck for spans between 30 m and 150 m.

Precast segmental construction usually produces savings in concrete volume when compared with in situ construction, due to thinner sections being more practical to achieve in the factory casting conditions. Incrementally launched box girder decks usually have slightly higher concrete volume than the other forms of deck construction due to the design requirements of the launching.

The overall reinforcement quantity per m² of deck is significantly less in precast segmental construction and precast beams, when compared to other forms of construction, due to reduced longitudinal

Figure 7.10 Typical quantities in prestressed concrete decks

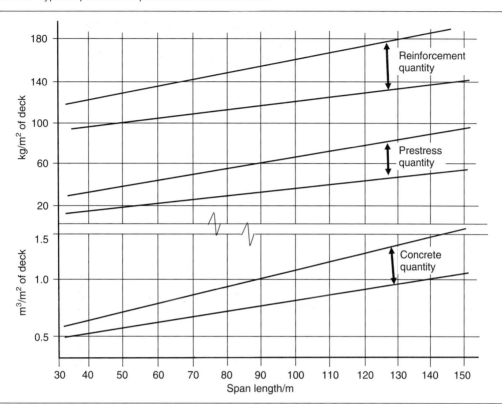

reinforcement, as early thermal effects and differential shrinkage are less; whereas launched box girders have more reinforcement to cater for the temporary loading and structural requirements during the launch.

Prestressed beams usually require the minimum prestress quantities, while precast segmental decks produce savings compared to in situ decks by reducing both the creep in the deck and the long-term losses in the tendons. Incrementally launched box girders need additional prestress to cater for the launch conditions. If external tendons are used instead of internal tendons the prestress quantity can increase due to a reduction in efficiency in resisting the bending effects, but this is offset to some extent by a reduction in friction losses.

The different design codes and loading criteria significantly affect the material quantities in a deck. Standard rail loading is heavier than highway loading, requiring more prestress, while the lighter loads on footbridges reduce the quantities. Bridge decks designed using BS 5400 (BSI, 1990) tend to result in higher reinforcement and prestress quantities when compared with equivalent structures designed to AASHTO (2011) or Eurocodes (BSI, 2004, 2005), although Eurocodes are heavily dependent on the choices in individual countries' national annexes.

REFERENCES

AASHTO (2011) Load and resistance factor design (LRFD) bridge design specifications, 4th edition. AASHTO, Washington, DC.

BSI (1990) BS 5400: Steel, concrete and composite bridges. Part 4. Code of practice for the design of concrete bridges. BSI, London.

BSI (2004) BS EN 1992–1-1:2004: Eurocode 2: Design of concrete structures. Part 1–1. General rules and rules for buildings. BSI, London.

BSI (2005) BS EN 1992–2:2005: Eurocode 2: Design of concrete structures. Part 2. Concrete bridges – design and detailing rules. BSI, London.

Ciria (1996) *Report 155. Bridges – design for improved buildability*. Ciria, London.

Highways Agency (1996) *The Appearance of Bridges and Other Highway Structures*. HMSO, London.

Prestressed Concrete Bridges, 2nd edition
ISBN: 978-0-7277-4113-4

ICE Publishing: All rights reserved
doi: 10.1680/pcb.41134.149

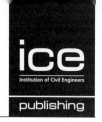

Chapter 8
Analysis of prestressed concrete bridges

Introduction

When analysing prestressed concrete bridges the different aspects discussed in Chapter 5 must all be taken into account. The staged build-up of self-weight, prestress and applied loading have to be combined in the overall structural behaviour, and the time-dependent effects, such as creep, shrinkage and loss of prestress force, included. Traditionally, these effects are considered separately and then combined to give the overall design requirements. In the 1990s, computer software programs evolved which combined the effects within a single analysis, while more recently specialist bridge design software packages have developed further to allow the most complex structures to be analysed.

The principles behind structural analysis and their application in bridge design are well documented and fully covered in numerous publications, such as Shanmugam and Narayanan (2008) and Hambly (1991). The following sections review the way that prestressed concrete bridge decks are usually analysed and the practical problems that may be encountered.

This chapter begins by describing the traditional approach to analysing prestressed concrete bridge decks, and it goes on to review one of the specialist software packages available to show how the approach to analysis is evolving.

Traditional approach to deriving forces, moments and shears

It has been traditional in bridge design to consider longitudinal and transverse designs, different loading conditions and long-term effects separately. They are then combined in order to check the design requirements. Forces and moments are derived from analysis using either 2D or 3D line beam or grillage models. Applied loading is directly applied within the model with prestress effects either generated by the use of equivalent loads or calculated independently.

In recent years, the availability of 3D finite-element specialist bridge software packages has enabled the longitudinal and transverse effects to be determined within a single analysis model, although it is still usual for the design in the two directions to be carried out separately and the requirements combined in the drawings.

Dead load and applied loading analysis

To analyse multiple beam structures a grillage model is established to derive the distribution of moments and shears under the different applied loading conditions. The deck is modelled as a series of discrete members both longitudinally and transversely, as illustrated in Figure 8.1. The moments and shears on each member are extracted from the grillage model and fed into the design process.

Figure 8.1 Grillage model for analysis

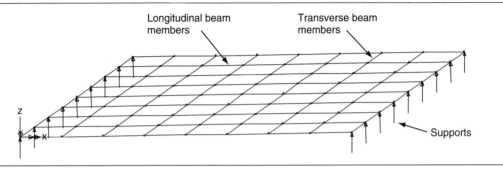

Longitudinally it is simplest to provide a grillage member along the centreline of each beam or web. The section properties of each member should include a portion of the top slab up to midway between adjacent beams or webs and, similarly, a portion of the bottom slab where 'U' beams or box beams are used. With box structures represented by individual grillage members for each web, the torsional stiffness of the box is calculated and a quarter of this is assigned to each web with the remaining half of the torsional stiffness assigned to the transverse members.

Transversely, grillage members are positioned at each diaphragm location with section properties to match the diaphragm beam and at regular intervals between, with section properties to match the length of top and bottom slabs represented. The location of the transverse members should give a relative spacing of longitudinal beams to transverse beams of between 1:1 and 1:2 to achieve reasonably accurate structural behaviour.

For single-beam-type structures the force and moment distributions in the longitudinal direction are derived from a line beam analysis. Figure 8.2 illustrates a typical 2D longitudinal analysis showing the dead load moments for a three-span frame structure. From this type of analysis the bending moments, shears and axial loads in each of the members are derived for the dead load, superimposed dead load and live load, and input into the design process. Differential settlement of the supports is

Figure 8.2 Longitudinal dead load moment from frame analysis

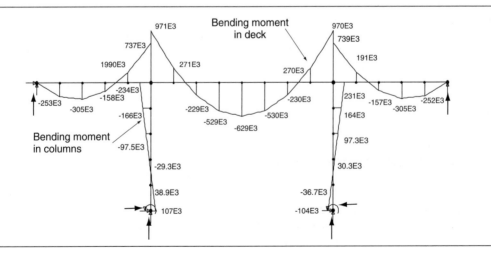

Figure 8.3 Frame model for transverse moment analysis

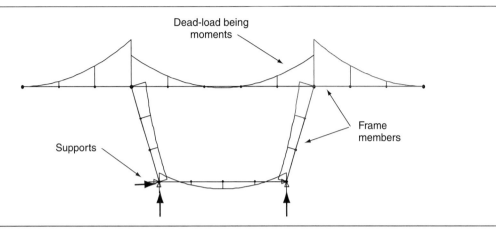

modelled by varying the support levels, and temperature variations are included by applying strain variations on to the model. Torsional effects are calculated manually using simple beam theory.

The transverse analysis for box-type structures is carried out on a 2D frame model, with a typical output for dead-load moment illustrated in Figure 8.3. By placing supports under the webs and applying the different loadings imposed on the section, the forces and moments around the box are obtained. Supports to the box section vary from 'rigid', where the webs sit directly on the bearings, to more flexible at mid-span, where the webs can deflect relatively to each other and are supported by the longitudinal structural action of the deck.

This difference in effective support stiffness results in variations to the distribution of forces and moments around the box. To represent this behaviour the analysis is carried out with rigid supports and again with 'spring' supports, with the worst case taken at each section around the box. The value of the spring support is derived from the longitudinal analysis model, and it is a function of the longitudinal deflection of the beam under a unit load.

Modern computer capacity and speed have made the use of 3D finite element analysis software more common when designing structures. A single model, as illustrated in Figure 8.4, can be used to analyse the complete structure and derive both the longitudinal and transverse effects. The advantage of a full finite element analysis is that the distribution of the load in both the longitudinal and transverse directions is more accurately modelled, resulting in a more refined and economic design. With this type of analysis the stresses or forces in the elements in all three directions are obtained directly. The stresses are input into the design to determine the prestress requirements, and the shears and moments are used to check the reinforcement requirements.

With grillage and line-beam analysis, effects such as shear lag and warping of the section are not included; however, a full-length 3D finite element model includes these effects automatically in the overall structural behaviour.

Deriving the prestress forces and moments

Design of the prestress is an iterative process that involves balancing the stresses generated by the self-weight and applied loading with those from the prestress. The secondary moments are usually a

Figure 8.4 Full-length 3D finite element model

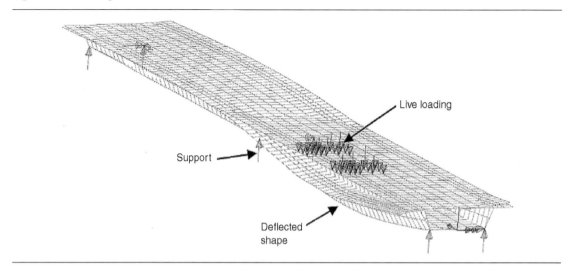

significant part of the prestress effect and the design process involves estimating the prestress required and then determining the primary and secondary effects so that these can be used to check the resulting stress levels. If these stresses are outside the allowable limits, further adjustments are made to the pre-stress arrangement until the final stresses are acceptable. It is important that the modelling and analysis of the prestress effects are set up to easily incorporate any subsequent adjustments to the prestress layout that may be necessary.

There are two stages in deriving the prestress effects. The first is to calculate the force in each of the tendons and the second is to establish the primary and secondary effects.

There are proprietary computer programs available that determine the force profile along a tendon or, alternatively, a simple spreadsheet can be established using the formula in Chapter 5, 'Friction losses and tendon extension'. The spreadsheet in Figure 8.5 illustrates the force profile for a single end stressed tendon during stressing, and after lock-off. The expected extension during stressing is also calculated. The force profiles for each tendon are combined to give the total prestress axial force and primary moment at any section along the deck.

The secondary effects can be derived using a number of different techniques, the most common being the equivalent load method where the forces from the tendons are applied directly to a grillage or simple line beam model, as illustrated in Figure 5.5(ii). The equivalent load approach is described in papers by Aalami (1990) and Catchick (1978).

The geometry of the model must closely match the structure's layout, and the prestress load must be precisely applied to ensure that the structural behaviour is accurately represented. If the tendon is inclined at the anchor, both a vertical and horizontal load is applied to the structure as well as the moment generated when the tendon is offset from the neutral axis. At each change in angle of the tendon an equivalent load equal to the force in the tendon multiplied by the angle change (in radians) is imposed on the structure. For a curved tendon the equivalent radial force per metre along the curve is equal to the force in the tendon divided by the radius of the curve.

Figure 8.5 Tendon friction-loss spreadsheet

SPREADSHEET FOR CALCULATION FOR POST-TENSIONING FORCE IN TENDONS

TENDON REF. **1** $Po =$ **202.5** kN

$\mu =$ **0.2**   Anchor Pull-in = **6** mm

STRESSING END NODE No. = **1** $k =$ **0.00700**

Area = **150** mm^2 E = **195000** N/mm^2

Node No.	Distance (m)	Vertical angle	Horiz. angle	Total angle(rad)	Force Px (kN)	Draw-in Dist. (mm)	Extension (mm)		Distance (m)	Reverse force	Total force after lock-off
1	0	0	0	0.0000	203	0.00	0.0		0.00	172	**172**
2	1	0	0	0.0000	202	0.010	6.9		1.0000	172	**172**
3	2	0	2	0.0349	201	0.183	6.9		2.0000	174	**174**
4	3	0	2	0.0698	199	0.469	6.8		3.0000	175	**175**
5	8	10.5	0	0.2532	190	3.664	33.3		8.0000	184	**184**
6	13	7	0	0.3754	184	7.895	32.0	187	10.7605	187	**184**
7	18	0	0	0.3754	183		31.4				**183**
8	20	0	0	0.3754	183		12.5				**183**
9	24	0	0	0.3754	182		24.9				**182**
10	28	3.5	0	0.4365	178		24.6				**178**
11	30	0	0	0.4365	178		12.2				**178**
12	33	0	2.2	0.4749	176		18.1				**176**
13	35	0	2.2	0.5133	174		12.0				**174**
14	38	6.8	0	0.6321	169		17.6				**169**
15	43	0	0	0.6321	168		28.8				**168**
16	48	0	0	0.6321	167		28.6				**167**
17	57.5	0	0	0.6321	165		53.8				**165**

Lock-off profile

Maximum force in tendon

TOTAL EXTENSION **351** mm

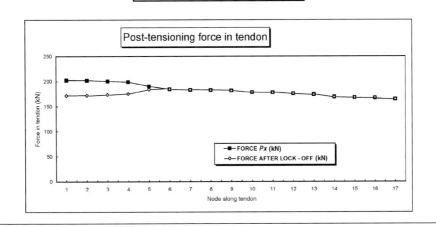

Post-tensioning force in tendon

FORCE Px (kN)
FORCE AFTER LOCK - OFF (kN)

The member layout in the model must accurately follow the neutral axis of the structure to ensure that the change in prestress primary moment is generated when the deck section changes. Friction losses in the tendons are incorporated by applying 'equivalent' forces and moments on the members along the model.

The advantages of the equivalent force method are that it is suitable for a wide range of structural arrangements and that it gives the combined primary and secondary effects. The disadvantages are that it is not always straightforward to establish accurately the analysis model and equivalent loads, and modelling the prestress losses can be complex.

An alternative approach to deriving the secondary effects is the influence or flexibility coefficient method, which readily takes into account changes in section and prestress losses. With a prestress

Figure 8.6 Prestress moments from influence coefficients

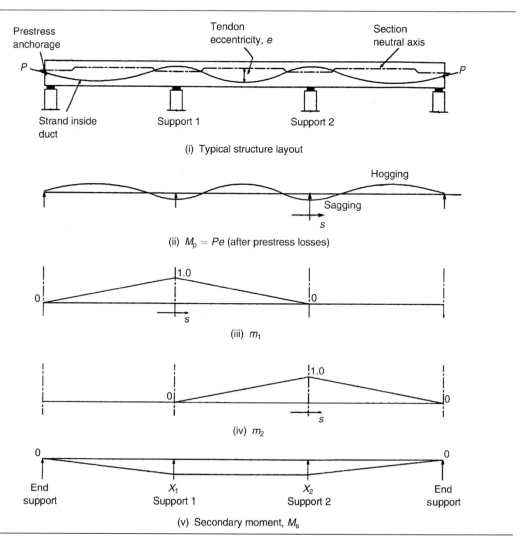

(i) Typical structure layout

(ii) $M_p = Pe$ (after prestress losses)

(iii) m_1

(iv) m_2

(v) Secondary moment, M_s

layout as shown in Figure 8.6(i), and by using the theory of least work, the following equations can be established:

$$f_{11}X_1 + f_{12}X_2 = -U_1$$

$$f_{21}X_1 + f_{22}X_2 = -U_2$$

where

$$f_{11} = \int_s \frac{m_1^2}{EI}\, ds$$

$$f_{12} = \int_s \frac{m_1 m_2}{EI}\, ds = f_{21}$$

$$f_{22} = \int_s \frac{m_2^2}{EI}\, ds$$

$$U_1 = \int_s \frac{m_1 M_p}{EI}\, ds$$

$$U_2 = \int_s \frac{m_2 M_p}{EI}\, ds$$

In Figure 8.6(ii), M_p is the free moment generated by the prestress on the beam. The moments for unit release at the intermediate supports are m_1 and m_2. If the deck is considered as a series of individual simply supported spans, when M_p is applied the beams deflect, causing relative rotations on either side of the supports. The moments required at the ends of the spans to counter the relative rotations are the secondary moments in the beam and are derived from the above equations.

By solving the equations X_1 and X_2 are determined: these are the secondary moments generated at their respective support. This approach can be extended to derive secondary moments for structures with many spans and is simplified by the use of spreadsheets. The 'secondary' shear forces and reactions at the supports are determined by considering the changes in the secondary moments along the structure. Once the spreadsheet is set up it is easy to input the necessary data, making this a simple, yet effective approach where more sophisticated computer software programs are not available.

After the derivation of both primary and secondary effects they are combined to obtain the total prestress effect. These are initially calculated with no long-term losses in the prestressing force. The losses due to elastic shortening, relaxation, creep and shrinkage are taken into account by estimating the percentage loss and reducing the primary and secondary effects accordingly.

Stage-by-stage and creep effects
Prestressed concrete bridges are usually built in stages with the resulting stage-by-stage forces and moments being locked into the structure and subsequently modified by creep behaviour, as discussed in 'Construction sequence and creep analysis', Chapter 5. Both dead-load and prestress moments and forces are affected by the stage-by-stage build-up of the structure.

Traditionally, each stage of the construction is modelled and analysed separately, using the techniques described above, to give the dead-load and prestress effects for that stage. The results from each stage analysis are combined with the results from all the previous stages to give the total moment and force distribution in the structure at the stage being considered. The moments, shears and prestress force from earlier stages are modified to take into account the long-term losses in the prestress and the creep redistribution of moments and shears.

In the final condition, the 'instantaneous' dead-load and prestress secondary moments are calculated for the completed structure after all long-term prestress losses have occurred, and these values are used in the creep assessment to determine the final design.

Combining effects

Once all the moments and forces have been derived for dead load, imposed loads and the prestress effects, they are combined to enable the design to be checked for the serviceability and ultimate-limit state requirements.

When using the traditional design approach, as described above, the effects are usually combined using spreadsheets to organise the data and output from each of the different analyses carried out at each stage of the construction.

Using standard structural analysis software and the traditional approach is time-consuming and complex, although it does give the designer a good understanding of the structural behaviour under the different effects. An alternative is to use specialist software as described below.

Specialist software for the analysis of prestressed concrete bridges

Many firms have developed their own specialist in-house software over the years to deal with analysing prestress effects, but frequently these only cater for some of the effects, and the designer is left to combine the results from other parts of the design process. A number of specialist proprietary software packages are now available which will combine most of the aspects to be considered when designing prestressed concrete bridges. The following sections show how these proprietary packages can be utilised. The specialist software program Midas Civil is used as an example to demonstrate their versatility. Midas Civil is a 3D Finite Element programme that models the structure as a series of beams or plate elements, and it includes the prestress as special line elements that can be input to follow the alignment of each individual prestressing tendon used.

General description

The main characteristic of specialist software programs like Midas Civil is their ability to combine the basic structural analysis with the time-dependent and load-history analysis that is essential in the design of prestressed concrete bridges. Many programs model each individual prestress tendon and will analyse the construction of composite structures with precast and in situ members. They also include the segmental and stage-by-stage construction of both pre-tensioned and post-tensioned concrete decks.

Bridges that are built by the balanced cantilever method, launched into position, built span by span or cabled-stayed can all be analysed at each stage in their construction with the prestress losses, creep, shrinkage, concrete ageing and creep recovery taken into account for the time that has elapsed since the start of construction. It is possible to model the temporary supports, temporary prestress, launching nose or other temporary works in the construction sequence to closely match the actual conditions that will be experienced on site and imposed on the structure in its final state.

By taking all the different aspects into consideration, and using the ability of the computer to handle large quantities of information and complex structural behaviour, it is possible with programs like Midas Civil to analyse a bridge in greater detail than has been possible in the past. More accurate modelling of the concrete behaviour and its long-term properties is possible than when using the traditional techniques discussed above. The rapid analysis provided by specialist software allows the designer to easily look at the sensitivity of the design to changes in the different parameters, which is useful during both the design and construction phases of a project.

The structure is usually modelled as a system of beam or plate elements connected at nodes with supports and other restraints imposed as appropriate. Prestressing tendon elements and stay-cable elements are included in the model and connected to the frame elements at the nodes. Time is divided into a number of steps to reflect the construction sequence and schedule. The program models the build-up of the structure and calculates the structural response at the end of each time step, or stage, incorporating the changes that have occurred since the previous step.

In common with other similar software programs, Midas Civil has a comprehensive set of input data. The Base model enables all the different parts of the structure and materials to be defined. Grouping of elements, nodes and tendons allows for ease of inputting and outputting data to assist in the design process. The loading can be defined in terms of individual load cases, as combinations of previously defined load cases and as envelopes of load effects. A moving load facility permits a user-defined train of loads to be run across the structure to extract the maximum effects or to generate influence lines. Forces and extensions of the prestress tendons are extracted from the analysis with the friction and time-dependent losses included.

Midas Civil input and output data and the graphics interface are all accessed through the display, which integrates the different operations into a single screen, as shown in Figure 8.7. The menu

Figure 8.7 Midas screen layout and modelling example

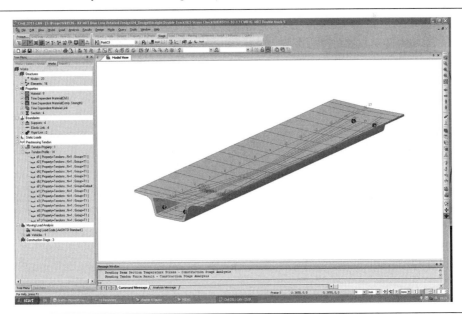

gives access to create and edit the different input files and to run the different stages of the analysis. The graphical display is used to view and check the input model and to see the different types of output. The output from Midas Civil combines the self-weight and applied loading with the prestress, including concrete creep and shrinkage effects, prestress losses and the stage-by-stage build-up of forces and moments. The moments, forces, stresses or deflections are presented in graphical form or in a text file ready for use in the design.

Input data

The input data for the Basic Midas Civil module is arranged in a series of sections as described below and as highlighted in the figures in this section of the chapter. Many of these are self-explanatory and demonstrate the different factors that are taken into account in the analysis.

ANALYSIS CONTROL input sets the type of analysis to be carried out and allows the time-dependent effects to be included, while geometric and material non-linear behaviour can be instigated. Different types of dynamic analysis can also be set up, while stage-by-stage analysis parameters are established.

STRUCTURES define the nodes and elements used in the analysis. The nodes establish the points that are used in the structural model. The elements range from beam and plates to trusses and stays. Complex 3D geometry can be introduced by importing CAD (computer-aided design) files, linking the design and drawing process.

PROPERTIES define the sections and materials used in the analysis. The different material properties can be called up from a library of standard materials or a user-defined material can be established. In this part the time-dependent behaviour of materials is defined, including the creep and shrinkage of the concrete and the variation of concrete strength and E-value with time, as shown on the screen in Figure 8.8. Again, standard properties can be called up from an established library relating to the

Figure 8.8 Concrete time-dependent properties input

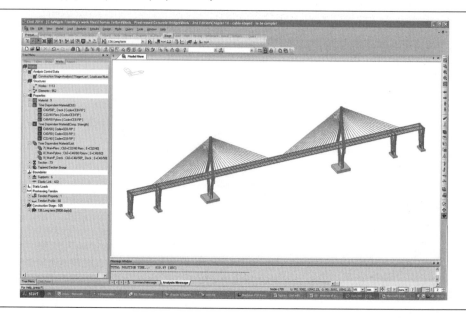

different design codes used around the world, or a user-defined behaviour can be assigned to different materials. This part is also used for setting up the section properties of each beam, plate or truss. The section properties can be developed from standard shapes, including box girder and 'I' sections, as well as the usual rectangle and round sections. A useful facility in Midas Civil is to be able to create a range of tapered sections along a member. It is possible to establish the section properties of a haunched box girder deck by simply establishing the deck sections at the pier and again at mid-span, with a soffit profile defined to create a series of tapered sections between the nodes along the deck.

BOUNDARIES include the supports and any links between the different parts of the structure. The supports are set with the appropriate stiffness in all six degrees of freedom, covering both the linear and rotational restraints. Rigid and elastic links can be used to link nodes in the structure, which is particularly useful when connecting the bridge deck to the piers, or the piers to the supports, where the axis of the two is offset but they are connected through the structural behaviour.

STATIC LOADS includes all the usual loading applied to the structure. The self-weight, surfacing, parapets and other permanent loads are all defined here. Other common standard static load cases include wind loading, temperature and settlement. The prestress loading is also defined, based on the tendon profiles and properties defined elsewhere. Figure 8.9 shows a three-span box girder deck with integral piers, with the parapet loading shown on the screen providing a simple check to ensure the loading is applied correctly.

PRESTRESSING TENDON input defines the tendon profiles and the tendon properties. The profiles are assigned to the adjacent structural members and are either internal bonded tendons or external unbounded tendons. The tendon profile has to be input carefully to ensure that the correct location of each tendon is achieved at the critical points and also to accurately model the friction losses along the length of the tendon.

Figure 8.9 Loading input and graphic display

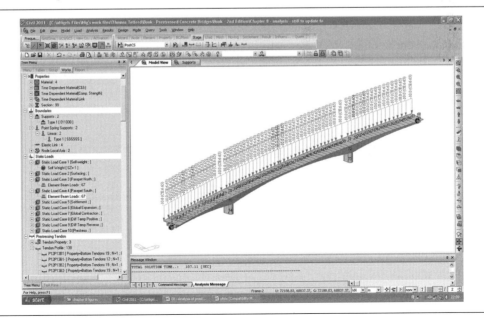

Figure 8.10 Staged construction viewed on the screen

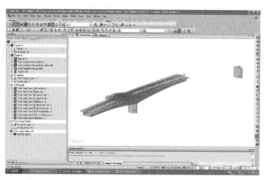

(a) First section of deck built as cantilever, with prestress tendons installed

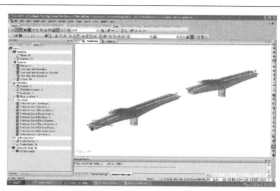

(c) Deck cantilevers continued out to mid-span

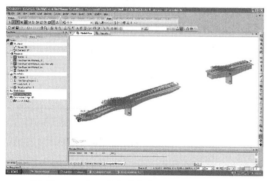

(b) First section of deck built as cantilever, with prestress tendons installed

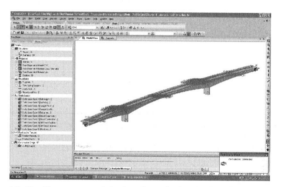

(d) Cantilevers completed and continuity tendons installed

MOVING LOAD ANALYSIS allows live load vehicles to be moved over the structure in small steps to generate the worst-case moments and shears at every element in the structure. BS 5400, Aashto and Eurocode loading patterns are automatically generated to cover the traffic lanes defined along the structure.

CONSTRUCTION STAGES allow every step of the construction to be modelled in the analysis, with the timeline established through the stage-by-stage build-up of the structure. This allows the creep, shrinkage and the prestress secondary moments to act on the partially completed structure, and to be modified as the structure is completed and matures in the long term. Figure 8.10 shows a series of construction stages for a concrete box girder deck built by the balanced cantilever technique. By showing the tendons and loading on the screen it simplifies the checking process, while the deflections and stresses may be displayed to ensure the overall structural behaviour is as expected.

The input data is input through the screen and dropdown menus, or by tabular input as shown in Figure 8.11. The tabular input may be generated in a spreadsheet such as Excel, and imported into Midas Civil.

Figure 8.11 Midas tabular input facility

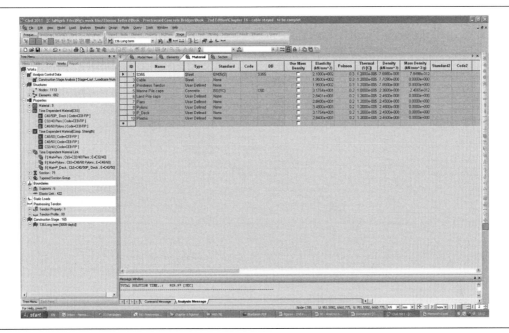

Analysis of the structure

The structure is analysed in a stage-by-stage process with the program building up the structure in the defined sequence and time. Each element is built or installed, and the tendons are stressed in the same order and period as expected on site. At the beginning of each time-step the stress, strain and displacement within the structure is known and this equilibrium state is used as the starting point for carrying out the analysis of the next time-step. The changes in the structure that occur during the next step are then superimposed onto the structure with the time-lapse taken into account when the time-dependent effects of the concrete and prestress losses are included. Because of the non-linearity of the time-dependent solutions, an iterative procedure is adopted in finding the solution to each time-step, with convergence criteria set to achieve sufficient accuracy. At the end of the time-step the new stress, strain and displacement distribution are stored ready for use in the next step of the analysis. The concrete parameters are also updated, including the current 'E' value and residual creep, shrinkage and relaxation.

For composite construction, such as a precast beam with in situ deck slab, the section is modelled as a single element with two parts. In the analysis, the separate parts are installed in the structure independently at the appropriate time. The first part acts independently until the second part is cast and starts to harden, at which time the two parts act together compositely. The redistribution of forces and moments due to differential shrinkage and creep between the two elements are automatically taken into account at each time step.

The boundary conditions at any node can be defined or changed at any time during the step-by-step analysis. Displacements can be applied to the structure to match those that might occur on site. In this way, the supports and restraints to the structure are installed, removed or adjusted as the construction proceeds.

Figure 8.12 Graphical output

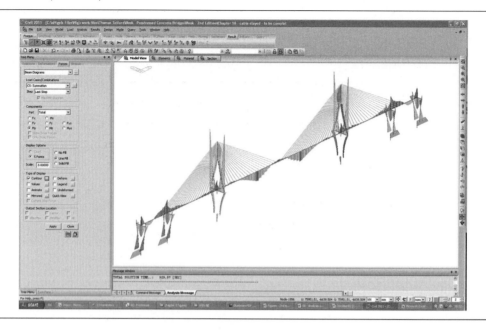

The differential temperature effects on the structure are analysed by defining the ambient temperature and temperature gradient across each element. The expansion or contraction and bending of the element is then imposed onto the structural model.

The program takes into account the friction losses and anchor lock-off effect when calculating the prestress force at the time of installing the tendon. When subsequent tendons are stressed the forces in the previously stressed tendons are adjusted due to the elastic deformations within the structure. Subsequent losses in the tendon force due to creep and shrinkage of the concrete are automatically incorporated, while the relaxation within the tendon over the time-step elapsed is calculated within the analysis and imposed onto the structural system.

Output

Midas Civil prepares the output in both graphical and tabular formats. At every step in the analysis it is possible to output the nodal displacements, element forces, support reactions and spring forces, element stresses and tendon forces. It is also possible to output the required camber to which the structure must be cast so that the structure achieves its required profile at the end of the construction. A typical example of the graphical output showing the bending moments in the members of a cable-stayed bridge is shown in Figure 8.12. By grouping elements it is possible to just show the elements currently being designed, simplifying the extraction of the results for each section. As an alternative to the graphical display, the output may be extracted in tabular format, as shown in Figure 8.13. The tables are transferred into prepared spreadsheets for the post-processing of the results and checking against the codified requirements.

Summary

Proprietary software, such as Midas Civil, bring together all the different aspects that are considered in the analysis of prestress concrete bridges and simplify the design process. This has also made the

Figure 8.13 Tabular output

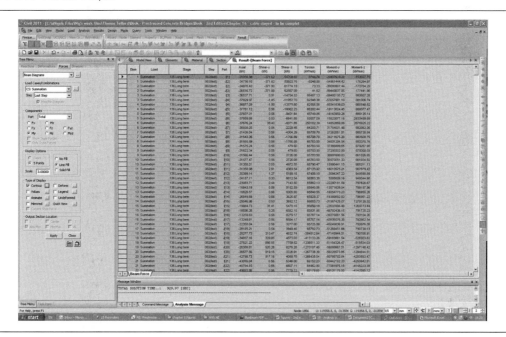

analysis and design more accurate when the time-dependent effects such as creep, shrinkage and pre-stress loss are considered; previously this had only been an informed estimate at best.

REFERENCES

Aalami BO (1990) Load Balancing: A comprehensive solution to post-tensioning. *ACI Structural Journal* **87(6):** 162–170.

Catchick BK (1978) Prestress analysis for continuous beams: some developments in the equivalent load method. *The Structural Engineer* **2(56B):** 29–36.

Hambly EC (1991) *Bridge Deck Analysis*, 2nd edition. Chapman and Hall, London.

Shanmugam NE and Narayanan R (2008) Structural analysis. In *ICE Manual of Bridge Engineering*, 2nd edition. Thomas Telford, London.

Prestressed Concrete Bridges, 2nd edition
ISBN: 978-0-7277-4113-4

ICE Publishing: All rights reserved
doi: 10.1680/pcb.41134.165

Chapter 9
Slab bridges

Introduction

Solid or voided prestressed concrete slabs are used for short spans and where good access is available for their construction. Using slabs simplifies the formwork and concreting operations, although the decks are relatively heavy and this makes prestressing inefficient. Slab decks were regularly used in the early period of prestressed bridges and are still used today on suitable projects. Figure 9.1 shows the Balbriggan Bridge in Ireland, a recent example of a bridge where this type of deck has been used, with the prestressing allowing a slender deck profile.

Solid slabs are used only for short spans. With spans above 15 m the deck is likely to be voided to reduce the deck section and dead load. For short-span lengths either plain reinforced concrete decks or prestressed beam-and-slab arrangements usually result in a more efficient and economical solution. For these reasons prestressed solid and voided slabs are not widely used in bridgeworks. However, prestressing of the concrete improves the structure's durability by reducing cracking, and it results in a more slender deck when compared to a reinforced concrete design – while many consider a slab

Figure 9.1 Balbriggan Bridge, Ireland (reproduced courtesy of Roughan & O'Donovan, Dublin, copyright reserved)

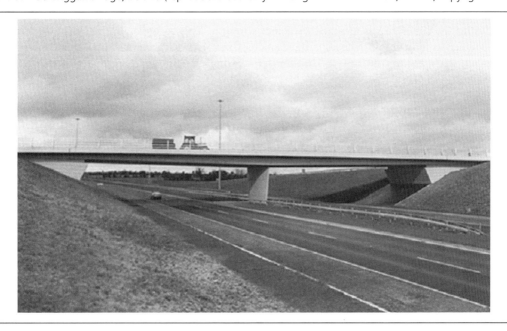

Figure 9.2 M4 Overbridge, Ireland (reproduced courtesy of Roughan & O'Donovan, Dublin, copyright reserved)

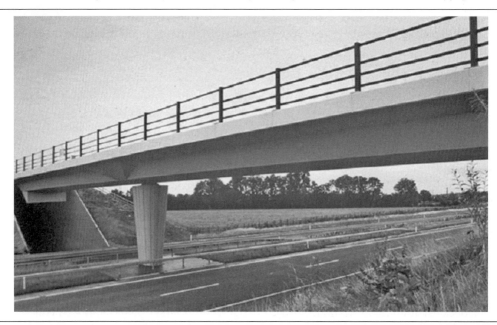

arrangement to be better aesthetically as compared to a beam-and-slab deck. Figure 9.2, showing an overbridge on the M4 in Ireland, illustrates a typical arrangement for this type of construction and demonstrates the pleasing appearance that is achievable.

Solid-slab bridges

Solid-slab in situ concrete decks are cast on full-height falsework with a simple formwork, reinforcement and prestress layout. The slabs are post-tensioned with either bar or multi-strand tendons that are installed after the concrete has attained sufficient strength. This arrangement is used for short spans where good access is available. The advantages include straightforward construction and simple formwork layout, while the disadvantages are that prestressing a solid concrete element is inefficient and dead load becomes excessive as span lengths get longer.

Prestressing of solid concrete slab decks has also been used to achieve aesthetic structures such as the proposed Marlborough Street Bridge in Dublin, shown in Figure 9.3. The deck is a slender prestressed concrete solid slab arrangement with a curved profiled soffit and integral abutment giving a partial arch behaviour. With a span of approximately 40 m and a minimum depth of 400 mm at midspan the span to minimum depth ratio of 100 is exceptional, while the profiled concrete soffit forms an elegant addition to the array of bridges crossing the Liffey River in the heart of Dublin.

Several short-span bridges have used sections of precast concrete slabs, lifted into place onto temporary supports and then prestressed together with post-tensioned tendons. The connections between the sections are made either with a narrow mortar or concrete stitch, or as a glued match-cast joint.

Precast pre-tensioned beams placed closely together and topped with in situ concrete to form a solid-slab construction are suitable for very short spans. By placing the beams side-by-side no formwork is required for the in situ concrete, resulting in a rapid and robust construction.

Figure 9.3 Marlborough Street Bridge montage, Dublin (reproduced courtesy of Roughan & O'Donovan, Dublin, copyright reserved)

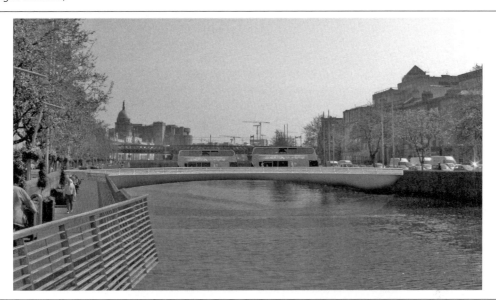

Voided-slab bridges

In situ voided-slab decks were commonly used during the early period of prestressed concrete bridge development within the span range of 25 m to 35 m and with span-to-depth ratios of up to 20:1. A typical cross-section is illustrated in Figure 9.4. The voids usually extend over only the centre portion of the span with a solid section used over the supports. In this way, the dead-load moments are reduced while the full shear capacity is maintained at each end of the span.

The prestress tendons usually extend the full length of the deck, anchored on the end faces. For continuous decks extending over several spans, multi-strand tendons follow a draped profile within the concrete 'webs' between the voids.

Figure 9.4 Voided-slab deck section

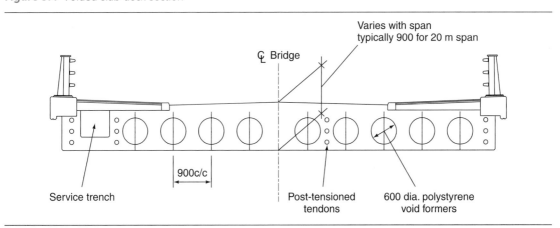

Figure 9.5 Void formers in deck

The concrete is cast within formwork supported by falsework built up from the ground. The voids are formed using either round or rectangular-shaped polystyrene blocks with bevelled corners, as seen in Figure 9.5. The void former shape allows the concrete to flow around the polystyrene and to fill the space underneath. A minimum concrete thickness of 150 mm is provided above and below the void for the reinforcement and concrete cover. Drainage pipes are provided at the low point in each of the voids to prevent any build-up of water during and after the curing of the concrete.

As the polystyrene void formers weigh less than the fluid concrete, they tend to float up as the concrete is placed and vibrated. To keep the void formers in place they are held down by ties through the soft formwork to the falsework below or by placing large weights above the deck supported on the polystyrene, as in Figure 9.6. The weights are removed after the initial set of the concrete beneath the voids, allowing the top concrete surface to be completed and finished off.

The advantages of voided-slab decks include simple construction and straightforward formwork arrangement, while the disadvantages include the requirement to hold down the voids during concreting and a heavy deck section. In recent years this has resulted in beam-and-slab or multi-cell boxes being preferred on most projects.

Figure 9.6 Weights holding void formers down

Design of slab bridges

Analysis of the slab, with or without voids, is carried out with sufficient accuracy using a beam grillage model. The longitudinal beam grillage members match the layout of the prestress tendons, or groups of tendons, and the transverse beam members are placed over the supports and at suitable spacing between to give an evenly balanced grid pattern. Hambly (1991) gives a detailed description of the behaviour and analysis of slab bridges using this method. The slab arrangement is efficient in distributing the applied live load transversely across the section.

The section properties of both the longitudinal and transverse beam members are based on the deck width to mid-way between the members with the resultant voided section taken longitudinally, and the section at the centre of the void taken for the full width transversely. This is normally found to give sufficiently accurate results for undertaking the design. When assigning the torsional stiffness to the grillage beam members it is usually sufficient to give both longitudinal and transverse members the same value of twice the 'I' value of the longitudinal members. Figure 9.7 shows a 3D beam grillage model used for a single-span slab bridge for the analysis using the Midas Civil software. In this example, by inputting the prestressing tendons and including the creep, shrinkage and other long-term effects, the full analysis was carried out within the one model.

In recent years it has become more common to use 3D plate elements to analyse solid slabs or voided slab structures. Plate analysis models are used to analyse both the longitudinal and transverse behaviour of a voided slab deck. The advantage of this type of analysis model is that the distribution of forces within the slabs and webs is more accurately derived in both the longitudinal and transverse directions. 3D FE (three-dimensional finite element) solid model software can also be used for analysing this type of structure, but it is usually more cumbersome, and more difficult to interpret the output, and this limits its popularity.

Figure 9.7 Grillage analysis model of slab bridge

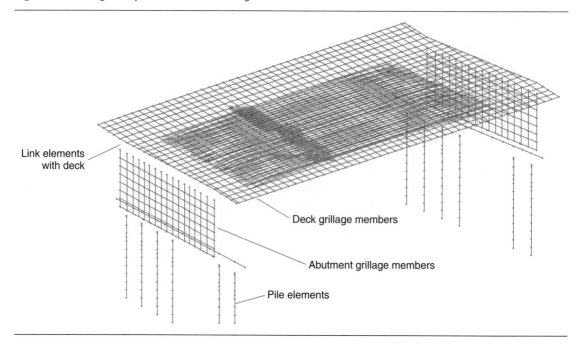

Link elements with deck

Deck grillage members

Abutment grillage members

Pile elements

The longitudinal prestress design is based on the bending moments and shear forces extracted from the grillage or plate model analysis output. The prestress is designed to balance the serviceability limit state stresses, with checks carried out to ensure adequate ultimate limit state strength as described in Chapter 5.

Transversely, the slab is designed as a reinforced concrete element, with reinforcement placed in the top and bottom of the section to take the transverse bending moments and shears derived from the grillage analysis, combined with the local effects caused by the live loading applied on the top slab. Additional reinforcement is placed around the void to tie together the top and bottom sections of concrete and to control cracking in the thinner sections of the slab.

REFERENCE
Hambly EC (1991) *Bridge Deck Analysis*, 2nd edition. Chapman and Hall, London.

Prestressed Concrete Bridges, 2nd edition
ISBN: 978-0-7277-4113-4

ICE Publishing: All rights reserved
doi: 10.1680/pcb.41134.171

Chapter 10
Beam-and-slab bridges

Introduction

Beam-and-slab arrangements are one of the most common forms of prestressed concrete bridge deck and utilise either precast or in situ concrete beams with the deck slab usually cast in situ. This type of deck is well suited for small- or medium-span bridges where the beam weights are small enough to use readily available cranes, and for projects where repetition in the construction gives cost savings. However, beam-and-slab arrangements are generally considered less attractive than the box girder form of construction for the longer spans.

The bridge in Figure 10.1 illustrates a typical beam-and-slab deck arrangement used. The beams span between piers or crossheads with transverse diaphragms at the ends and along the span to provide support and help distribute the loads, while the beams support a thin deck slab which provides the platform for the users above. The deck slab and the beams work together as a composite section to resist the superimposed and live loading. The efficient arrangement of the concrete and prestressing results in less overall material being used than in most other types of prestressed concrete deck construction.

With cast in situ beams, haunched arrangements are possible, as illustrated in the Analee Bridge, Ireland, in Figure 10.2. The haunched profile of the beams provides an efficient structural arrangement and an elegant short-span crossing over the river. In situ concrete beams are more suited for smaller, 'one-off' bridges, where it is difficult to justify setting up a pre-casting yard, if one does not already exist. They are also well suited for bridges in developing countries where the simple construction techniques often give the best results.

Precast beams are usually of constant depth along their length and are frequently used for underpasses or overbridges on highway schemes, as shown in Figure 10.3. In countries where precasting yards are already established, beam-and-slab decks provide a quick and economical form of construction. Many different beam shapes have been developed, with the choice of which to use left to the designer. On large projects, with a long length of viaduct or with many bridges to construct, the cost savings from the repetition associated with precast beams may justify the construction of a special casting yard if there is not an existing one close by.

When building over obstructions, such as a river, the use of precast beams reduces the need for temporary works. Figure 10.4 shows a bridge under construction, on the A55 in Anglesey, where the precast beams spanning the river were placed by crane. The resulting reduction in work carried out at the site, and the shorter construction period, minimised the impact on the environment.

Bridges constructed over railways have their own set of requirements, including restricted possession periods in which the deck is built. Rail operation safety considerations limit the construction options,

Figure 10.1 Typical precast beam deck

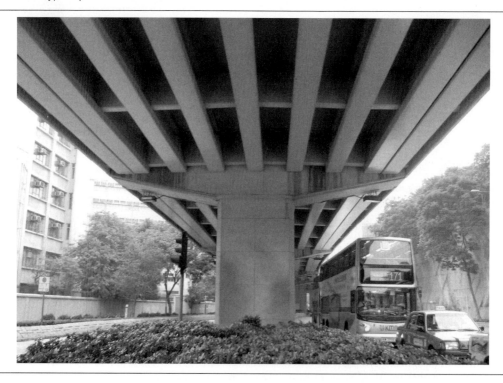

Figure 10.2 Analee Bridge, Ireland (reproduced courtesy of Roughan & O'Donovan, Dublin, copyright reserved)

Figure 10.3 Ghantoot Interchange Overbridge, Abu Dhabi, UAE (reproduced courtesy of Hyder Consulting (UK) Ltd, copyright reserved)

Figure 10.4 Precast beam bridge under construction over river (reproduced courtesy of Hyder Consulting (UK) Ltd, copyright reserved)

and the use of precast beams is an established solution for a deck in this situation. The single-span road-over-rail crossing shown in Figure 10.5 has precast beams lifted into place by a crane during a short possession period, with the deck completed by casting an in situ concrete deck slab over the top.

The span lengths of deck using beam-and-slab arrangements range from 25 m up to 40 m. In exceptional cases, spans of up to 60 m have been constructed, although for spans greater than 40 m the box girder form of construction is usually more efficient. For short spans, in situ reinforced concrete decks are more economic, although over rivers or railways precast beams are usually preferred in order to overcome the access problems.

Precast beams are more common than in situ beams, especially where existing pre-casting facilities exist. Precast 'M' beams and box beams are used for shorter spans and minimise the area of deck slab formwork needed. 'I' and 'Y' beams are the most common type for medium-length spans, while the 'U' and 'T' beams are used for the longer spans. In the UK, 'super Y' or 'SY' beams have been developed for the longer span range and they are similar to the original 'Y' beams except for being deeper and more slender. Figure 10.6 illustrates some of the different beam shapes. The history of precast beams in the UK is described in a paper by Taylor (1998).

The advantages of using precast beams include economies from repetition and rapid construction, as well as minimising the on-site work. The disadvantages include the need for good access to site and for heavy-lifting equipment for the longer beams.

General arrangement

A typical section through a deck with precast 'Y' beams is illustrated in Figure 10.7. The precast beams span longitudinally and support an in situ concrete deck slab, usually cast using permanent formwork

Figure 10.6 Typical precast beam arrangements

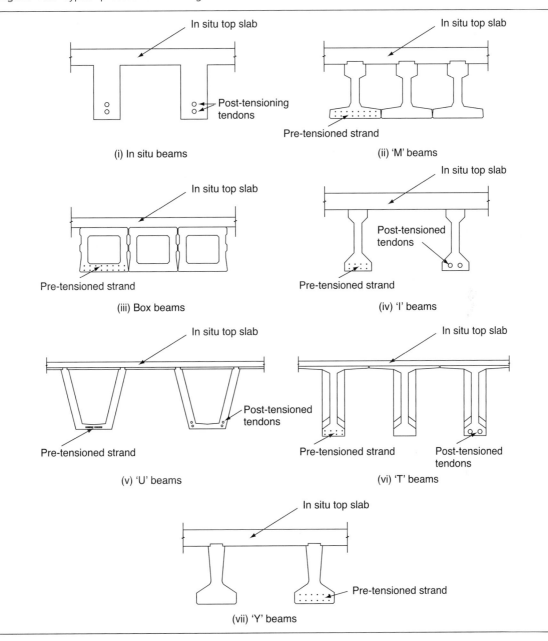

(i) In situ beams

(ii) 'M' beams

(iii) Box beams

(iv) 'I' beams

(v) 'U' beams

(vi) 'T' beams

(vii) 'Y' beams

supported from the beams themselves. Special edge beams with a plain outer face are used to improve the appearance of the deck. The deck slab is provided with a concrete upstand along both sides of the deck to support the parapet and finish off the verge and waterproofing detail.

The spacing of the beams depends on the type and capacity of the beam used. The spacing is usually less than 3 m to allow standard systems of permanent formwork to be used for the deck slab

Figure 10.7 Typical precast beam deck section

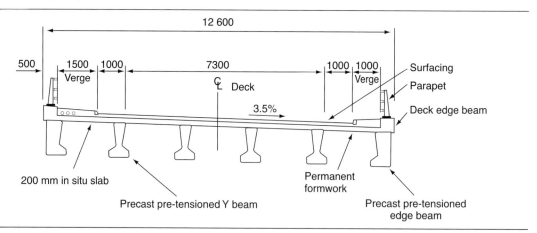

200 mm in situ slab

Precast pre-tensioned Y beam

Permanent formwork

Precast pre-tensioned edge beam

construction, although it may be wider than this if necessary to suit the design and construction requirements. The deck slab is typically between 200 mm and 300 mm thick which is governed by the beam spacing and the live loading carried. Problems have been encountered when the deck slabs are too thin, with the deck concrete deteriorating under traffic loading. When the overall width of the deck slab varies along the span length the beams are 'fanned' out, giving a variation in beam spacing over the span.

Standard beam shapes have developed in countries where precast beams are established and permanent casting yards exist. In the UK, 'M', 'Y' and 'SY' beams are the most widely used, with 'inverted T', 'box' and 'U' beams also available. The USA tend to use widely spaced 'I' beams or 'bulb Ts'. In Australia the 'I', 'U', 'T' and 'super T' beams are common arrangements. The 'super T' is shaped similar to a deep 'U' beam with side cantilevers extending out from the top of the webs on either side. It was first used in southern Europe in the 1980s. Where dedicated precasting facilities are set up for a particular project the beam type and shape is developed to suit the project's requirements.

The choice of which precast beam shape to use is partly governed by the span length and loading, but is often based on the designer's preference.

The standard precast beam types consist of families covering a range of different depths and strand arrangements. The length that each beam is able to span depends on its depth and the beam spacing. The deeper and more closely spaced beams are able to support heavier loads over longer spans. The typical span ranges for the different types of precast beams are given in Figure 10.8, with the typical span-to-depth ratios of the deck usually between 16:1 and 20:1.

The layout of cast in situ beams depends on the deck arrangement required. For short- or medium-span decks the beams are often cast closely spaced, as illustrated in Figure 10.6(i). Where in situ beams are used on longer spans and wide decks an efficient solution is achieved with a 'ladder-beam' arrangement, where larger longitudinal edge beams and smaller transverse crossbeams support the deck slab. The arrangement illustrated in Figure 10.9 spanned 60 m, with the 3.5 m-deep edge beams supporting shallower crossbeams at 5 m centres and with a 250 mm-thick deck slab. Both the edge beams and crossbeams were post-tensioned with multi-strand tendons.

Figure 10.8 Precast beam span range

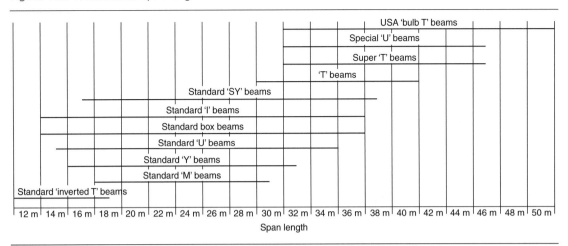

For precast beams the prestressing is achieved with pre-tensioned strand or post-tensioned tendons positioned to take the sagging moments, as indicated in Figure 5.1. Most standard precast beams, produced in established casting yards, use pre-tensioned strand, while post-tensioning is often adopted for the very long beams and when a project-specific casting yard is set up.

At abutments the beams are either supported on bearings or built integral with the abutment wall. At intermediate piers the beams are simply supported or made continuous, as illustrated in Figure 10.10. With a fully continuous arrangement, as shown in Figure 10.10(ii), the beams are connected together longitudinally with an in situ diaphragm over the pier. An arrangement often used is to make the deck slab continuous and the beams simply supported, as shown in Figure 10.10(iii). This arrangement simplifies the construction while reducing the need for expansion joints in the road surface.

Figure 10.9 In situ ladder beam arrangement (reproduced courtesy of Cowi, copyright reserved)

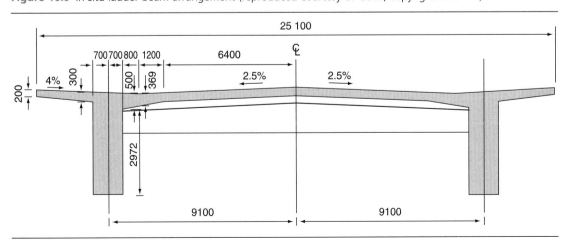

Figure 10.10 Beam continuity at pier

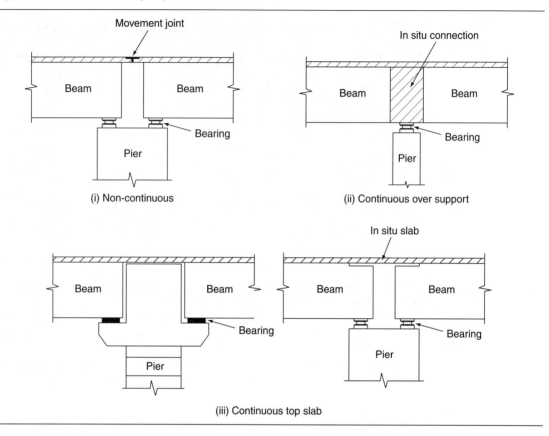

(i) Non-continuous

(ii) Continuous over support

(iii) Continuous top slab

The non-continuous arrangement shown in Figure 10.10(i) is generally not preferred, due to the susceptibility of movement joints to leaking, and the difficulties involved in inspection and maintenance.

At the supports the beams are connected transversely by an in situ concrete diaphragm to provide rigidity to the deck structure. The beams either are individually supported by bearings on a crosshead arrangement at the pier, as shown in Figure 10.11, or use the transverse diaphragms to transfer the beam load to the bearings, as illustrated in Figure 6.12(iii). The height of the transverse diaphragms may be reduced to provide additional space above the crosshead to insert temporary jacks, should it become necessary to replace the bearings in the future.

The beams in Figure 10.11 have a half-joint arrangement at the ends to reduce the overall construction depth of the deck and crosshead at the pier. This reduces the visual impact of the crosshead, although the beam construction is made more complex. Where half-joints are used, sufficient space must be provided around the bearings to allow for inspections and maintenance. In Figure 10.11, it can be seen that the transverse diaphragm under construction at the end of the beams is only over part of the beam depth to allow for access around the bearing.

Transverse diaphragms are sometimes incorporated in the span to improve the distribution of live load between the individual beams. These intermediate diaphragms have become less common in recent

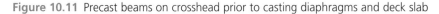
Figure 10.11 Precast beams on crosshead prior to casting diaphragms and deck slab

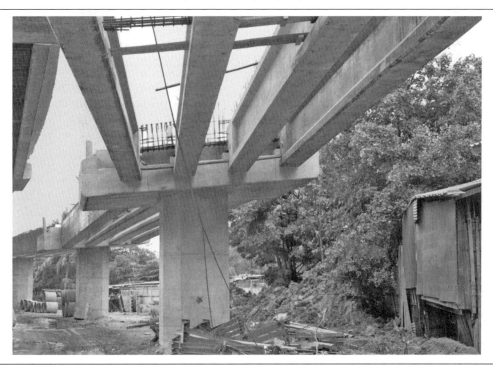

years as designers usually prefer to simplify construction by omitting them and designing the beams accordingly.

Construction of in situ beam-and-slab decks

Where the beams are cast in situ the formwork is supported on full height falsework or trusses spanning between the piers. The full height falsework shown in Figure 10.12 was used where good access and suitable ground conditions were available. Although it is time-consuming and labour intensive to erect and dismantle the full-height scaffolding, the equipment is usually available locally, and the system is versatile, being able to cope with a wide variety of different deck arrangements and site conditions. If sufficient scaffolding and formwork is available several spans may be constructed at the same time, which speeds up the overall rate of construction.

Span-by-span construction using a self-launching truss, as shown in Figure 10.13, is used where access is difficult and a sufficient length of deck justifies the high cost of the truss system. In this example, the formwork is hung from the truss, which is supported off the piers and the previously completed deck. Alternatively, the truss may be placed underneath the deck and directly support the formwork when sufficient headroom is available. Construction usually starts at one end of the bridge and progresses one span at a time, using the completed deck for access. A typical construction cycle would achieve one span every four to six weeks.

The beams and deck slab are usually cast in one pour to give a monolithic construction and to eliminate differential shrinkage effects. After the concrete has hardened the post-tensioning tendons are threaded

Figure 10.12 Falsework for in situ beam-and-slab deck

Figure 10.13 Overhead gantry for in situ beam-and-slab deck (reproduced courtesy of Hyder Consulting (UK) Ltd, copyright reserved)

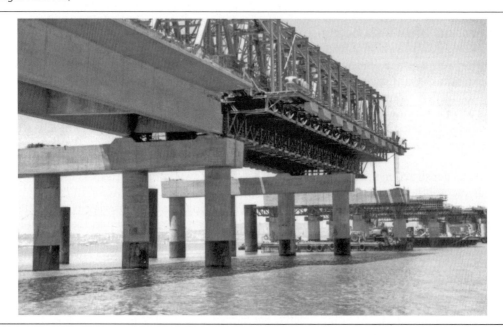

through the ducts cast into the concrete and stressed, lifting the deck off the formwork. The falsework and formwork are then removed or the truss launched forward to cast the next section of deck.

Casting and transportation of precast beams

Precast beams are either pre-tensioned or post-tensioned and are made in special casting yards set up to produce large quantities of beams in factory conditions.

Casting beds consist of a soffit form, side shutters and internal forms. With pre-tensioning it is usual for a bed to be long enough to cast several beams at a time, while with post-tensioning each beam is cast in an individual bed.

Figure 10.14 shows a casting yard for pre-tensioned precast 'U' beams where up to five beams were cast on the bed during each cycle. The reinforcement is seen placed on the soffit shutter ready to be fixed in position. Pre-tensioned strands extend between anchor frames at each end of the bed. The side shutters, shown in the insert, and the internal shutter move down the bed to cast each beam on a daily cycle. The completed beams are seen at the top of the picture in the storage area.

With pre-tensioned beams, jacking frames, as seen in Figure 10.15, are located at either end of the casting bed to provide an anchorage for the strands to be stressed against. These frames are subjected to high forces from the prestressing strand, especially for the larger and longer beams, and have to be anchored into the ground to stop them sliding forward or overturning.

The strands are individually stressed using a mono-jack, as seen in the insert to Figure 10.15. After the beams have been concreted the strands are released using a large jack placed within the frame. This jack

Figure 10.14 Casting bed for pre-tensioned beams

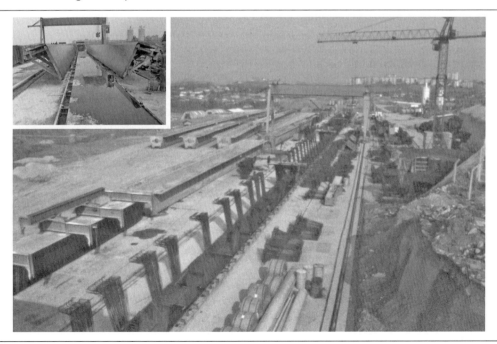

Figure 10.15 Jacking frame at end of casting bed with inset of strand jack

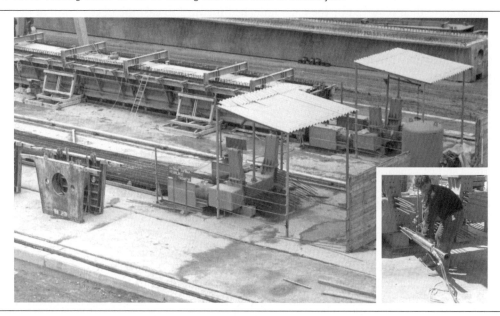

contracts to allow the back section of the frame to move forward, releasing all the strands at the same time. Where several beams are cast on a single bed the strands between the beams are cut and trimmed back flush with the concrete surface.

An alternative to using jacking frames is to anchor the pre-tensioned strand onto the shutters, as seen in Figure 10.16. The strands are tensioned against a steel frame at each end of the beam, with the forces transmitted through the shutter to balance each other. The advantage of this is that the jacking frames and foundations are dispensed with; however, the individual shutters have to be designed to take the high compressive forces, and this adds to their cost.

After installation of the strand, the reinforcement and shutters are assembled and the concrete cast. The side shutters are usually stripped when the concrete has reached a strength of $12\,\mathrm{N/mm^2}$. A concrete strength of at least $35\,\mathrm{N/mm^2}$ is normally required before the strands are released from the jacking frames and the force transferred into the beam. The release of the force in the strands should be done gradually as a sudden release increases the draw-in and anchorage transmission length of the strand.

Steam curing of the concrete is often used to accelerate the strength gain for earlier transfer of the pre-stress and to shorten the cycle time between casting and removing the beams.

To complete the beams the pre-tensioned strands are cut flush with the concrete surface at the ends of the beam, and a water resistant sealant is applied over the strand ends to keep moisture out and reduce the risk of corrosion occurring.

For post-tensioned precast beams the casting cell layout is simpler, as a jacking frame is not needed. A typical arrangement for a precast 'I' beam is shown in Figure 10.17, where the shutters are set up ready

Figure 10.16 Shutter with strands anchored at ends

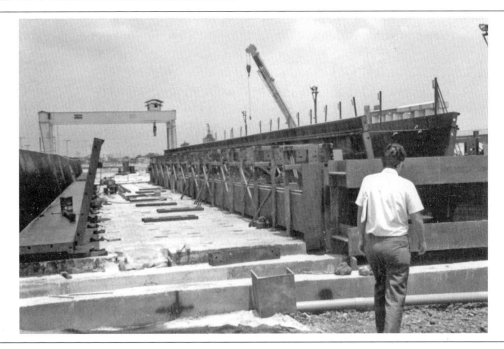

for concreting. The beams are assembled with ducts and tendon anchorages cast in. This allows the tendon to be threaded through and stressed after the concrete has gained sufficient strength. The ducts are stiffened and protected from the possible ingress of concrete by inserting and inflating a pneumatic tube.

Figure 10.17 Shutter for post-tensioned precast beam (reproduced courtesy of Hyder Consulting (UK) Ltd, copyright reserved)

Figure 10.18 Stressing post-tensioned beam (reproduced courtesy of Hyder Consulting (UK) Ltd, copyright reserved)

The stressing of the post-tensioned tendons is carried out with standard stressing equipment, as shown in Figure 10.18. Tensioning all the tendons with only the beam weight present may overstress the concrete, and with post-tensioning the stressing is often carried out in several stages. Sufficient prestress is initially applied to take the beam weight and later the deck slab weight. After casting the deck slab the remaining prestress is applied to cater for the imposed and live loading.

Post-tensioned tendons are normally more expensive than pre-tensioned strand when taking into account all the costs of installation and grouting. For this reason pre-tensioning is generally preferred unless other factors influence the choice, such as when existing precasting facilities are not available and only a small number of beams are needed. In this situation, the cheaper casting set-up makes post-tensioning economic.

For simply supported beams the applied dead-load moment reduces away from the centre to zero at the supports. If a constant prestress force is applied along the bottom of the beam throughout its full length this can generate high compression in the bottom and tension in the top over a significant length at the beam ends. This is overcome in post-tensioned beams by draping the tendons, with the anchors raised up at the ends.

With pre-tensioned beams the prestress force at the ends is reduced by debonding some of the strands with a plastic tube. Alternatively, some of the pre-tensioned strands may be deflected upwards at the ends by means of an anchor frame positioned at quarter or third points along the beam length. Deflecting the strand complicates the casting bed arrangement and is not commonly done.

Figure 10.19 Beam transporter (reproduced courtesy of Hyder Consulting (UK) Ltd, copyright reserved)

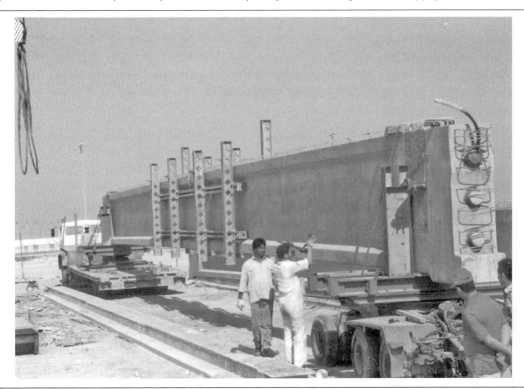

During the transfer of prestress to a beam, the beam tends to lift up along its length as the prestress counteracts the dead load. As the beam lifts off the bottom shutter, as seen in the insert in Figure 10.18, its weight is shed back to the ends of the beam. If this weight acts through a short length at each end of the beam the concentrated load can cause local overstressing and spalling of the concrete. To prevent this a short length of soft bearing material is placed under the beam soffit at the ends to help spread out the load.

After the beam has been prestressed it is lifted off the casting bed using a crane or gantry and moved to a storage area. Moving the beams short distances around the casting or storage areas is often done by using specialist equipment such as overhead lifting frames or transporters. When moving beams long distances and over public highways, specialist bogey arrangements, as seen in Figure 10.19, are used. Smaller beams are often placed on a standard low-loader for transporting.

The longer and more slender beams, such as the 'SY' type, have little transverse strength and when lifted or transported they are susceptible to lateral bending and buckling, and they may require temporary support frames.

Beam weights vary greatly depending on the type of beam and its length. A typical 16 m-long 'Y' beam weighs approximately 12 tonnes, while similar beams spanning 30 m would weigh approximately 40 tonnes. For the long-span 'U' beams the weights can exceed 150 tonnes and specialist heavy lifting and transporting equipment is required.

Beams are lifted using hooks cast into the concrete or by inserting a lifting bar through holes formed in the ends of the beam. The lifting points are placed as near to the permanent support positions as possible to set up similar bending moments in the beam.

Erection of precast beams

The placing of precast beams into position is normally done by crane; however, for long multi-span bridges or where ground access is poor an erection gantry is used to lift the beams and carry them into place.

The precast beams are lifted individually and placed on to the bearings or supports. Figure 10.20 shows an edge beam being lifted into position by a crane. The beams are spanning a small river and the crane was able to reach the deck from a position behind the abutment. When selecting the beams to be used the weight of each should be considered, to ensure that there is a suitable crane available with the capacity to lift the beams at the radius required.

With long and heavy beams it is sometimes necessary to use two cranes, although in some countries this is discouraged due to the high safety risks involved. With a crane at each end of the beam careful control is needed to ensure that they are evenly and correctly loaded at all times.

Precast beams are usually placed directly onto the bearings at the supports, although where they are built into a diaphragm or made continuous over the support they are placed on temporary supports while the in situ connection is made.

Figure 10.20 Beam placed by crane (reproduced courtesy of Hyder Consulting (UK) Ltd, copyright reserved)

Figure 10.21 Gantry for placing beams

Where bearings are used under each beam they are installed on the bridge piers or abutments prior to lifting in the precast beams. Rubber or laminated bearings are either placed directly on to the prepared concrete support or bedded on a thin layer of cement mortar. Pot or mechanical bearings are positioned on a cement grout bedding with dowels fixing the bearing to the concrete below.

With rubber bearings the beams rest directly on to the bearing surface or are bedded on a thin layer of cement mortar. If pot bearings are used the beams are held above the bearing, leaving a small gap which is filled with a cement grout.

Erection gantries are used on long viaducts where there are many spans, and the cost of the gantry is spread out over many uses. The gantry sits on the pier crossheads or abutment, as shown in Figure 10.21, and is self-launching, moving from pier to pier as the deck progresses. Gantries operate in one of two ways depending on how the beams are delivered. If the beams are delivered along the completed deck they are picked up by the back of the gantry and moved out over the span before being lowered into position. If the beams are delivered at ground level, the gantry reaches down and raises them up to deck level, placing them on the bearings.

When the beams are placed on the bearings they are held in position by temporary frames or struts, which prevent them from sliding off or from toppling over. The beams may also need holding in place during the concreting of the deck slab, although the deck formwork often provides sufficient support.

Casting of deck slab

After positioning the precast beams the formwork is placed to allow the top slab and diaphragms to be cast. With closely spaced beams only minimal formwork is needed, while with widely spaced beams the formwork and falsework may be significant.

Where there is good access from the ground the deck slab formwork may be supported by full-height scaffolding. As the deck slab weight is not transferred to the beams until the slab concrete has gained strength, full-height scaffolding has the advantage of improving the stress distribution across the composite section in the completed structure.

It is common to use permanent formwork between the beams, while for the deck-side cantilevers and parapet beams the formwork is supported off the outer beam or from the deck slab between the beams if this is completed first.

Permanent formwork is supported on the edges of the beams, spanning across the gap. Glass-fibre reinforced cement (GRC) panels are used where the beams are closely spaced, with ribbed glass reinforced polymer (GRP) panels used for spans up to 4 m. Precast concrete planks, either ribbed or a constant depth, are used for a wide range of beam spacing and act compositely with the deck slab. The panels are bedded down on the edge of the beam and sealed to prevent grout loss.

The formwork arrangement for the edge cantilevers is simplified by casting the deck width in stages, although this needs to be considered during the design to cater for the differential shrinkage effects. In the first stage, the deck slab is cast across the beams omitting the edge cantilever section. The formwork for the edge cantilever is then supported off the completed deck slab. Alternatively, the edge cantilevers may be precast and lifted into position before being connected to the deck with an in situ stitch.

With the beams and deck formwork in place the deck slab reinforcement is fixed, as seen in Figure 10.22. The bars are detailed with a simple layout for easy positioning and fixing. Holes are left through

Figure 10.22 Formwork and reinforcement being placed (reproduced courtesy of Hyder Consulting (UK) Ltd, copyright reserved)

Figure 10.23 Deck slab being cast (reproduced courtesy of Hyder Consulting (UK) Ltd, copyright reserved)

the webs of the beams at the transverse diaphragm positions to feed bars through and provide the transverse connection.

The deck slab concrete is placed by skip or with a pump, as shown in Figure 10.23. The deck is divided into bays to help control the concreting and to establish the finished levels required. On large areas of deck a finishing machine is often used to help level and compact the concrete surface and to give access across the concrete to apply the surface finish and curing. A typical finishing machine is shown in Figure 11.9.

Design of beam-and-slab decks

Design of beam-and-slab decks involves the superimposing of many different effects to build up the overall design state, as described in Chapter 5. The structural behaviour of beam-and-slab decks is described by Hambly (1991).

General design

Longitudinal and transverse effects in the deck slab under superimposed dead load and live load are derived from a grillage analysis. This gives satisfactory results for most standard beam-and-slab arrangements. Recommendations for setting up grillage models are given by West (1973).

Where unusual beam-and-slab arrangements occur or secondary effects, such as distortions or transverse bending in the beams, become significant, a 3D finite element model should be used to derive the load effects, as seen in Figure 10.24. Plate elements are used to model the webs, slabs and diaphragms of the 3D structure. The dead, superimposed and live loads are applied directly to the model to give the forces and moments at the critical sections in both the longitudinal and the transverse directions for the beam-and-slab design.

Figure 10.24 3D FE model of beam-and-slab deck

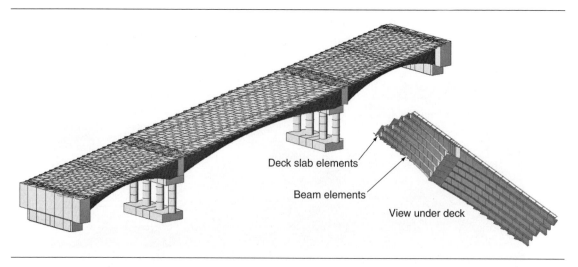

Deck slab elements

Beam elements

View under deck

Serviceability design considering the stresses in the deck is described in the next section. The design of the deck for ultimate moment, shear and torsion is carried out as described in Chapter 5. The check of the longitudinal shear along the interface between the slab-and-beam must include the forces generated by the differential shrinkage and creep effects.

During lifting and transporting, precast beams are subjected to different loading conditions causing a change to the bending moments and shears, which must be considered in the design of the beam. The beams must be handled carefully to minimise any impact or dynamic loading, while lifting and temporary support positions should be located as close to the permanent support position as possible. With long slender beams, transverse instability may occur either during lifting or when placing the concrete for the top slab, and suitable restraints may be needed. This effect is described in the paper by Burgoyne and Stratford (2001).

Stress distribution through section

The prestress and dead-load stress distribution through the beam-and-slab and the time-dependent effects of creep, shrinkage and relaxation of the prestress are all taken into account in the design. Differential shrinkage between the precast beams and the in situ deck slab also affects the stress distribution in the section.

The effect of the construction and stressing sequence and the time-dependent behaviour are considered in the design by building up the stress profile through the deck, superimposing the different effects at each stage in the construction and during service.

When the precast beams are cast they are supported by the formwork along their full length, and on the application of the prestress they tend to hog upwards between the ends. On moving to the storage yard they are supported only at the ends of the beams, giving a typical stress profile at mid-span as indicated in Figure 10.25(i). At this stage the beams are subject to their own dead weight and the applied prestress, with the resulting stresses at the top and bottom of the beam derived from simple bending theory. With the full prestress present and only the self-weight of the beam the difficulty is to ensure

Figure 10.25 Stresses in precast beam deck

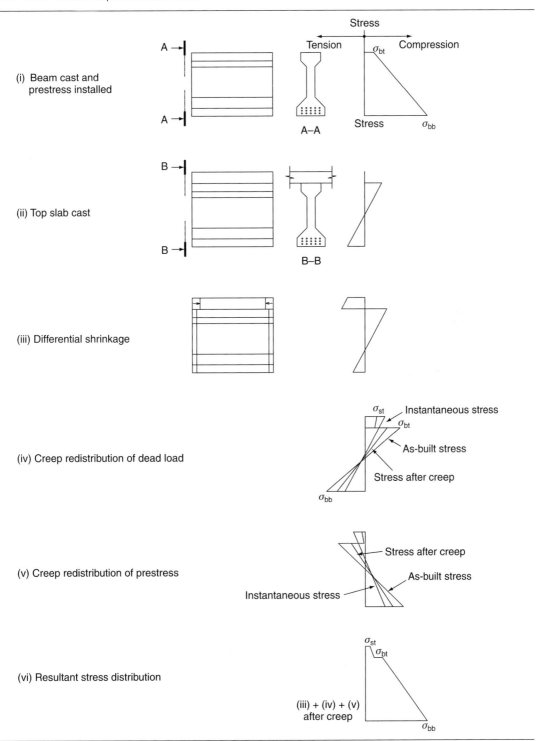

that the compression in the bottom and tension in the top of the beam do not exceed the allowable limits.

After erecting the precast beams, if the deck slab is cast on permanent formwork the weight of the slab is carried by the beam section alone to give a stress distribution, as indicated in Figure 10.25(ii). If the deck slab is cast on full-height scaffolding when the scaffold is lowered the slab's weight is carried by the beam-and-slab's combined section.

After the deck slab concrete has hardened, creep of the concrete redistributes the dead load and pre-stress stresses within the section, while differential shrinkage between the beam and deck concrete sets up additional stresses in the section. The superimposed and live loads are carried by the combined beam and deck slab section.

The creep redistribution of the dead load and prestress stresses is estimated by comparing the as-built condition with the 'instantaneous' state. In the as-built condition the entire dead load and prestress force is carried on the beam and the stresses are a combination of Figure 10.25(i) and (ii). In the theoretical 'instantaneous' condition the full dead load and prestress force is assumed to be applied to the composite section. The stresses will creep from the as-built condition towards the instantaneous condition, with the final stress at any level in the section being given by:

$$\sigma_{\text{final}} = \sigma_{\text{as-built}} + (1 - e^{-\phi})(\sigma_{\text{inst}} - \sigma_{\text{as-built}})$$

The loss of prestress forces due to relaxation in the strand, and the concrete creep and shrinkage, affect the stresses in both the slab and beam. These long-term prestress losses should be considered in two phases. For the first phase, before the deck slab is cast, the prestress losses are estimated and the pre-stress force in the beam reduced accordingly. In the second phase, the remaining losses that occur after casting the slab are estimated and applied as a 'tensile' force to the composite section at the position of the centroid of the strand or tendons. The resulting stresses are then combined with the other stresses in the section.

When the top slab is cast, the concrete in the beam will have already completed a large proportion of its shrinkage, and differential shrinkage occurs between the two. The deck slab tries to shrink more than the beam and is restrained by it, creating tension in the slab and a combined compressive force and sagging moment in the beam. The equivalent differential shrinkage force assuming the top slab was fully restrained is estimated from:

$$\text{force, } F_{\text{s}} = \Delta_{\text{s}} \cdot A_{\text{s}} \cdot E_{\text{c}} \frac{1 - e^{-\phi}}{\phi}$$

with $(1 - e^{-\phi})/\phi$ reducing the effect due to creep of the concrete.

The stresses in the composite section are derived by adding the stress of the 'fully restrained' slab to the stress due to the released force applied to the composite section, that is:

(a) tensile stress from force F_{s} acting on slab alone applied at the centroid of slab, plus
(b) stresses generated from a compressive force F_{s} acting on the composite section, applied at the centroid of the slab.

This gives a stress profile similar to Figure 10.25(iii).

The stresses generated at the top and bottom of the slab and top and bottom of the beam are checked at each stage of construction, as well as in the permanent condition after opening and again after the long-term effects have occurred.

Modern computer software packages, such as Midas Civil described in Chapter 8, are able to incorporate the long-term effects such as creep, shrinkage and prestress losses with the staged construction of the beams as well as the applied loading to combine all of the above requirements for checking the stresses in the beams and deck slab at each stage of construction and in service.

Precast beams in continuous and integral decks

When the beams are fully connected longitudinally over the piers to make the deck continuous, as illustrated in Figure 10.10(ii), the dead load and prestress effects are redistributed due to creep of the concrete. The dead-load sagging moment is reduced at mid-span and a hogging moment set up at the pier. The superimposed and live loads will also cause a hogging moment at the connection. The creep of the concrete under the prestress force causes the beam to hog at mid-span, generating a sagging restraint over the pier. This may lead to cracks in the deck soffit above the pier if the effect is not fully considered and suitable reinforcement provided.

The final moment at the pier due to the creep of the dead load and prestress may be derived by a similar approach to that described in Chapter 5, with:

$$M_{final} = M_{as\text{-}built} + (1 - e^{-\phi})(M_{inst} - M_{as\text{-}built})$$

ϕ is based on the residual creep left in the beam at the time of casting the connection and $M_{as\text{-}built}$ is zero where the connections are made after erecting the beam and casting the top slab. The moment, M_{inst}, calculated as if the deck was built continuous 'instantaneously', includes the prestress secondary moment as well as the prestress primary moment and dead-load moment.

The differential shrinkage between the top slab and beam generates a secondary hogging moment in the deck at the pier after the connection has been made, which must also be considered in the design.

Reinforcement is provided across the connection to cater for the hogging and sagging moments in the deck above the pier. The section is designed as a reinforced concrete element, with reinforcement placed in the deck slab and between the bottom of the beams. With larger beams it is possible to provide post-tensioned tendons in the webs which run through the connection to prestress the in situ concrete.

A similar behaviour occurs when precast beams are built integral with the abutments. Reinforcement is provided to connect the deck slab and beam with the abutment to take the moments' set up. Depending on the overall arrangement and the construction sequence of the bridge it may be possible for further hogging to be imposed on the beam and deck slab along the span. The hogging moments may also set up tensile stresses in the top of the precast beam and extra reinforcement is provided to control any cracking that may occur. In these circumstances the top of the beam and the in situ slab above may be considered as a reinforced rather than prestressed concrete section, with sufficient reinforcement to take the applied bending and shear and to control the crack widths under serviceability loading conditions.

Prestress and reinforcement

Post-tensioned tendons usually follow a draped profile and need additional reinforcement at the beam ends around the anchorages, as described in Chapter 6 and shown in Figure 6.3. Reinforcement is provided along the beam to cater for shear and torsion, and the early thermal effects within the concrete.

With pre-tensioned beams longitudinally the strands are sufficient with only nominal additional reinforcement required in the top of the beam, as illustrated in Figure 10.26. The main prestressing strands are spread out evenly over the bottom of the beam with a minimum spacing of 50 mm and a concrete cover similar to that normally provided for reinforcement. Additional strands are sometimes provided in the top of the beam to overcome tension in the temporary condition before the beam is fully loaded.

The reinforcement in the deck slab is designed to cater for the moments, shears and torsions acting on the elements, as derived from the grillage or plate analysis of the structure. The design moments used to calculate the reinforcement requirements are derived by combining the direct bending and torsion, as described in the paper by Wood (1968).

The quantity of reinforcement required depends on the depth of deck slab, beam spacing and the live load carried. A typical arrangement for the deck slab reinforcement is illustrated in Figure 10.27.

Details of the standard precast beams are found in manufacturers' literature and on the appropriate websites noted in Appendix C.

Figure 10.26 Typical reinforcement and prestress arrangement

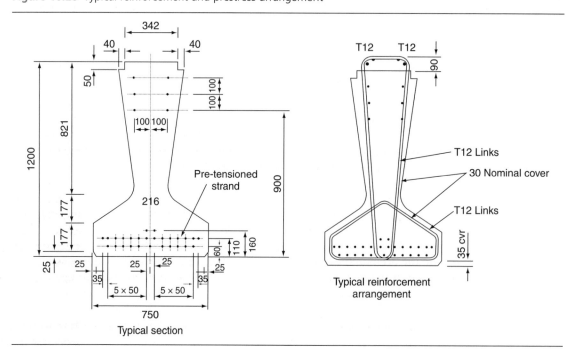

Typical section

Typical reinforcement arrangement

Figure 10.27 Typical deck slab reinforcement

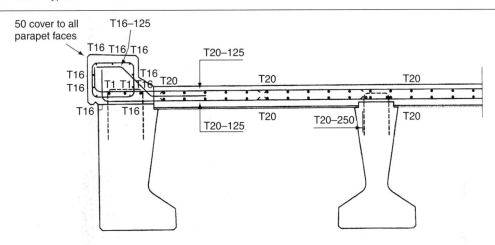

REFERENCES

Burgoyne CJ and Stratford TJ (2001) Lateral instability of long-span prestressed concrete beams on flexible bearings. *The Structural Engineer* **79(6)**: 23–26.

Hambly EC (1991) *Bridge Deck Analysis*, 2nd edition. Chapman and Hall, London.

Taylor HPJ (1998) The precast concrete bridge beam: the first 50 years. *The Structural Engineer* **76(21)**: 407–414.

West R (1973) *Recommendations on the use of grillage analysis for slab and pseudo-slab bridge decks.* C&CA/Ciria, London.

Wood RH (1968) The reinforcement of slabs in accordance with a pre-determined field of moments. *Concrete* **February**: 69–76.

Prestressed Concrete Bridges, 2nd edition
ISBN: 978-0-7277-4113-4

ICE Publishing: All rights reserved
doi: 10.1680/pcb.41134.197

Chapter 11
In situ multi-cell box girder decks

Introduction

In situ multi-cell concrete box girder bridges are frequently used in the medium span range from 30 m up to 50 m. They provide a versatile arrangement suitable for simple highway bridges or longer viaducts, while able to cope with complex geometry, tight curvature and variable deck widths. The box shape gives a good aesthetic appearance as well as being efficient in its design. Figure 11.1 shows an example of a bridge with this type of deck built as part of a highway interchange scheme in Hong Kong. Below spans of 30 m precast beams or voided slab decks are more suitable while above 50 m a single-cell box arrangement is usually more economic.

The advantages of this type of construction include the efficient use of the concrete and prestress, simple construction where access is easy and the ability to cope with complex geometry. The techniques used are well established and experienced labour and equipment are, usually, readily available locally.

The disadvantages include the need for extensive labour activities on site, slow construction and complex falsework if access is difficult.

A variation of this type of deck is the two- or three-cell box girder which is also used for longer spans, utilising the techniques as described in Chapters 12 and 13. This chapter concentrates on the multi-cell arrangements used for the medium-length spans when cast on full-height falsework or temporary support trusses.

General arrangement

The deck arrangement is similar to a voided slab, but with the voids occupying a larger proportion of the deck area and usually being rectangular in section. A typical deck section is illustrated in Figure 11.2. The outer webs are often sloped and the side cantilevers made longer to improve the appearance. The web thickness is governed by the shear requirements, but they must be wide enough to provide space for the reinforcement and concrete to be placed around the prestressing ducts. This usually requires a minimum web thickness of 300 mm, but may need to be wider if large tendons are used. The deck slab size is governed by the web spacing and the live loading carried and is typically between 200 mm and 300 mm thick. There is very little loading on the bottom slab, with a thickness between 150 mm and 200 mm being sufficient.

Transverse diaphragms are provided across the full width of the box at each of the support locations. The diaphragms provide rigidity to the box and assist in transferring the loads in the webs to the supports. Intermediate diaphragms are often placed at quarter or third points along the span to stiffen up the box and to help distribute the loading between the webs.

Figure 11.1 Kwun Tong Bypass, Hong Kong

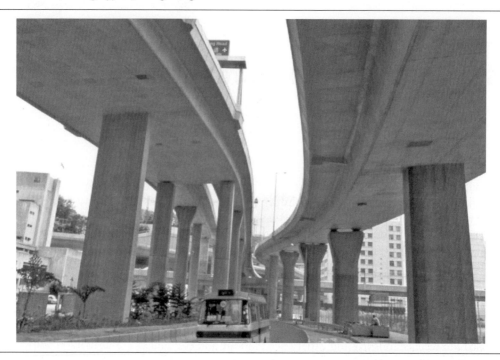

Access into the box cells is achieved through soffit access holes of a minimum of 600 mm diameter, which are usually located near the abutments. Similar-sized holes are provided through each of the diaphragms and webs, as required, to give access into each section inside of the deck. Small drainage holes, typically 50 mm diameter, are provided through the bottom slab at the low point in each section of deck to ensure that water cannot collect inside the box cells.

Concreting and construction restraints usually dictate a minimum deck depth of 1200 mm; although for reasonable inspection and maintenance access a depth of at least 1800 mm is needed. With an

Figure 11.2 Typical box cross-section

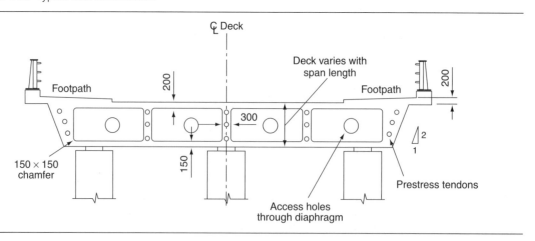

Figure 11.3 Multi-cell box girder bridge

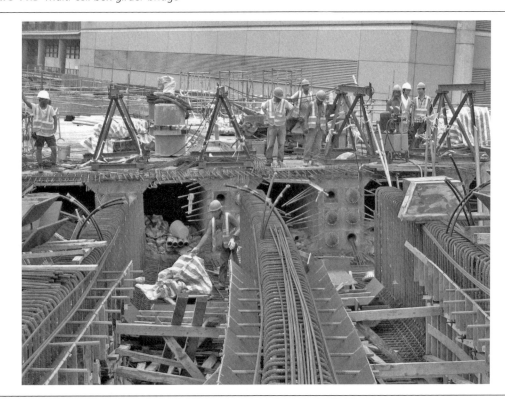

optimum span-to-depth ratio of between 18:1 and 25:1, the preferred span lengths are usually greater than 30 m.

Multi-strand tendons are used following a draped profile, and are located in the bottom of the webs in mid-span and at the top of the webs over the supports. For decks with an overall length less than 80 m and fully cast before applying the prestress, the tendons would usually extend over the full deck length and be anchored on the end diaphragms. Longer decks are cast in stages on a span-by-span basis, with the prestress tendons anchored on the webs at the construction joint, as seen in Figure 11.3. The tendons are then continued into the next stage of deck by the use of couplers.

Construction of in situ multi-cell box girders

Most in situ multi-cell box girders are cast on full-height scaffolding built up from the ground, as seen in Figure 11.4. Where good access exists this form of construction provides flexibility in the construction sequence and deck layout. Obstructions under the deck, such as live roads, railways or small rivers, are overcome by spanning with temporary works to support the falsework. Guidance on the design and construction of falsework is given in the reports by Bragg *et al.* (1975) and the Concrete Society and IStructE (1971).

After erecting the scaffolding the formwork is placed to the required shape and profile. Timber formwork, consisting of a plywood facing supported by timber studding, is shown being installed in Figure 11.5. Steel forms are used when long lengths of deck are to be cast in stages and the shutters

Figure 11.4 Full-height scaffolding

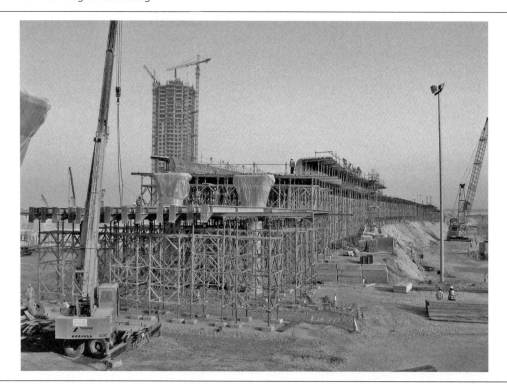

Figure 11.5 Formwork being installed

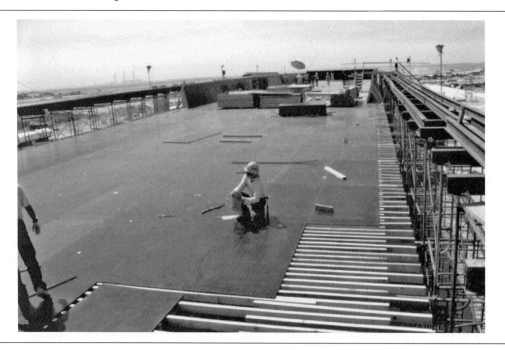

Figure 11.6 Reinforcement and tendon ducts

are re-used many times. With timber forms it is easier to have square or angled corners and flat faces while steel forms are able to more easily incorporate curved corners and sides.

Casting the deck section in several stages simplifies the formwork. This also makes the concreting operations much simpler and easier to control. The bottom slab, outer webs and diaphragms are cast first, followed by the inner webs and top slab soon after. The time delay between castings should be kept to a minimum to reduce any early thermal and differential shrinkage effects. It is preferable to cast the outer webs with the bottom slab so that any construction joint is at the top of the web and hidden in the corner with the top slab. A construction joint between the bottom slab and webs is difficult to hide on the concrete surface and, although this is not important for the inner webs, it mars the appearance on the outer webs. The formwork for the inner webs and top slab is supported off the bottom slab concrete, simplifying the overall arrangement.

With the formwork in position, the next activity is the fixing of the reinforcement, prestressing ducts and anchorages. Figure 11.6 shows the reinforcement and ducts in place for concreting the bottom slab and outer webs. Short shutters are being installed along the bottom of the webs to form kickers when the webs are cast in the next stage.

Without the inner web and top slab formwork in place the access for placing, compacting and finishing the concrete in the bottom slab is improved. The subsequent concreting of the inner webs and top slab is done from above the deck without needing access to the void. Figure 11.7 shows a multi-cell box after the bottom slab and webs have been cast and before installing the top slab formwork and reinforcement. At this stage the deck is still fully supported by the falsework, which remains in place until the concreting is completed and the tendons installed.

Figure 11.7 Bottom slab, webs and diaphragm cast

Either permanent formwork panels or removable table forms are used between the webs to support the wet deck slab concrete. The removal of formwork from inside the voids, after the deck is completed, requires it to be broken down into small sections and passed out through the access holes in the diaphragms and bottom slab. Alternatively, a larger temporary access hole is left in the top slab at one end of the deck which is concreted after the rest of the formwork has been removed.

Figure 11.8 Stage-by-stage construction

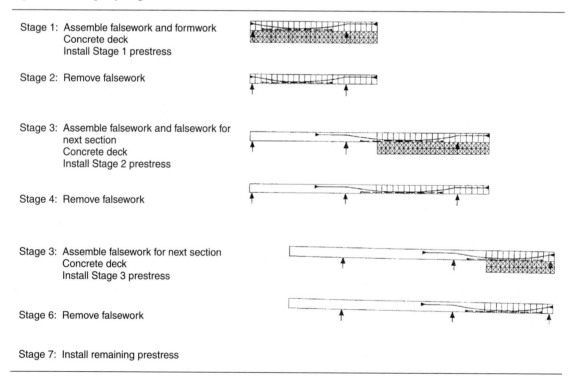

Stage 1: Assemble falsework and formwork
Concrete deck
Install Stage 1 prestress

Stage 2: Remove falsework

Stage 3: Assemble falsework and falsework for next section
Concrete deck
Install Stage 2 prestress

Stage 4: Remove falsework

Stage 3: Assemble falsework for next section
Concrete deck
Install Stage 3 prestress

Stage 6: Remove falsework

Stage 7: Install remaining prestress

Longer bridge decks, extending over several spans, are usually cast in sections on a span-by-span basis, as illustrated in Figure 11.8. This has several benefits including reducing the size of concrete pours to a more manageable quantity, optimising the length of the prestress tendons and permitting the maximum re-use of the falsework and formwork. The first section cast is a complete span plus part of the adjacent spans to give short cantilevers. This moves the construction joints away from the highly stressed region at the pier and helps to balance the moments in the deck in the temporary and permanent situations. To optimise the overall moment distribution the construction joint is placed between the quarter or third points of a span. Subsequent sections of deck extend out from the construction joint over the next pier with a short cantilever, as before. This process is continued until the end of the deck is reached.

During concreting of the deck slab the level and finishing of the top surface has to be carefully controlled. On smaller decks this is achieved by placing levelling timbers on the reinforcement and screeding the concrete to the top of these. For larger areas of slab, a finishing machine, as seen in Figure 11.9, is used to assist in accurately levelling the top surface. The finishing machine runs along rails preset to the correct levels down either side of the deck. The machine is used to compact and float off the top surface of the concrete as well as to provide a working platform to give access over the completed deck for curing and covering of the concrete.

Figure 11.9 Deck finishing machine

Figure 11.10 Balanced cantilever construction of multi-cell box (reproduced courtesy of DYWIDAG-Systems International GmbH, copyright reserved)

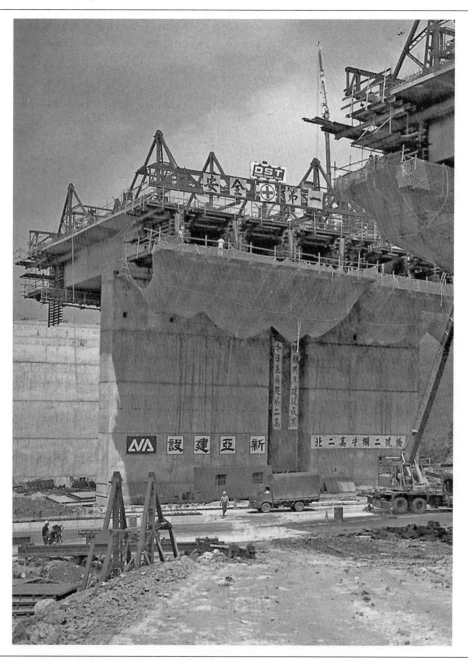

When the concrete has attained the required strength the prestress tendons are installed and stressed. The deck tends to lift up along its span and reduce the load on the falsework as the prestress is applied. The falsework is removed after sufficient tendons have been stressed to carry the deck dead load.

If some parts of the deck deflect downwards when the prestress is applied this causes additional load on the falsework underneath. This situation occurs when the deck is cast in stages using the span-by-span approach as described above. The cantilever extending beyond the pier in the direction of construction tends to deflect down as the previous span is prestressed and lifts up. The falsework must be designed to take the increased load. The increase may be minimised by suitably arranging the prestress layout and stressing only the minimum number of tendons required to support the deck dead load before removing the falsework.

Where access is difficult, or the deck is at a high level, the formwork is supported from a truss spanning between the piers, similar to Figure 10.13, or the deck built with a form traveller as a balanced canti-lever as shown in Figure 11.10. A truss system is usually self-launching to move itself between the piers. To simplify the truss design, deck construction is on a span-by-span basis with the construction joint just beyond the pier diaphragm.

Design of in situ multi-cell box girders

Multi-cell box structures are often analysed using a grillage model to derive the forces and moments in the webs and slabs, as discussed in Chapter 8. The local bending moments in the deck slab from wheel loads are then derived using Pucher charts or similar design guides. The structural behaviour of this type of deck is described by Hambly (1991).

Thin webs and slabs can significantly deform and distort under the imposed loading, and in this case the deck analysis is most accurately carried out using a 3D finite element model with plate elements for each web and slab member, as illustrated in Figure 11.11. The moments and shears in each element in each direction are taken directly from the analysis, and are used for both the longitudinal prestress design and the transverse reinforcement design.

For the longitudinal design, the deck section is divided up into a series of 'I' beams consisting of a web and associated top and bottom slabs up to the mid-point of each cell. Prestress layouts are designed for each web to balance the stresses derived from the finite element or grillage analysis and to provide sufficient ultimate strength.

The prestress for long multi-cell box girder decks constructed in stages is often arranged as illustrated in Figure 11.12, with similar tendon profiles and arrangements in each web. In this example, the deck

Figure 11.11 Three-dimensional finite element model of deck

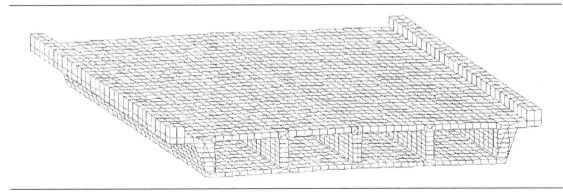

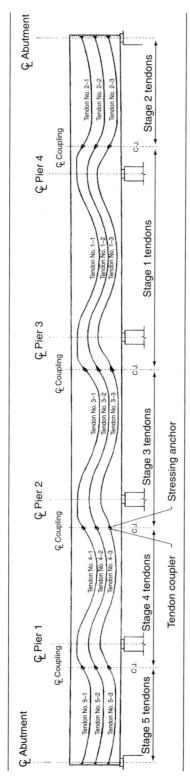

Figure 11.12 Longitudinal tendon profile

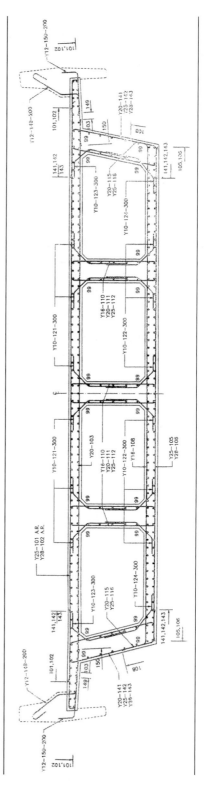

Figure 11.13 Typical reinforcement arrangement

construction is shown starting with the span between piers 3 and 4, progressing outwards. Construction joints are approximately at the quarter points along the spans.

The tendons at each stage are anchored in the webs at the 'free' end of the deck. Where the new section of deck meets the previously cast section at the construction joint the prestress is joined together using couplers, as described in Chapter 2. Alternatively, the prestress tendons are 'lapped' and anchored in blisters inside the deck.

Transverse reinforcement is provided in the concrete section to cater for the transverse moments and shears expected in the webs and slabs. Additional reinforcement is required for the longitudinal shear in the webs and torsion in the box section. A typical layout of reinforcement for a multi-cell box is illustrated in Figure 11.13.

With the deck fully prestressed longitudinally only nominal longitudinal reinforcement is required in the permanent condition. Early thermal effects within the concrete and differential shrinkage between the concrete pours may require additional reinforcement to control cracking in the temporary condition before the prestress is applied. It can be beneficial to install and tension part of the prestress as early as possible to counter the early thermal and differential shrinkage effects.

The diaphragms over the supports are designed as described in Chapter 6, while the intermediate diaphragm design is based on the forces and moments generated from the grillage or finite element analysis, with reinforcement provided based on a reinforced concrete beam section.

REFERENCES

Bragg SL, Ahm P, Bowen FM, Champion S, Kemp LC, Mott JCS *et al.* (1975) *Final Report of the Advisory Committee on Falsework*. London: HMSO.

Concrete Society and IStructE. (1971) *Technical Report TRCS 4, Falsework*. Concrete Society, London.

Hambly EC (1991) *Bridge Deck Analysis*. 2nd Edition. Chapman and Hall, London.

Pucher A (1976) *Influence Surfaces of Elastic Plates*. 5th Edition. Springer-Verlag, Wien.

Prestressed Concrete Bridges, 2nd edition
ISBN: 978-0-7277-4113-4

ICE Publishing: All rights reserved
doi: 10.1680/pcb.41134.209

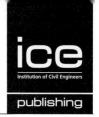

Chapter 12
In situ single-cell box girder bridges

Introduction

Single-cell box girders cast in situ are used for spans ranging from 40 m up to 300 m. The box arrangement is considered to give a good aesthetic appearance, with the web of the box in shadow producing a slender appearance when combined with a slim parapet profile.

Single box arrangements are efficient for both the longitudinal and transverse designs, and they produce an economical solution for most medium and long-span structures. The box structure is well suited for curved decks, and it is able to cope with complex highway geometry.

This type of deck is constructed span by span using full-height scaffolding or trusses, or as a balanced cantilever using form travellers. The post-tensioned concrete box girder spans for the Hong Kong MTR Island Line, shown in Figure 12.1, were cast in situ on falsework with span-by-span construction.

Balanced cantilever construction is suitable for the longer spans and where access is restricted. For the Malaysia–Singapore Second Crossing, shown in Figure 12.2, the 165 m long main navigational span and adjacent side spans were cast in situ as balanced cantilevers while the shorter spans, on the approach viaduct, utilised precast segments. Precast segmental construction produced the most economical solution for the repetitive spans of the approach viaduct, whereas the 'one-off' nature of the longer main span favoured in situ construction.

The advantages of in situ single-cell box girders are their efficient use of concrete, reinforcement and prestress, flexibility in span arrangements, and good aesthetics. Their disadvantages include the labour-intensive site activities and the long construction times needed for the larger structures.

General arrangement

Constant depth sections are adopted on spans up to 70 m and with a span-to-depth ratio of approximately 20:1. For spans above 50 m, a haunched profile is common with a typical span-to-depth ratio of 16:1 at the piers, increasing to 45:1 at mid-span. Between 50 m and 70 m, the deck is either haunched or a constant depth throughout, depending on the site constraints and the designer's preference. A publication by Podolny and Muller (1982) contains many early examples of concrete box girder arrangements used around the world.

Using a constant depth section throughout allows the deck details to be standardised, simplifying the construction. With a haunched profile the shallower deck at mid-span reduces the dead-load bending moments and shears in the deck, while the deeper section at the pier is more efficient resulting in an overall reduction in concrete and prestress quantities.

Figure 12.1 MTR Island Line, Hong Kong

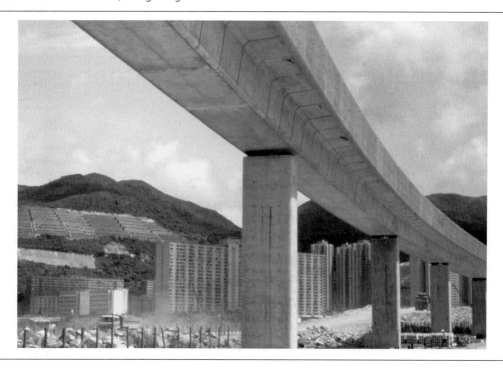

Figure 12.3 illustrates a typical section for a single-cell box girder. The width of the top slab is governed by the user's requirement in terms of highway, rail or footway layout. The side cantilevers are usually between 2 m and 3.5 m long, above which the slab design becomes less economically viable. It is preferable to slope the webs to improve the appearance of the box, while the width of the bottom slab should be sufficient to provide transverse stability to the deck when overturning about the bearings is being considered.

Figure 12.2 Malaysia–Singapore Second Crossing (reproduced courtesy of Hyder Consulting (UK) Ltd, copyright reserved)

Figure 12.3 Typical single-cell box cross-section

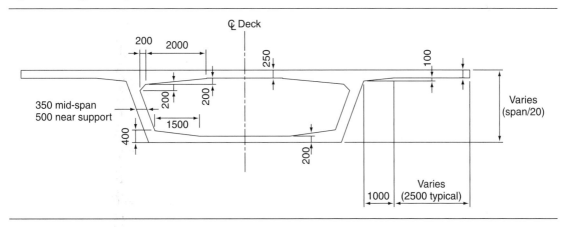

Local bending and punching shear design under live load governs the top slab thickness. The punching shear from wheel loads generally requires a minimum slab thickness of 200 mm while transverse bending requirements dictate a deeper slab near the webs.

The web thickness is governed by the shear design from the longitudinal analysis. A minimum thickness of 350 mm is usually adopted for ease of construction, with the webs thickening adjacent to the piers to suit the ultimate shear requirements.

The bottom slab is usually a nominal thickness of between 160 mm and 200 mm over the mid-span region, but it may need to be thicker near the supports to cater for the compressive stresses generated by the longitudinal hogging moments and the torsional stresses in the box. Where large pre-stressing tendons are placed in the top or bottom slabs sufficient concrete depth is needed to provide adequate cover to the ducts.

With a single-cell box the deck width is typically in the range of 6 m to 16 m. Below a width of 6 m the resulting narrow box is difficult to construct and causes problems with stability of the deck at the supports. Above a width of 16 m the top slab arrangement becomes less efficient. For wider decks, a multi-cell box, as described in Chapter 11, or a twin-box arrangement with the inner cantilevers joined together, as illustrated in Figure 12.4, is often used.

Wider decks are achievable with a single-cell box by introducing additional support to the deck slab. Providing transverse ribs under the deck slab allows longer side cantilevers and wider spacing of the webs but requires a more complex shutter arrangement. The arrangement illustrated in Figure 12.5, where struts are used to support the deck slab and transfer the load back to the webs, has been used successfully on a number of projects in the UK and worldwide.

In the UK, it is usual to design the top slab as a reinforced concrete element, while in Europe and elsewhere the use of transverse prestressing is common. Flat slab-type tendons are anchored in the cantilevers and extend over the full width of the top slab. This improves the durability of the structure by reducing the crack widths associated with a reinforced concrete design. However, transverse prestressing is usually only an economical choice for wider decks.

Figure 12.4 Twin-box arrangement

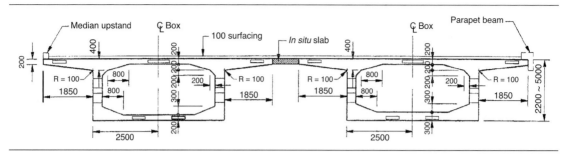

Longitudinal prestressing consists of post-tensioned tendons, usually with multi-strand systems, but occasionally with bar systems as well. Both internal and external prestress tendons are widely used, with the choice dependent on particular project requirements and the designer's preference. The advantages and disadvantages of internal and external tendons are discussed in 'Post-tensioning with internal or external tendons', Chapter 7.

With both span-by-span and balanced cantilever construction the deck is either supported on bearings at the supports or built integrally with the piers. Building the deck integrally eliminates the need for bearings and in the case of balanced cantilever construction reduces the need for temporary supports. Integral construction is not always possible with long decks as the movements in the deck due to creep, shrinkage and temperature change may generate excessive forces in the substructure at the ends of the structure.

Figure 12.5 Single-cell box with struts supporting side cantilevers (reproduced courtesy of C.S. Park, copyright reserved)

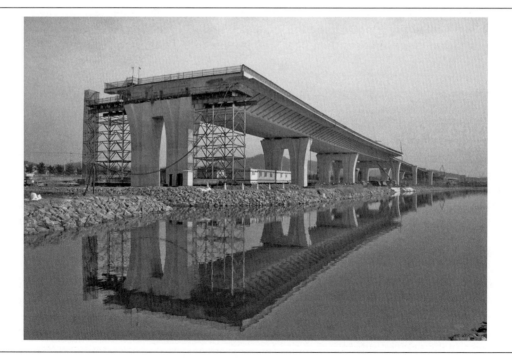

Many of the early balanced cantilever bridges built had hinges at the mid-span points to leave the deck acting as a series of propped cantilevers in its final form. However, several of these structures developed problems, with significant sag occurring in the spans. The use of hinges is seldom seen in modern bridges, with designers preferring to join the cantilevers with in situ stitches.

Construction, span by span

Simply supported spans are constructed full length before applying the prestress. For continuous decks, span-by-span construction involves casting a length of deck between vertical construction joints positioned in each span.

A typical span-by-span construction sequence for a continuous deck is similar to that for a multi-cell box illustrated in Figure 11.8. The first stage cast consists of a full span plus a short cantilever, with subsequent stages extending out over the adjacent piers. The length cast is usually between the quarter or third points in adjacent spans to optimise the bending moments and prestress requirements during construction and in the final structure. It is also preferable to place transverse construction joints away from the piers where the maximum moments and shears occur. After concreting a span the prestress is installed allowing the falsework to be lowered and moved forward ready for the next span to be constructed.

Where good access is available beneath the deck, full-height scaffolding is used to support the form-work, as shown in Figure 12.6, although this becomes expensive if the deck is at a high level. With full-height scaffolding it is possible to work on several spans at the same time to speed up the rate of construction. The formwork and reinforcement can be installed for the next span while the concreting and prestressing progresses on the previous span.

For high-level decks, or where access is difficult, the formwork is supported on trestles or by a truss or gantry spanning between the piers, as shown in Figure 12.7. Trestle and steel beam arrangements are able to span over small obstructions such as rivers, roads and rail lines, and reduce the amount of scaffolding needed. They are usually quicker to erect and dismantle than full-height scaffolding.

Gantry systems are normally self-launching, moving forward across the tops of the piers after each deck section is completed. This results in a more rapid construction than either full-height scaffolding or trestle arrangements, and gantries are popular when long lengths of viaducts are being constructed.

The deck section is usually cast in two stages with the bottom slab, webs and diaphragms cast first and the top slab cast soon after. After setting up the outer formwork, the reinforcement and prestressing tendon ducts are fixed, as seen in Figure 12.8. The internal formwork for the webs and diaphragms is then placed. A short width of shutter is placed along the top of the bottom slab adjacent to the webs to prevent the web concrete from slumping. However, it is not usually necessary to provide shutters over the rest of the bottom slab.

A separate internal form is used for the top slab between the webs. With the bottom slab and webs already cast, the top slab shutter is usually arranged as a 'table' form supported on the hardened concrete, as seen in Figure 12.9. After the top slab concrete has gained strength the table form is lowered and moved forward for the next section of deck.

The time lapse between casting the webs and the top slab should be kept to a minimum to reduce any potential early thermal and differential shrinkage effects between the two concrete pours.

Figure 12.6 Full-height scaffolding from the ground

A typical construction cycle takes approximately four to six weeks per span, depending on the span length, the rate of strength gain in the concrete and the time to fix the reinforcement and prepare for concreting.

The advantages of span-by-span construction are the ability to arrange the prestress in an efficient manner and the relatively fast construction that is achievable when compared to the balanced canti-lever method. The disadvantages are that it requires a lot of falsework, formwork and labour resources to achieve a good construction rate.

Figure 12.7 Gantry support for formwork (reproduced courtesy of C.S. Park, copyright reserved)

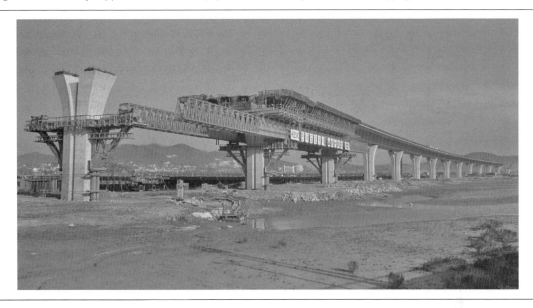

Figure 12.8 Reinforcement and ducts being placed

Figure 12.9 Shutter for top slab between webs

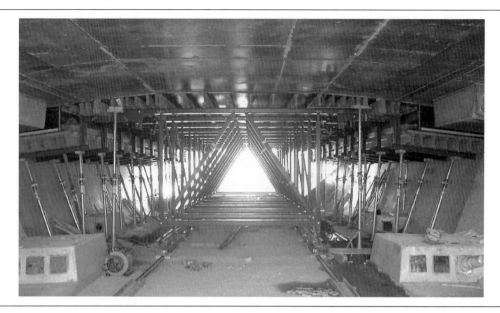

Construction by balanced cantilever

With this technique the deck is cast in short lengths within a pair of form travellers as a balanced cantilever about a pier, as shown in Figure 12.10. This form of construction is particularly suited to long spans and where access beneath the deck is difficult.

The sequence of construction is illustrated in Figure 12.11. The deck is built out from each side of each pier in a balanced sequence to minimise the out-of-balance forces on the piers. When the cantilevers from adjacent piers have been completed and face each other at mid-span the remaining gap is concreted and the prestress installed to make the deck continuous. The end span length is usually

Figure 12.10 Balanced cantilever construction (reproduced courtesy of Hyder Consulting (UK) Ltd, copyright reserved)

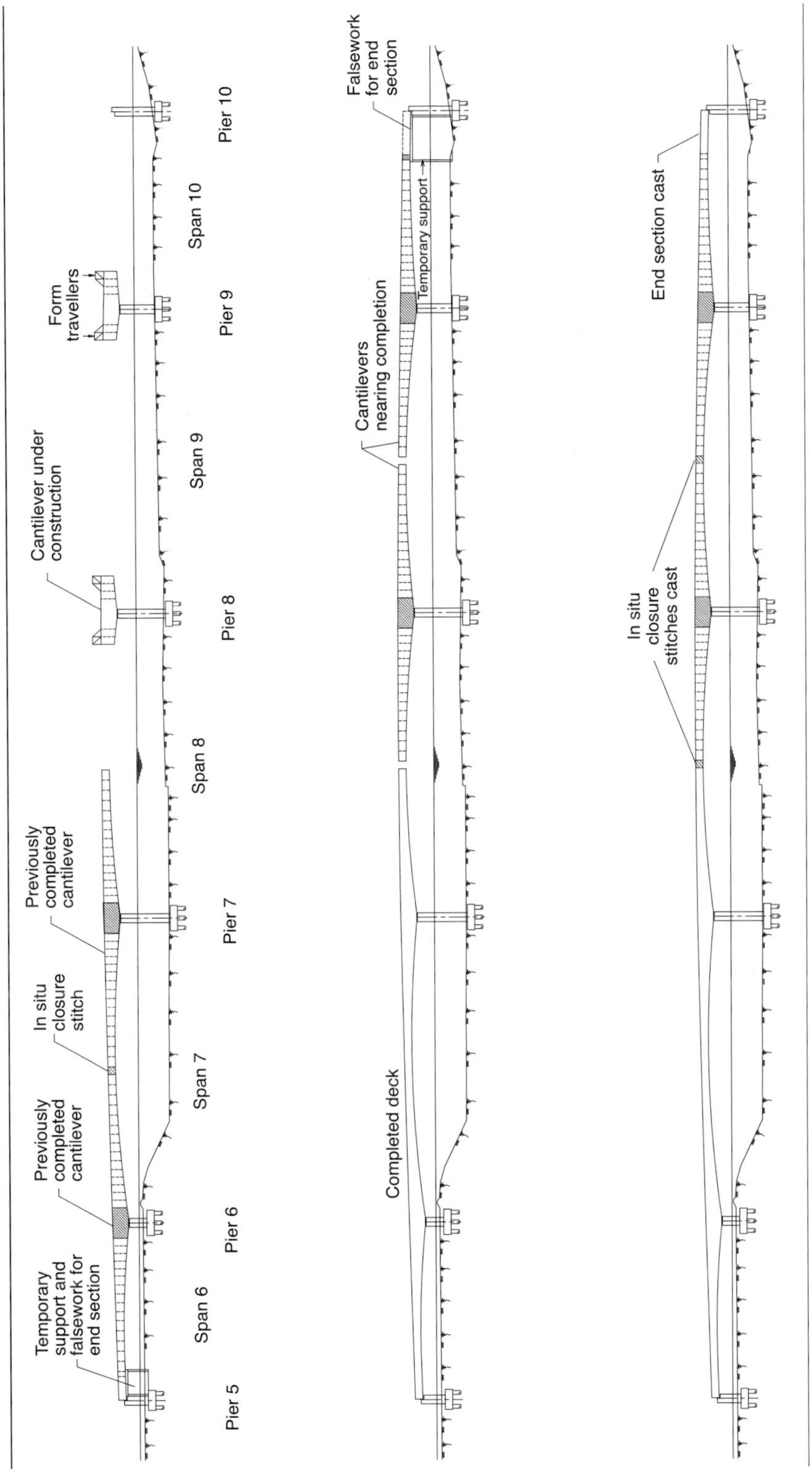

Figure **12.11** Balanced cantilever construction of deck (reproduced courtesy of Hyder Consulting (UK) Ltd, copyright reserved)

Figure 12.12 Pierhead and setting up traveller

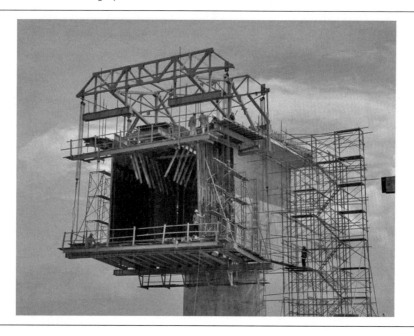

between 60% and 65% of the adjacent span length and the 'balanced' cantilever from the last pier is extended out until it reaches the abutment. Alternatively, this out-of-balance length of deck is supported by temporary props in front of the abutment, or the end section of deck is cast on falsework to support the concrete until the prestress is installed and the span completed.

A section of deck is initially constructed above each pier to enable the form traveller to be installed. This pierhead section is cast on falsework or temporary supports fixed to the pier. Figure 12.12 shows the pierhead section in the background completed and the frame for the form traveller being assembled. In this example the deck is built integral with the main piers which resist the out-of-balance moments from the cantilever construction without the need for any further temporary works. A twin-leaf pier arrangement provides the necessary moment restraint during construction without being too stiff along the line of the deck, as this would set up large moments and shears in the permanent structure due to the deck movement.

The form traveller consists of a series of shutters that hang from the support frame, as illustrated in Figure 12.13. The soffit shutter and inner web shutters are adjustable to cater for a varying depth section. The webs are shown vertical which simplifies the shutter arrangement. Sloping webs are also used, although this complicates the formwork design with the soffit shutter varying in width as the segment height changes. Temporary access is provided between the top of the deck and the void, and working platforms incorporated around the traveller. The frame of the traveller is bolted down to the previously completed deck. Shown in Figure 12.14, the support frame carries the weight of the shutters and the wet concrete until the concrete has hardened and the prestress 'cantilever' tendons are installed. The shutters are then stripped from the concrete face and the frame released from the deck to allow it to be slid or rolled forward to construct the next segment. With the form traveller set up, the reinforcement and tendon ducts are placed and the concrete cast.

Figure 12.13 Travelling form arrangement (reproduced courtesy of VSL International, copyright reserved)

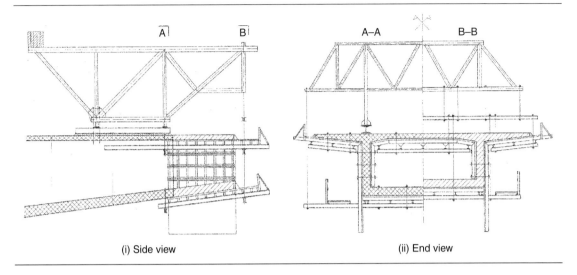

(i) Side view   (ii) End view

Figure 12.14 Travelling form on deck

Cantilever segments are typically between 3 m and 5 m long with the complete cross-section cast at one time. When the concrete has reached sufficient strength, prestressing tendons are installed in the top slab to support the cantilevering deck. Concrete with an early high strength is needed to enable the prestress to be installed as soon as possible and to achieve a typical cycle time of between seven or eight days per segment.

Construction times are reduced by prefabricating sections of the reinforcement cages and lifting them into the travelling form, or by precasting the prestress anchor region so that the prestress may be installed sooner. Using this approach, deck construction cycle times have been reduced to four days per segment.

As an alternative to using a form traveller, several projects have utilised a temporary overhead truss to support the formwork. Spanning between the piers and the previously completed deck, it moves the formwork forward in stages. When the cantilevers are completed, the truss launches forward to the next pier. This is usually suitable only for spans up to 100 m and is generally less economical than using form travellers in most situations.

Figure 12.15 Balanced cantilever construction with bearings (reproduced courtesy of DYWIDAG-Systems International GmbH, copyright reserved)

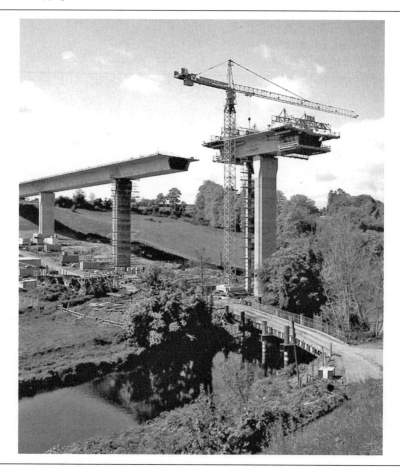

At the pier head, the deck is either built into the pier or is placed on bearings to allow rotational and horizontal movements. If the deck is built in at the pier, as shown in Figure 12.12, this provides the support during the construction on the cantilevers and greatly simplifies the temporary works required. When placed on bearings the deck is held by temporary supports or a tie-down arrangement between the deck and the pier to maintain stability while the deck is being cantilevered. Figure 12.15 shows a balanced cantilever under construction with a prop supporting the deck adjacent to the pier. When the cantilever is completed and joined to the adjacent deck the temporary supports are dismantled to allow the deck to move relative to the pier as it expands, contracts or rotates.

The advantages of the balanced cantilever method are the minimal formwork used with a high degree of repetition, while propping and other temporary works are also reduced or eliminated. The disadvantage is its slow rate of construction.

Design of in situ single-cell box girders
For a single-cell box girder the longitudinal and transverse designs have traditionally been considered separately although it is now more usual to establish three-dimensional finite element models that combine the effects.

The general design, including the creep effects of the dead load and prestress, and the stage-by-stage construction is described in Chapters 5 and 8. The design of the details associated with box girders is given in Chapter 6.

Box behaviour
Thin-walled box girders distort or warp when subjected to torsion and applied loading. This causes a redistribution of both the longitudinal and transverse forces and stresses in the box. These effects are fully described in the C&CA Technical Report on concrete box beams by Maisel and Roll (1974), and they are most accurately determined by using a full-length three-dimensional finite element model for the design.

Shear lag is not usually a problem for normally proportioned boxes, but it may become significant for wider or slimmer decks. In these cases, the shear lag will modify the stress distribution across the section and the peak stresses are used in the serviceability longitudinal stress check.

Prestress layout
The prestress arrangement is governed by the construction method and the stresses generated in the deck during both the construction stages and in the permanent structure.

With span-by-span construction the prestress is also arranged on a span-by-span basis, with a layout similar to that for multi-cell boxes, as illustrated in Figure 11.12. The tendons are anchored at each construction joint and either coupled or overlapped for subsequent stages. The prestress in each span is either fully installed as each stage is completed, or only the minimum number of tendons needed for construction are stressed, with the remainder installed at the end.

With balanced cantilever construction there are three distinct sets of prestressing tendons, as illustrated in Figure 12.16. First are the 'cantilever tendons' that are placed in the top slab and installed as the deck is being cantilevered out. The second are the 'bottom tendons' placed in the bottom slab to take the sagging moments over the mid-span region, and the third are the 'continuity tendons' which usually follow a draped profile and extend over several spans. In Figure 12.16, the continuity

Figure 12.16 Tendon layout for balanced cantilever construction

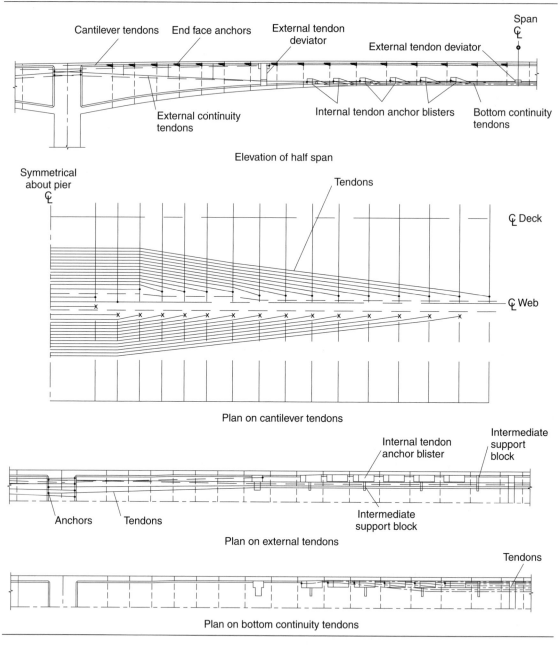

Elevation of half span

Plan on cantilever tendons

Plan on external tendons

Plan on bottom continuity tendons

tendons are shown as external tendons, but they could be internal tendons if preferred by the designer.

With both span-by-span and cantilever construction the prestress tendons are often anchored on the end face of a section. Anchored on either the slab or the webs, the design follows the procedure described in 'Post-tensioned tendons', Chapter 6, with bursting, spalling and equilibrium reinforce-

Figure 12.17 Transverse tendons profile and anchorage (courtesy of Tony Gee and Partners LLP, copyright reserved)

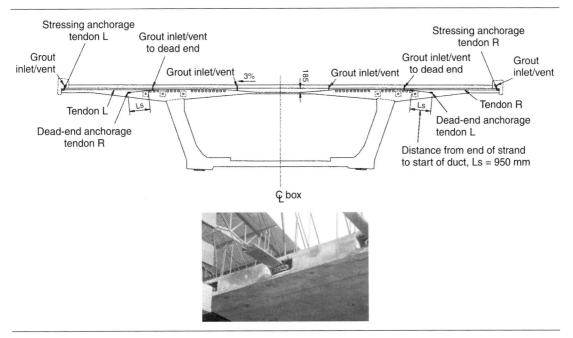

ment being provided. When the next section of deck is cast the anchors become fully encased in the concrete.

Transverse prestressing of top slab

Prestressing the slab transversely as well as longitudinally improves the durability of the structure by reducing crack widths. However, it is usually more costly than the reinforced alternative for all but the wider decks. A typical transverse tendon profile is illustrated in Figure 12.17, with the tendons anchored at the end of the box side cantilevers or with dead-end anchors within the slab. The transverse tendons pass over the top of the longitudinal tendons and flat slab tendon systems are used to take up the minimum space in the top slab.

The top slab is designed as a normal prestressed element with the primary and secondary effects taken into account. Additional tendons are needed over the diaphragms to counter the tension forces caused by the truss action from the bearing loads and diaphragm behaviour.

Deck articulation

The choice of supporting the deck on bearings or building it integral with the substructure is a function of the design and construction method used. Placing the deck on bearings simplifies the design of the permanent structure, while building the deck into the pier provides stability during balanced cantilever construction, eliminates the bearings and reduces the future maintenance of the bridge.

An integral connection is usually feasible only for piers where the deck movements from shrinkage, creep and temperature are small. Where large deck movements occur the moments and shears in the piers become excessive, although making the pier more flexible reduces the effect. A twin-wall pier arrangement with a gap in between provides the necessary moment restraint to the deck while minimising the stiffness in the longitudinal direction. This type of arrangement is often used on long-span bridges where the scale of the structure suits the appearance of the twin-wall pier.

A similar effect is achieved using two rows of bearings on each pier. The bearings provide rotational restraint while allowing horizontal movement. Either pot or rubber bearings are used with the stiffness chosen to optimise the design effects.

Deck construction

The stresses in the deck and substructure are checked at each stage of construction and with the most critical loading arrangement applied. The stability and strength of the deck and substructure are also checked to ensure that adequate factors of safety are achieved. Temporary props or tie-down arrangement at the piers have to cater for all the out-of-balance forces likely to occur. This is particularly relevant for balanced cantilever construction, but span-by-span construction must also be checked.

For decks constructed in situ by the balanced cantilever technique, the temporary loading on the deck includes the following:

- form traveller at the end of the cantilever (typically between 400–1200 kN for spans between 50 m and 200 m)
- construction equipment (typically taken as 10–20 kN on the cantilever tip)
- construction live load (typically taken as 0.5 kN/m^2) applied to the deck to give most adverse effect
- wind loading on the deck including 'upward' wind under one of the cantilevers

In addition, it is common to consider an out-of-balance dead load, with the concrete 2.5% heavier on one side of the balanced cantilever.

REFERENCES

Maisel BI and Roll F (1974) *Methods of analysis and design of concrete box beams with side cantilevers*. C&CA, London.

Podolny W and Muller JM (1982) *Construction and Design of Prestressed Concrete Segmental Bridges*. Wiley, New York.

Prestressed Concrete Bridges, 2nd edition
ISBN: 978-0-7277-4113-4

ICE Publishing: All rights reserved
doi: 10.1680/pcb.41134.225

Chapter 13
Precast segmental match-cast decks

Introduction

The technique of precasting concrete box girders in short segments and transporting them to site to be lifted into position is well established, and it offers many benefits on suitable projects. On early bridges constructed using this technique the precast segments were connected with a narrow in situ concrete or mortar joint, while since the 1980s it has been more common for the segments to be match-cast in a factory and joined on site with only a thin layer of epoxy between them. Several projects in the USA and Asia have used match-cast segments erected without any epoxy in the joints, a technique known as 'dry-jointed'.

The UK's first use of precast match-cast segments with epoxy joints was the Byker Viaduct, shown in Figure 13.1. Completed in 1978 the viaduct carries the Tyne and Wear metro over the Byker valley, and

Figure 13.1 Byker Viaduct, Newcastle, UK

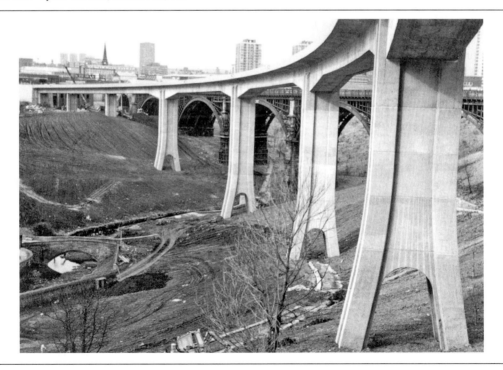

225

Figure 13.2 Belfast Cross-Harbour Links, Northern Ireland (reproduced courtesy of Hyder Consulting (UK) Ltd, copyright reserved)

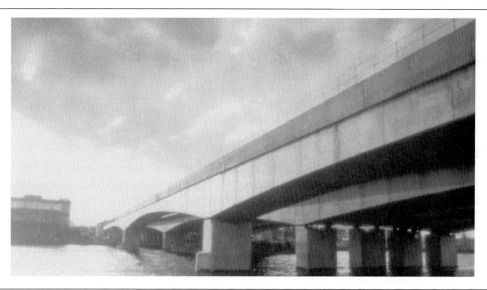

was built by a combination of balanced cantilevering and progressive placing. The segments were placed using either a crane or a lifting frame.

The Belfast Cross-Harbour Links, shown in Figure 13.2, carries both road and rail traffic through the heart of Belfast; it was built with precast segments using the balanced cantilever technique. The design was developed during a competitive design-and-build tender and provided an aesthetically pleasing as well as economic solution. The box girder segments were lifted into place by crane before being glued together.

Dry joints were pioneered in the USA, and their first use in Asia was on the Bangkok Second Expressway System (BSES), built in 1990 and shown in Figure 13.3. Since then dry-jointed decks have been used on many projects in the region. The first section of viaduct constructed on the BSES included over 60 km of precast segmental deck designed and built in a 30-month period. Large predicted differential settlements between the piers led to a simply supported deck arrangement with the segments erected on a span-by-span basis using either overhead or underslung self-launching trusses. External prestressing tendons and dry joints were used to achieve a very rapid rate of erection.

Hung Hom Bypass and Princess Margaret Road Links in Hong Kong, shown in Figure 13.4, includes over 6 km of viaduct built with complex geometry over some of the most congested roads in the city. The viaduct passes over part of the harbour and a busy mainline railway as well as the existing highway network. Using precast segments minimised disruption to the existing infrastructure and allowed the construction to be standardised throughout with all the spans built as balanced cantilevers using a combination of an overhead erection gantry and cranes to position the segments.

In recent years, several U-trough-shaped decks have been constructed using the precast match-cast technique. These have been used on Light Rail Transit (LRT) viaducts, where the U-trough results

Figure 13.3 Bangkok Second Expressway, Bangkok (reproduced courtesy of Hyder Consulting (UK) Ltd, copyright reserved)

Figure 13.4 Hung Hom Bypass and Princess Margaret Road Links, Hong Kong (reproduced courtesy of Hyder Consulting (UK) Ltd, copyright reserved)

Figure 13.5 Dubai LRT U-trough deck

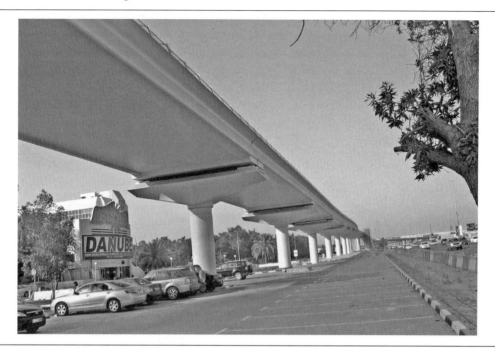

in an reduced depth of deck overall to improve the appearance of the structure and lower the platform level in the station, which has benefits in the urban setting where these viaducts are usually located. The trains run on rails and trackforms resting on the lower slab of the U-trough, while the edge beams also serve as the walkways with only a handrail required above that. Figure 13.5 shows the U-trough used on the Dubai LRT which has a typical span length of 36 m and a total structure depth when viewed from the side of only 2 m.

There have been many other notable bridges built around the world using precast segments which have become a popular form of construction for long urban viaducts as well as major river and estuary crossings. Several cable-stayed bridges have utilised precast segmental concrete decks, such as the Sunshine Skyway Bridge in Florida with a 366 m main span and shown in Figure 16.1.

The advantages of precast segmental construction include the economies from repetition in the construction and the high quality achieved through the factory production, as well as the rapid construction and the minimal disruption to site. The disadvantages include the high initial cost of the erection equipment and casting yard, and the high level of technology needed for both the design and construction.

General arrangement

Precast segmental box girder decks usually have single-cell arrangements with side cantilevers and are similar to the in situ box girder decks described in Chapter 12, with a typical section shown in Figure 12.3. Top slab widths range typically from 6 m to 16 m with box widths between 3.5 m and 8 m. Wider decks are possible using a twin-cell box arrangement; although with three webs this complicates the casting cell design and operation as well as the lifting and transportation activities. Wider decks are

Figure 13.6 Jahra Ghazali Viaduct, Kuwait

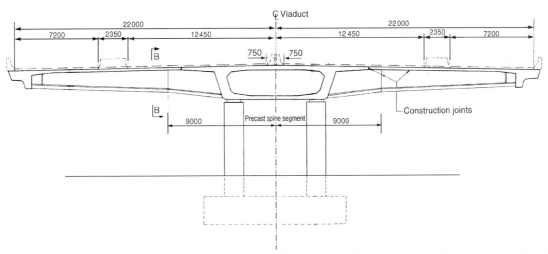

also achieved by supporting the top slab transversely, either by struts, as seen in Figure 12.5, or by transverse beams extending out from the webs. The Jahra Ghazali Viaduct in Kuwait, with the section illustrated in Figure 13.6, used a single 8 m-wide box spine with transverse beams supporting the deck slab as it widened out to 40 m to carry the diverging carriageway at the access ramps.

Precast segments are used on box girder decks with span lengths of 40 m to 150 m. Below 40 m other forms of construction are likely to be more economical, while for spans above 150 m the segment weights become excessive and the use of in situ balanced cantilever construction is usually preferred. For spans up to 70 m a constant-depth section is used with a span-to-depth ratio of approximately 20:1, although between 50 m and 70 m the deck is sometimes haunched in profile. Spans above 70 m are usually haunched with typical span-to-depth ratios of 16:1 at the pier, increasing to 35:1 at mid-span.

Web and slab thickness are similar to the in situ single-cell box girders described in 'General arrangement', Chapter 12. The factory conditions for casting the segments make it practical to construct thinner webs and bottom slabs where the design allows.

Diaphragms are used in the deck only at the pier and abutment positions, to stiffen the box and transfer the deck load onto the bearings and substructure. A typical diaphragm arrangement for a box is shown in Figure 6.16. A publication by Podolny and Muller (1982) contains many early examples of precast segmental box girder projects around the world, illustrating the deck arrangements and concrete details used.

U-trough decks, as seen in Figures 13.5 and 13.45, have been used with spans of up to 44 m when made continuous and 36 m when simply supported. The edgebeams down the sides provide the longitudinal stiffness and shear resistance, while the bottom slab spans transversely between the edgebeams. The bottom slab thickness is governed by the space requirements for the longitudinal prestress and is thickened at the ends to accommodate for the prestressing anchorages and stiffen up the deck above the bearings. The shape and size of the edgebeams are dictated by the shear and the bending moments along the deck. The beams are thinnest at the mid-span, thickening out as the shear increases near the supports. The bending moments at mid-span cause large compressive forces in the top surface of the edgebeam which may require widening of the top, with this widening being thicker at midspan and reducing towards the piers.

Segment lengths are governed by the handling and transportation constraints. Lengths of up to 3.6 m are usually transportable on public roads without excessive restrictions being imposed by the highway authorities. Where no restraints exist, the segments are made as long as practically possible to reduce the overall number of segments to a minimum.

Segments weighing up to 65 tonnes are usually within the capacity of cranes readily available locally, while heavier segments may require special cranes and lifting equipment. The heaviest segment is usually the diaphragm segment and this is often reduced in length to match the weight of the other segments.

Shear keys are formed on the concrete faces of the segment joints. They are detailed as either large or small keys as discussed later in this chapter ('Shear keys at joint'). The shear keys assist in lining up adjacent segments when they are erected to achieve the required horizontal and vertical alignment. The keys also transfer the shear across the joint during the erection process and until the epoxy in the joints has set. For dry-jointed decks, the shear keys assist in transferring the shear across the joint in the permanent condition.

Casting of segments

The segments are made in a purpose-built casting yard with special casting cells designed to allow a daily production of segments. These yards are often covered to protect the casting cells against the weather. A typical casting yard, as seen in Figure 13.7 with a short-line casting cell, is divided into a

Figure 13.7 General view inside casting yard

Figure 13.8 Casting cell schematic layout (reproduced courtesy of VSL International, copyright reserved)

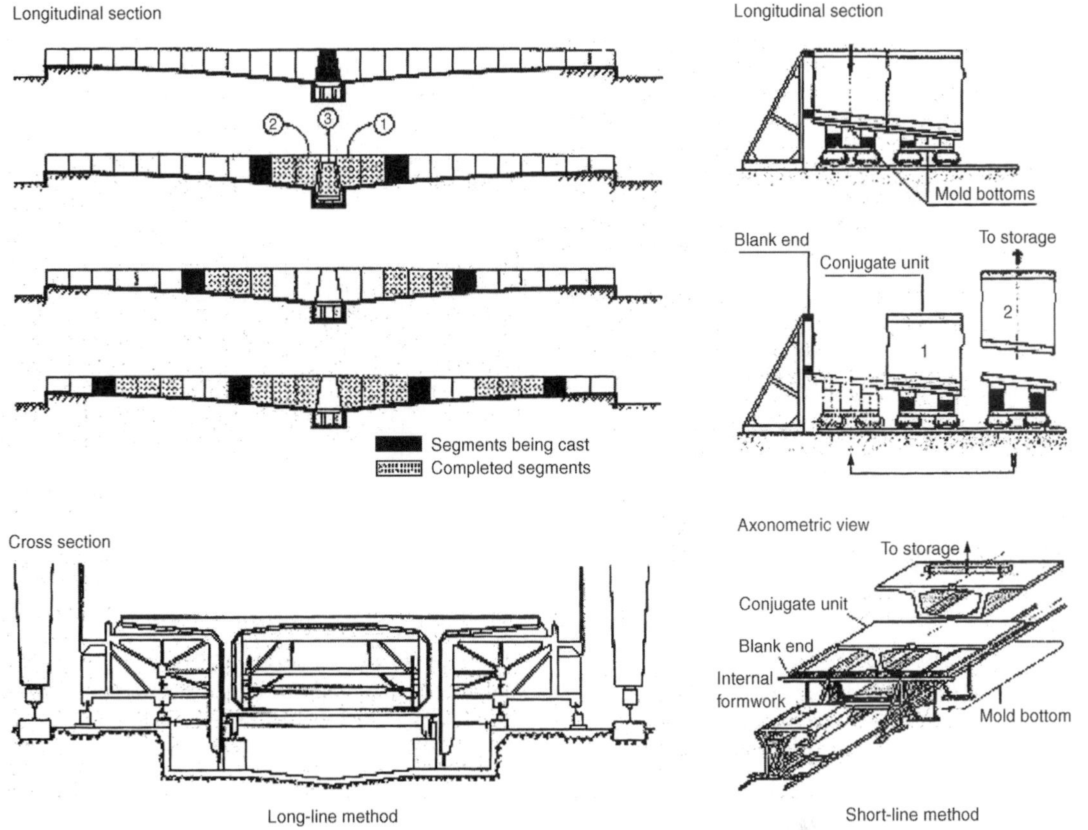

number of areas for the different activities undertaken to construct the segments. A separate area is established for the assembly of the reinforcement in jigs, seen on the left side of the picture. The casting cell is shown in the right foreground, while the segment storage is usually to one side of the casting area. Concrete is often batched on site to ensure good quality control and an uninterrupted supply.

Segments are cast in specially designed steel formwork, using either the short-line or the long-line methods, as illustrated in Figure 13.8. In the short-line method, the segment is cast against its neighbour, and when hardened it is moved into the counter-cast, or conjugate, position to be cast against, after which it is moved to a storage area. With the long-line method the segments are cast one at a time in a long bed until all the segments in a section of span are finished. The short-line casting method is by far the more common of the two due to its need for less space and its greater flexibility in achieving complex geometric alignments, while the long-line method can make the casting operation simpler where sufficient land for the casting yard is available and the deck alignment is straighter.

For the short-line method the casting cell consists of a bulkhead, side and cantilever shutters, a soffit form, and an internal shutter, as seen in Figure 13.9. The bulkhead is usually rigidly fixed at the far end of the cell, while the side shutters are able to drop away from the segment to assist in stripping. The

Figure 13.9 Short-line casting bed

soffit shutter supports the reinforcement cage during the concreting of the new segment and needs to be adjustable to set up the counter-cast segment to its required vertical and horizontal alignment.

The inner shutter, shown in Figure 13.10 in its withdrawn position, forms the central void in the box and has to cater for the formation of the prestress anchor blisters, diaphragms and deviators. The operation of the casting cell requires the inner shutter to be moved back out of the way to allow the segment just cast to be moved into the counter-cast position and the reinforcement for the next segment to be lifted into the casting cell. The inner shutter is designed to allow it to be removed, adjusted and repositioned on a daily cycle. It often incorporates hydraulic supports and adjusting systems to speed up its operating and setting up.

At the bulkhead shutter, seen in the top view of Figure 13.10, a series of protrusions form shear keys across the end face of the newly cast concrete segment. When the new segment is moved back and cast against, the concrete match-cast face of the new segment follows the shape of the shear keys which are mirrored across the match-cast joint.

Figure 13.11 shows the segment moved back into the counter-cast position while still on the soffit shutter. The counter-cast segment is set to the required vertical alignment by adjusting a series of jacks under each leg which move the corners of the segment to the required level. The horizontal alignment is set with a slide mechanism under the shutter which allows the front or back of the counter-cast segment to be moved sideways. With the counter-cast segment in position the reinforcement and tendon ducts are placed and the shutters set up ready to cast the new segment.

In the alternative long-line method the segments are cast against each other on a long casting bed, as shown in Figure 13.12. The advantages of the long-line method are the simpler shutter design, operation, and easier geometry control systems, while the disadvantages are the extra cost of the longer soffit shutters and the extra space needed for the casting area. In the example shown in Figure 13.12, the decks are straight where the long-line casting method was used, allowing a simple stationary concrete soffit shutter to be used throughout. On projects where the alignment geometry varies, a long steel

Figure 13.10 Inner shutter ready to slide into position

soffit shutter is used, designed and assembled so that it can be adjusted to the required alignment for each span.

The full length of soffit shutter is set up in position, while the side shutters and internal formwork, as seen in Figure 13.13, move down the line for each segment casting. Automated systems allow the rapid stripping and moving of the shutters to achieve the required daily casting cycle. After each line of

Figure 13.11 Counter-cast segment positioned

Figure 13.12 Segments on long line casting bed

Figure 13.13 Long line formwork

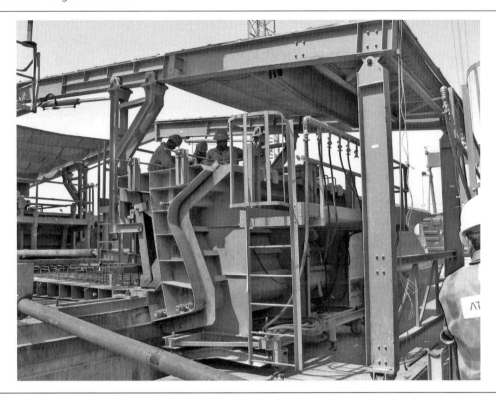

segments has been cast they are individually removed from the bed to a storage area, freeing up the bed for the next set of segments.

With either the short-line or the long-line method, a casting cycle of one segment per day per casting cell is usually achieved for the typical segments in a span. Diaphragm segments and other more complex segments may take two or three days each. To achieve these cycle times, high-early-strength concrete reaching 12 to 14 N/mm^2 after 12 hours is needed to allow the internal, side and cantilever shutters to be stripped in preparation for producing the next segment.

The reinforcement cage for each segment, including the prestress anchorages and ducts, is assembled in a jig which is exactly the same shape as the segment to be made. A typical jig is shown in Figure 13.14 with steel sections used to define the outline of the segment and to support the reinforcement during its assembly. The reinforcement and prestress components are accurately fixed to ensure that the cage will fit into the segment shutters with the required cover and so that the prestress ducts match up with the ducts cast into the adjacent segments. There are usually two or three jigs per casting cell allowing the reinforcement to be assembled over a two or three day period to match the daily concrete casting cycle for each cell.

After the assembly of the reinforcement and prestressing ducts, the cage is lifted and transported to the casting cell using a lifting frame, as seen in Figure 13.15. The lifting frame supports the different parts of the cage and prevents it from bending or distorting as it is moved. An overhead gantry or special lifting equipment is used to pick up the frame and cage and move them around the casting yard.

Figure 13.14 Reinforcement assembly jig

Figure 13.15 Reinforcement cage being lifted and moved to casting cell

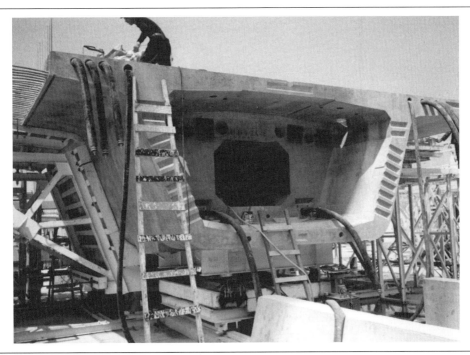

One problem that occurs in segmental construction with internal tendons is angular misalignment between the ducts across the joints. When the ducts are installed and set up over the short length of a single segment it is difficult to get them aligned exactly with the corresponding duct in the adjacent segment. During concreting the ducts are sometimes dislodged or tend to float up in the concrete if they are not rigidly supported. Misaligned ducts lead to an increase in friction during the stressing of the tendons, resulting in a lower achieved prestress force than required. To hold the ducts in position during the concreting and to prevent any concrete getting into them they are filled with a pneumatic tube, as seen extruding from the countercast segment in Figure 13.16. The tube is inflated to seal the duct and cover any gaps across the segment joints.

By casting a segment directly against its neighbour a perfect fit is obtained when they are erected. Complex alignments are achieved by carefully controlling the relative position of adjacent segments when they are being cast. For the long-line method, the soffit shutter is set up to the required alignment before casting begins, while with the short-line method the alignment is controlled by adjusting the position of the segment in the counter-cast position. Decks with a 6% gradient, 10% crossfall and horizontal radii down to 45 m have been successfully built with the short-line casting system.

Figure 13.17 illustrates the alignment control points on a segment. The horizontal alignment is controlled by a series of centreline pins placed at either end of the segments on the top slab mid-way between webs, as indicated in Figure 13.17(i). A permanent horizontal control line is established along the centreline of the casting cell, with a theodolite left in place at one end and a target permanently placed at the other. These are often supported on steel piles and protected by an enclosure from the effects of the weather to minimise any disturbance to the established setting-out line.

Figure 13.17 Segment setting out

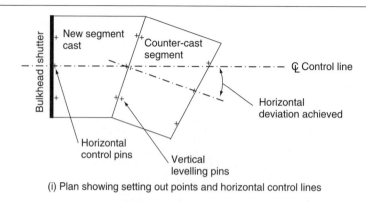

(i) Plan showing setting out points and horizontal control lines

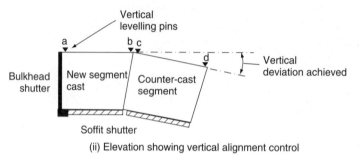

(ii) Elevation showing vertical alignment control

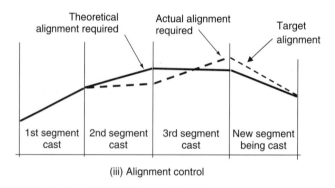

(iii) Alignment control

The vertical alignment is controlled by four levelling bolts per segment placed above each web and again at each end of the segments, as indicated in Figure 13.17(i) and (ii). Precise levelling equipment is used to obtain an accurate measurement of the level of each bolt. The vertical alignment of each web is monitored and used to control the deck crossfall as well as the longitudinal profile.

By measuring the relative horizontal and vertical deviations between adjacent segments while they are against each other in the casting cell, it is possible to predict the alignment that will be achieved when all the segments are erected. Figure 13.17(iii) illustrates a typical alignment control curve. The achieved deviation between adjacent segments is plotted against the theoretical alignment curve from which the target alignment for the next segment is derived.

Construction tolerances and movements in the shutters during casting mean that the exact target alignment is seldom achieved, but by precise surveying, accurately setting up the segments and careful construction the errors are kept to acceptable values and within normal construction tolerances. It is possible to compensate for small misalignments in the casting of each segment by adjusting the target alignment for the next segments to be cast. In this way, the theoretical alignment points are continually 'aimed' at and any misalignments compensated for.

The centreline pins are short lengths of bar inserted into the wet concrete, after casting the segment, with the centreline inscribed into the top surface to line up with the horizontal control line. The levelling pins are small round-headed bolts pushed into the fresh concrete. Typical centreline and levelling pins are shown in Figure 13.18.

Figure 13.18 Survey pins

Centreline pins

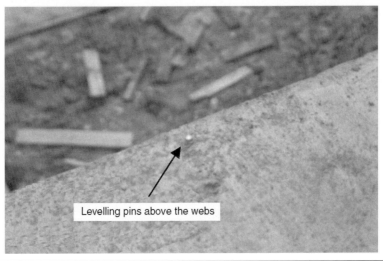

Levelling pins above the webs

Storage and transportation of segments

After leaving the casting area the segments are kept in a storage yard, as seen in Figure 13.19, until they have gained their full strength and the deck erection has reached the stage when they are required. In order not to warp the segments during their storage, a simple three-point support arrangement is used, with two support points under each end of one web and the third point under the mid-point of the other web. Should any one of the supports settle it should not cause twisting or warping of the segment. Sandbags to spread the load out are often placed on concrete beams or pads to provide a robust support base. Where there is limited space to store the segments it is possible to stack them on top of each other; although the strength of the segments and stability of the arrangement need to be checked. Segment production normally starts while the substructure is being constructed and it is usual to accumulate segments over five or six months of production before segment erection commences. The storage yards are sized to cater for this number of segments, although any delay in segment erection may require additional storage areas to be sourced. Once segment erection has commenced, it is common to create secondary storage areas close to where the segments are to be erected. If the segments are transported along the completed deck, this secondary storage area is often on the completed length of deck.

Segments are moved around the casting yard and storage areas using a wheeled lifting frame as seen in Figure 13.20. These segment transporters are able to move segments quickly from one place to another in the casting and storage cycle. The height of the transporter seen in Figure 13.20 allows it to place segments on top of each other in the storage yard permitting double storage to save space, although this particular transporter was not tall enough for triple storage. These transporters need a good ground surface to run over efficiently and safely, and layers of sub-base and granular surfacing are needed to support the wheels and keep the area free from surface water.

Figure 13.19 Segments in storage yard

Figure 13.20 Segment transporter

A segment carrier, such as the type shown in Figure 13.21, is often used to move the segment from the casting area to the storage yard and from the storage yard to the erection site. Segments transported over a public highway are usually limited in length to 3.6 m so that when they are placed lengthwise on the carrier they fit within a typical highway lane to minimise disruption to the traffic. Segments deeper

Figure 13.21 Segment on special carrier

Figure 13.22 Segments lifted by crane at storage yard

than 4.0 m, when placed on a carrier, may exceed the vertical clearance under normal highway bridges en route from the casting yard to the erection site, and either special routes are required avoiding any bridges or the segments need to be cast at the bridge site to avoid the problem.

The segments are lifted on and off the carriers by transporters such as the one shown in Figure 13.20, or with a crane, as shown in Figure 13.22. The crane ropes are passed under the box side cantilever as near to the web as possible to minimise any bending in the top slab, while rubber protection guides are used to protect the concrete surface and end faces of the segment. Each segment is likely to be lifted and stored several times between the casting bed and its final erection position, and it may be stored for several months before being used.

Although it is sometimes possible to lift the segments with ropes slung under the segment cantilevers, as seen in Figure 13.22, lifting devices are usually built into the top slab of the segments. Figure 13.23 shows an arrangement with a steel anchor cast into the web so that a lifting hook can be fitted when required. An alternative arrangement is to leave small holes through the top slab and to use these to place stressbars and clamp on a removable lifting assembly.

For decks built with an erection gantry, the segments are usually delivered along the completed deck with the transporter driving up under the back of the gantry. With erection by crane or lifting frame the

Figure 13.23 Cast-in lifting hooks

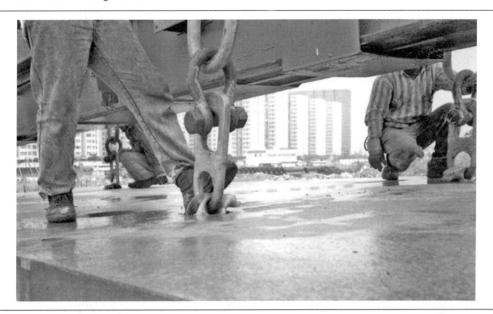

segments are first delivered and positioned at ground level ready to be lifted into position. Some lifting frames are designed to pick the segments up off the deck behind and move the segments into position at deck level. This results in a more complex lifting frame but is useful to avoid any obstructions under the deck. Where access at the erection site is not possible for wheel-based transporters, rail-mounted bogeys, as seen in Figure 13.24, may be used to move the segment under the deck ready to be lifted into place. This is a slow process for the erection and is generally limited to projects where access is difficult, and lifting frames on the deck are being used to raise the segments into position.

Barges, as shown in Figure 13.25, are used to deliver the segments when crossing a navigable river. Used in conjunction with barge-mounted cranes or lifting frames on the deck the barges are usually able to transport several segments at a time. This method of transportation requires facilities for loading the segments onto the barges, which could be by crane or special transfer gantries. The water depth at point of loading and at the landing place for the erection site needs to be deep enough for the barges to travel without restriction. Where there is fast-flowing water at the site some means of holding the barges stationary is required while the segments are being lifted.

Segment erection
General
The segments are positioned in the deck either by a self-launching gantry supported on the piers, by a crane or by a ground level gantry if access is suitable, or by lifting frames fixed to the previously erected deck.

Three different techniques are used for erecting the segments, as illustrated in Figure 13.26. These are the 'balanced cantilever' approach, the 'span-by-span' approach, and the 'progressive placing' of the segments. The balanced cantilever method involves erecting the segments as a pair of cantilevers about each pier. For the span-by-span method all the segments for a span are positioned before the prestress is installed. With progressive placing the segments are built out in one direction from a

Figure 13.24 Segment positioned on rail mounted bogey

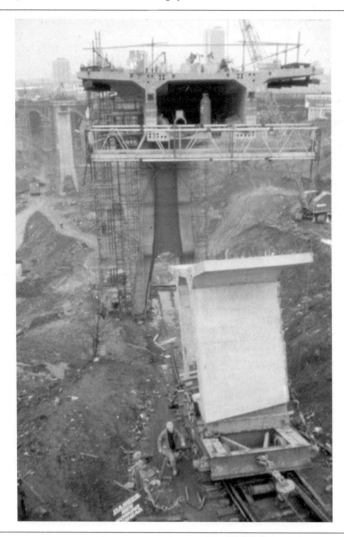

starting point passing over the piers in the process. These three different techniques are described in more detail in the next three sections.

The first segment to be erected is usually a diaphragm segment, sitting over a pier on temporary supports while it is bedded onto the bearings. If the bearing has fixing bolts protruding up from the top plate, box-outs are formed in the underside of the segment to match the bolt location. With the segment held above the bearing the gap between the soffit and top plate, including the box-outs, is filled with cement mortar. If the top plate has no fixing bolts, the bearing top plate is grooved and, with the segment held above the bearing, the gap is filled with epoxy mortar.

Segments to be erected with epoxy in the joints are lightly sandblasted on their end faces before being erected. This removes any deleterious material, such as debonding agents used in the casting or curing

Figure 13.25 Segment transported by barge (reproduced courtesy of Hyder Consulting (UK) Ltd, copyright reserved)

compounds sprayed on during storage, and provides a good bond with the epoxy glue. The sand-blasting should only partially remove the surface latents and not expose the aggregates, just creating a roughened clean surface. Cleaning of the joints by jet washing or wire brushing is used on some projects but tends to be less effective in fully cleaning the joints. Spread evenly over the surface, the epoxy helps to lubricate the joint during erection allowing the segment to slide into position. When

Figure 13.26 Precast segmental deck erection techniques

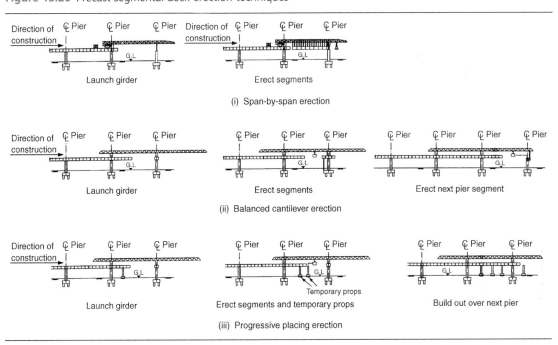

Figure 13.27 Epoxy being spread over joint (reproduced courtesy of Hyder Consulting (UK) Ltd, copyright reserved)

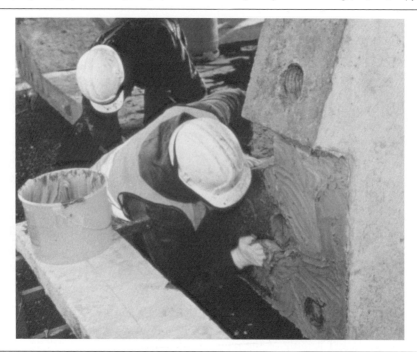

hardened the epoxy seals the joint and provides structural continuity between the segments. Application of the epoxy onto one face of each joint between the segments is usually 'by hand', as shown in Figure 13.27. A uniform thickness is achieved by combing across the joint with a serrated edge that leaves a 1–3 mm thick coating.

When the segments are brought together and the temporary prestress applied the epoxy is squeezed across the joint to fill in any gaps and provide a watertight seal. After stressing the deck the thickness of the epoxy in the joint is between 1 mm and 1.5 mm. The epoxy usually has a minimum compressive strength of 70 N/mm^2, which gives a connection stronger than the concrete either side. Compressible rubber or foam 'O' rings, as shown in Figure 13.28, are glued to the concrete around each hole to prevent the epoxy from being squeezed into the prestressing ducts.

The use of corrugated plastic ducts, as shown in Figure 2.8, to improve the protection of the tendons in the concrete has led to the development of new jointing systems between the ducts across a match-cast joint such as the Liaseal Coupler from Freyssinet, shown in Figure 13.29. In the UK, the use of internal prestressing tendons with precast matchcast segmental decks is prohibited as described in Chapter 3, 'Durability and detailing', due to concerns over the protection to the tendons at the match-cast joints. Previous experience has shown that the epoxy and 'O' ring arrangement does not always achieve a fully water-tight join between the ends of the segments. With corrugated steel ducting this also is not water-tight, especially at the segment joints where the steel ducting is not continuous. The use of water-tight plastic ducts requires a water-tight connection between the ends of the segments across the joints to provide effective protection to the tendons, and the Liaseal Coupler is intended to provide this.

Figure 13.28 'O' rings around ducts at segment joints

The epoxy used in the joints only has a short 'open' life and after being placed on the concrete surface the joint has to be closed with the temporary prestress applied within approximately one hour. The temporary prestress, as illustrated in Figure 13.30, normally consists of bars. These are quick to install, and hold the segments in place until the permanent tendons are stressed. To keep the epoxy thickness uniform over the joint, the temporary prestress is applied with an average compressive stress of between $0.2 \, \text{N/mm}^2$ and $0.3 \, \text{N/mm}^2$. Until the epoxy has hardened the local stress in the segment joint should not be less than $0.15 \, \text{N/mm}^2$ and the stress difference due to flexure not more than

Figure 13.29 Liaseal Coupler (reproduced courtesy of Freyssinet International, copyright reserved)

Figure 13.30 Temporary prestress for segments

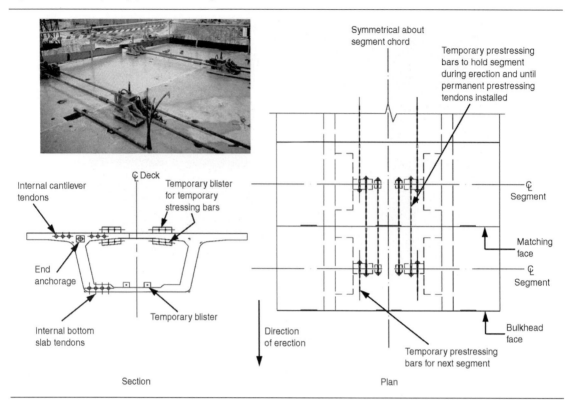

Section

Plan

$0.5\,N/mm^2$. When the epoxy is put under pressure by the prestress force it tends to squeeze out of the sides and bottom of the joint. A sheet or other cover is placed under the joint to catch the epoxy as it drops.

The temporary prestressing bars are either placed through the segment's top and bottom slabs or arranged outside the concrete, as shown in Figure 13.30. Bars inside the slabs are usually left in place and grouted to become part of the permanent prestress for the deck. Bars located outside the concrete section are positioned above the slabs and anchored on temporary steel or concrete blocks stressed down to the segment, or placed inside the segment and anchored to concrete nibs protruding from the inner surface of the box. The advantage of using bars outside the concrete is that they are easy to de-stress and re-use.

When segments are being erected they are carefully guided into position to ensure that no damage occurs to the concrete. Figure 13.31 shows the inside of the deck as the segment is being positioned. The segment is initially offered up to the end of the previously erected segment on a 'dry run' with the temporary prestressing in place to ensure that everything fits together and matches. It is then moved back by at least 300 mm, and the epoxy is applied to the joint surface, after which the segment is pulled into position and the temporary prestress fully installed.

After the epoxy has hardened it becomes very brittle and any material protruding from the joint is chipped off to leave a level surface. Figure 13.32 shows a close-up view of the outer surface of the

Figure 13.31 View inside box girder as segment is being positioned

joints after completion of the deck erection. It is not necessary to do any further work on the joints, which are noticeable only when viewed from near the deck.

Several segmental bridge decks, such as the Bangkok Second Expressway System, shown in Figure 13.33, have been built with dry joints where no epoxy is used between the segments. The design

Figure 13.32 Close up of joint after segment erected

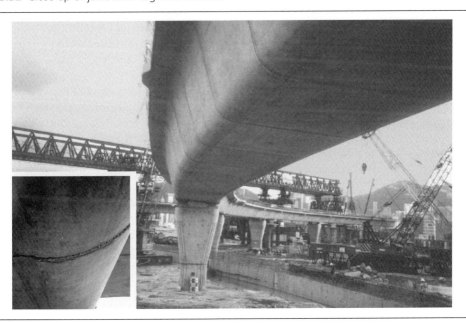

Figure 13.33 Dry-jointed segmental deck

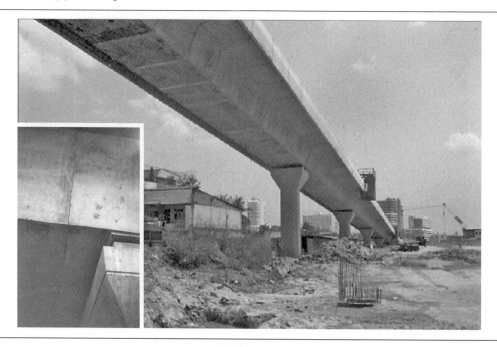

relies on shear friction over the surface and the shear resistance of the keys to transfer the loads across the joint in both permanent and temporary conditions. Omitting the epoxy speeds up the erection times, removes hazards associated with epoxy, and saves cost. Dry-jointed decks are used only in tropical climates where freeze–thaw cycles do not occur, while their effectiveness in more aggressive climates has yet to be proved. In all other respects, the construction of dry-jointed decks is similar to segmental bridge decks with epoxy joints. The casting techniques and erection procedures are the same except that the epoxy is omitted and external tendons are always used as the joints would not provide a water-tight protection to any internal tendons.

Before erecting segments for dry-jointed construction the joint faces are cleaned by wire brushing but are not sandblasted, as this may damage the mating surfaces. The wire brushing needs to be thorough and extensive so that debonding agents or curing compounds are removed, as these may lower the friction resistance of the joint. Without epoxy in the joints temporary prestress is not needed, although bars are sometimes used to pull the segments together and hold them in place until the permanent prestress is installed. When the deck is prestressed the joints fully close up, presenting a neat appearance.

Dry joints do not achieve a fully watertight seal between the segments, and this makes the grouting and protection of internal tendons difficult. For this reason, external tendons are always used with dry joints. To prevent water seepage through the top slab the top of the joint must be sealed in some way. The early techniques for this included inserting a rubber tube into a preformed groove running across the full width of the top slab. This did not prove to be 100% successful and water was able to seep through the joint and stain the underside of the segment or create 'stalactites'. On the Bangkok Second Expressway System viaducts the dry joints were made watertight with an epoxy seal poured into a narrow recess along the top surface, as illustrated in Figure 13.34.

Figure 13.34 Details of seal along top of dry joint

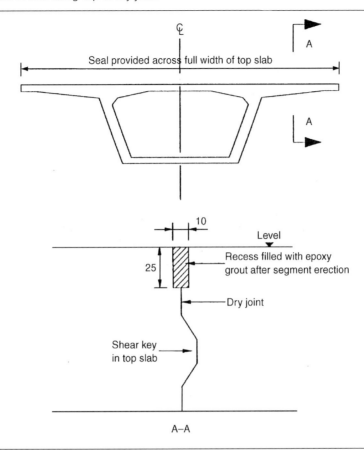

Seal provided across full width of top slab

A

A

10

Level

25

Recess filled with epoxy
grout after segment erection

Dry joint

Shear key
in top slab

A–A

The advantages of using dry joints with precast segmental construction are

- elimination of epoxy
- elimination of the temporary prestress required with epoxy joints
- rapid construction times.

The disadvantages of dry joints are

- more prestress is required to maintain the ultimate shear and moment capacities
- external tendons must be used
- top surface must be sealed to prevent moisture ingress.

When the segments are being transported and erected they must be protected from damage, especially the match-cast faces. They should not be stored where they could be knocked by passing vehicles and must be carefully handled when moved. Spalling of the concrete edges, as seen in Figure 13.35, may also occur when the segments are being stressed together due to segment misalignment or the presence of deleterious material in the joint. Any damage to the joint surfaces should not be repaired until after the segment has been erected to ensure the match-cast faces maintain a perfect fit.

Figure 13.35 Spalling of concrete edges at joint between segments

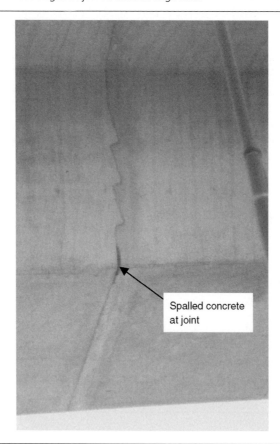

Spalled concrete
at joint

When the segments are erected they join up to the adjacent segments with the same horizontal and vertical angle deviation that existed when they were cast against each other. This is facilitated by the presence of the shear keys on the webs and slabs which guide the segment into position. The erected segments follow an 'erection alignment' that matches the 'cast alignment' modified by the self-weight and prestress deflections as the deck is built up.

Construction and surveying tolerances may require some correction to the alignment of the segments during the erection process. Small inaccuracies in individual segments or misalignment across joints accumulate when long lengths of deck are constructed. A small deviation at the first segment results in an increasing misalignment as more segments are erected. In decks erected by balanced cantilever any small misalignments are corrected within the in situ stitch at mid-span. Similarly for the span-by-span and progressive placing methods, misalignments are taken out by short in situ stitches introduced at suitable locations along the deck. If it is necessary to adjust the alignment of the individual segments during their erection this is achieved by using shims or fibre sheets to increase the thickness of the epoxy joint along one edge. Although this technique achieves only a small adjustment to the deviation at any one segment joint the effect is magnified as subsequent segments are built out.

Segment erection by the balanced cantilever method

The most common method of erecting the segments is by the balanced cantilever technique, either with an overhead or ground-level gantry, with a crane, or by a special lifting frame fixed to the deck.

The first segment over the pier is lifted into position and placed on temporary supports while it is set to the correct alignment. The matchcast segments are then transported into position and lifted either side of the pier segment, being fixed to it with epoxy in the joints and temporary prestress. Further segments are positioned with erection progressing out from both sides of the pier in a balanced manner, as illustrated in Figure 13.26(ii). When the cantilever reaches mid-span it is connected to the opposing cantilever from the adjacent pier by an in situ concrete stitch. Prestressing tendons are then installed along the length of the span and across the stitch to complete the connection.

A variation on this is to use a 'key' segment in mid-span with two short in situ stitches either side. After erecting the cantilevers out from both ends of the span, the gantry or lifting frame holds the 'key' segment in place and restrains the ends of the cantilevers so that 100 mm wide in situ stitches on either side of the segment can be cast to connect it to the rest of the deck. The longitudinal prestress is then applied to complete the continuity. The advantage of using a 'key' segment is that it provides greater scope for catering for any misalignment between the ends of the two cantilevers. If it is not possible to pull the ends of the segments into alignment the 'key' segment provides a sloped transition rather than having a sudden step across a short stitch.

The cantilevers on either side of the pier are kept as balanced as possible to minimise the temporary works needed to prevent the deck from overturning as it is being built out. The deck and temporary works are usually designed for a maximum of one segment out-of-balance although on some projects two segments are erected simultaneously, one on either side of the pier to reduce the out-of-balance forces.

Using the balanced cantilever technique it is possible to erect a typical 40 m span over a 2.5- to 3-day cycle, while spans of up to 100 m typically take between seven and 12 days to erect.

Overhead gantries are supported off the piers or the already completed deck, as seen in Figure 13.36. The gantries are designed to move backwards or forwards and to slide sideways at both the front and back supports so that they can traverse tight curves and position themselves to deliver segments to the correct location along the cantilevers. The segments are transported along the completed deck and lifted up at the back of the gantry by a bogey arrangement which runs along the top chord of the truss and moves the segment into its required position.

After the balanced cantilever at a pier is completed, the gantry moves forward to the next pier ready to erect the next pair of cantilevers. Most gantries are self-launching with the back section supported on rear legs running along the completed deck and with front legs moving onto the next pier. A typical sequence for erection by gantry is as follows.

(a) Gantry used to position pier segment onto next pier. At this stage the gantry's front leg is supported by a bracket fixed to the next pier, and the middle and back legs are on the previously completed deck.
(b) Middle leg moved onto pier segment.
(c) Gantry launched forward to sit 'symmetrically' above the pier where the balanced cantilever is to be erected. The gantry is supported with its middle leg on the pier segment, and its back leg on the previously completed deck, usually at the previous pier location.

(*d*) Segments delivered along completed deck, lifted up by the gantry, and moved into position, as seen in Figure 13.37.

(*e*) Segments erected and prestress installed until cantilevers completed.

(*f*) In situ stitch at mid-span cast and continuity prestress installed.

(*g*) Next pier segment positioned and cycle repeated.

Where a crane is used, as in Figure 13.38, good access is needed to the deck and surrounding area to allow the crane to be positioned and the segments to be transported nearby for lifting into place. Cranes usually give the simplest and most rapid erection procedures with the minimum of temporary works; however, where there are obstructions on the ground crane erection may not be possible. Where cranes are used several cantilevers may be erected at one time, with the erection sequence being more flexible compared to gantries which require decks to be erected consecutively to follow the gantry direction of travel.

Ground-level gantries running on tyres or rails are used to position segments where a level access is available. The segments are moved into position under the gantry, which lifts them up to be stressed onto the already completed section of deck. Figure 13.39 shows a ground-level gantry being used to erect the deck as a balanced cantilever. Less versatile than a crane, ground-level gantries are usually used only on long viaducts and when an existing gantry is readily available.

Lifting frames fixed to the end of the cantilever are used to pick up and position the next segment, as seen in Figure 13.40. At the beginning of the erection of each balanced cantilever the pier segment is

Figure 13.37 Segment being moved into position (reproduced courtesy of H&T Associates, copyright reserved)

Figure 13.38 Precast segments erected by crane (reproduced courtesy of Hyder Consulting (UK) Ltd, copyright reserved)

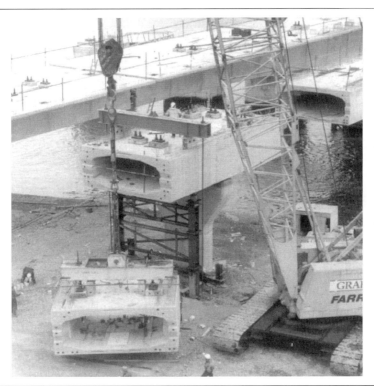

Figure 13.39 Precast segments erected by ground level gantry (reproduced courtesy of Freyssinet International, copyright reserved)

Figure 13.40 Precast segments erected by lifting frame

positioned by crane or cast in situ to establish a platform on which the lifting frame is assembled and secured. After picking up and placing the new segment the frame is moved forward and re-secured to the deck, ready to pick up the next segment. After a balanced cantilever is completed the lifting frame is removed from the completed section of deck and lifted onto the pierhead segment at the next pier to repeat the process.

Using a pair of lifting frames, one on each cantilever, simplifies the erection process, although using a single lifting frame is also possible. When using a single lifting frame it is moved along the deck from one side of the pier to the other to lift the segments in turn, which results in a relatively slow erection process.

The advantages of placing segments with a crane are that very rapid erection is achieved and the deck may be constructed in several places at once, but good access is required along the complete bridge length. Ground-level gantries also allow rapid erection of the segments but require good access. Overhead gantries achieve fast construction rates and minimise the disruption to the ground beneath the deck. They are well suited to building over rivers or other obstructions although they are usually limited to erecting the deck in a sequential manner and are delayed if problems occur at any one pier or in any one span. Lifting frames are the least used due to their relative slowness and the disruption when moving to the next pier. They also require good access underneath the deck although they can offer economic advantages if only a short length of deck is being constructed.

Where the deck is supported on bearings at the pier the stability of the deck during the cantilevering sequence is achieved by providing a temporary support. When the deck is at a low level a prop and tie-down arrangement is often used, as in Figure 13.41. It is convenient with this arrangement to support the prop and tie from the pier foundation.

Figure 13.41 Prop and tie to stabilise cantilever (reproduced courtesy of H&T Associates, copyright reserved)

Figure 13.42 Deck cantilever supported off bracket fixed to pier

For taller piers, or where there are obstructions beneath the deck, stability is achieved during canti-levering by a bracket clamped to the top of the pier (as seen in Figure 13.42), which is used to provide a support to prop or tie-down the deck. Alternatively, the deck may be temporarily fixed to the top of the pier with a tie-down arrangement. With this method the deck is supported on the pier top by a combination of the permanent bearings and temporary jacks, and prestress tendons are installed from the top of the deck, through the diaphragm and down into the pier to provide a clamping force.

The temporary prestress to hold the segments together is arranged as shown in Figure 13.30. As the cantilever extends, additional temporary prestress bars are added to support the extra dead load, or alternatively some of the permanent prestress tendons, commonly referred to as 'cantilever' tendons, are installed through the top slab.

At mid-span the in situ stitch connecting the adjacent cantilevers is concreted using a short length of formwork, as seen in Figure 13.43. The formwork is supported and clamped to the concrete surfaces either side of the stitch. The width of the stitch is typically between 200 mm and 1200 mm depending on the designer's preference. The advantage of a narrow stitch is that no reinforcement is needed within the short length, while a wide stitch gives more room to manoeuvre the segments into position and to stress cantilever tendons on the end face if required.

In addition to the stitch formwork, the ends of the cantilevers are held in place by a frame that spans across the stitch and is fixed to the deck on either side, as shown in Figure 13.44. If the cantilevers are slightly misaligned the frame is used to pull the ends of the deck to the correct position. It then holds

Figure 13.43 Shutter for in situ stitch at midspan

the deck in place to prevent any undesirable movements while the stitch is being concreted and until some of the permanent prestress is installed along the span.

With balanced cantilever erection the span lengths are arranged to suit the cantilevering out from each pier, keeping the two sides reasonably balanced. For end spans, which are typically 60% of the length of

Figure 13.44 Frame to align cantilevers

the adjacent span, the last few 'out-of-balance' segments are supported on temporary props or falsework until the span is completed and the deck made continuous through the installation of the prestress.

The deck is not continuous at movement joints over internal piers, and the spans on either side are 'end spans'. In this situation the deck is constructed either with the segments placed on falsework until the adjacent balanced cantilevers are completed, or with the first segments on either side of the movement joint clamped together with temporary prestress, and a balanced cantilever being constructed in the normal way. After the adjacent cantilevers have been constructed, and the deck connected, the temporary prestress across the movement joint is removed and the movement joint formed.

Segment erection by the span-by-span method

For span lengths of less than 50 m, a common erection method is to place the segments span-by-span, as illustrated in Figure 13.26(i). With this technique a gantry, either overhead or underslung, is used to support a complete span of segments which are pulled together by installing either temporary prestress or the permanent tendons. The gantry then releases the span of segments onto the bearings and launches itself forward ready to erect the next span. For spans above 50 m erection gantries become very heavy and tend to be uneconomical when compared to the balanced cantilever erection method.

Overhead gantries, as shown in Figure 13.45, are supported off the piers or the completed deck. They move forward by 'stepping' onto the next pier in front of the erection and by running along the top of the already completed section of deck.

Figure 13.45 Overhead gantry for span-by-span erection

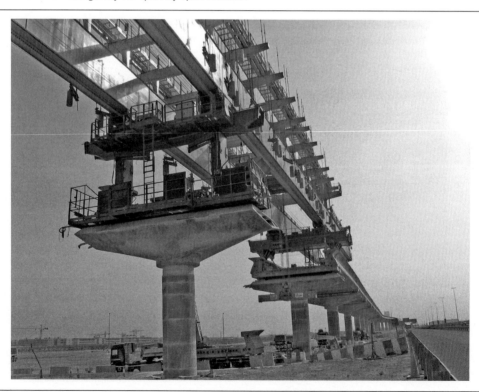

Figure 13.46 Segments hang from gantry

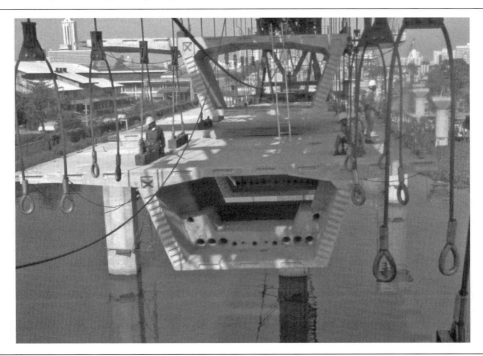

Segments are transported along the completed deck up to the back of the gantry, where they are lifted up and moved out along the span. The individual segments are placed on hangers, as shown in Figure 13.46, which are used to align the segment in position. Once the complete span of segments is in place, and the permanent prestress installed, the gantry lowers the deck onto the bearings on the piers. Overhead gantries are able to deal easily with variations in span lengths and deck geometry, and being above the deck they are not affected by ground-level constraints.

Underslung gantries, as seen in Figure 13.47, are positioned beneath the deck supporting the segments under their side cantilevers or soffit. The gantries are supported at the piers by brackets fixed to the column top or by props from the foundations. They move forward by launching themselves and stepping between the piers. Underslung gantries are not suitable for decks on a tight horizontal curve where the piers are significantly out of line and the segments are offsets from the straight chord between the piers. They also project beneath the deck which may cause temporary headroom problems when passing over existing roads or railways.

The segments are lifted onto the gantry with a crane or lifting frame, as seen on the deck behind the gantry in Figure 13.47. They are then slid along the top of the gantry into position, as shown in the inset. When all the segments for the span are in place the longitudinal prestress is installed and the span released from the gantry onto the bearings, allowing the gantry to be launched forward onto the next pier to repeat the erection cycle.

The deck in Figure 13.48 was erected using an underslung gantry with a lifting arm at the front to position the segments. The gantry spanned between the piers and supported the segments under the box soffit.

Figure 13.47 Underslung truss for span-by-span erection

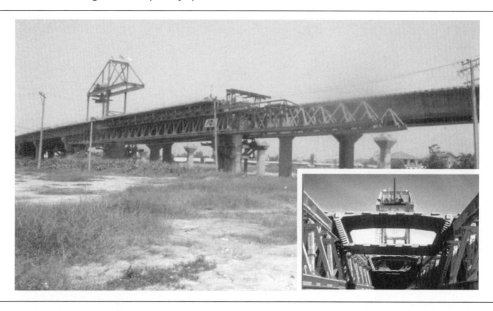

Figure 13.48 Lifting arm on underslung gantry (reproduced courtesy of Jonathon Hiscock, copyright reserved)

Most dry-jointed segmental decks are constructed using the span-by-span erection method. With the segments supported by the gantry, temporary prestress is not needed other than to help position the segments.

With epoxy in the segment joints temporary prestressing bars are used to hold the segments together while the epoxy is setting. The erection process is simplified by applying the epoxy and stressing several segments together at one time.

Span-by-span erection is used on both continuous decks and simply supported spans; the process of erecting the segments is the same for either arrangement. With a continuous deck adjacent 'spans' are joined together with a narrow in situ concrete stitch typically 100 mm or 150 mm wide. This stitch is positioned adjacent to the pier segment or further into the span to suit the design and construction requirements. After the stitch concrete has hardened, the continuity prestress tendons are installed to make the deck continuous.

The advantage of overhead gantries is that they are able to cope with a wide range of deck geometries and tight horizontal curves; however they are generally more complex to design and assemble, and the erection of the segments is slow when compared to the erection of segments for an underslung gantry.

The advantages of underslung gantries are that they are generally simpler to design, assemble and operate; it is also easier to erect the segments, but underslung gantries are used only with decks on a large horizontal radius and when headroom beneath the deck is not critical.

With the span-by-span placement method and epoxy in the segment joints, a typical 40 m span is usually erected every two or three days. With an underslung gantry and a dry-jointed deck an erection rate of up to one span a day is achievable.

Segment erection by the progressive placement method

With 'progressive' placing the erection starts at one end of the bridge and continues out progressively until the other end is reached, as illustrated in Figure 13.26(iii). The deck is usually unable to support itself fully as it cantilevers out towards the next pier and temporary stays or props are required along the span, as shown in Figure 13.49.

As the deck extends away from the pier it deflects under its self-weight and the props or stays are introduced to support the deck and reduce the bending moments and shears. The loads in the props or stays change every time another segment is erected or the lifting frame is moved forward. If the props are susceptible to settlement this is compensated for by incorporating a suitable jack into the arrangement.

When the deck reaches the next pier, sufficient prestress is installed to allow removal of all the intermediate props, just leaving in place the last prop nearest to the end of the deck. This last prop is used to jack the deck up to the correct level to allow the diaphragm segment to be placed over the bearings at the pier. Additional prestress is then installed to enable the deck to be lowered onto the bearings and the cycle to be repeated for the next span.

The segments are placed using a crane, a lifting frame, or a gantry, as with to the balanced cantilever method described above, and they are erected either with epoxy in the joints or with dry joints if appropriate. In situ concrete stitches are often introduced along the deck to assist in controlling the alignment.

Figure 13.49 Progressive placing of segments with props

The disadvantages of progressively placing the segments are the slow erection rate when compared to the other techniques, and the fact that this method is not well suited to high decks or to decks which pass over obstructions such as rivers and highways. The slow erection and the need for regular propping or staying of the deck tend to make this method more expensive compared to the balanced cantilever and span-by-span methods.

Design aspects associated with precast segmental decks

The design of precast segmental decks is similar to that of in situ box girders as described in Chapter 12, with the main difference being due to the presence of the joints. The joints create a discontinuity in the concrete and longitudinal reinforcement, resulting in a reduction in performance under both service-ability and ultimate limit states.

Prestress tendon layout

For precast segmental match-cast decks the prestressing tendon layout is arranged so that the ducts pass through standard duct positions at each segment joint. This is to simplify the construction of the bulkhead shutter used to cast the segments. The tendon profile should incorporate a straight section of duct for a minimum of 100 mm either side of the segment joint, which allows the segment manufacturer to use straight positioning devices to hold the ducts in position and to the correct angle on the bulkhead while also making it easier to align the ducts either side of the match-cast joints.

The tendon anchorage positions and arrangements are also standardised to simplify the shutter arrangement. Anchors located on the segment end face are commonly used with balanced cantilever construction, while top and bottom blisters located in the corners between the webs and the slabs are fixed at standard positions within box segments.

The prestress layout and tendon profile is dependent on the method of construction. A typical arrangement for a balanced cantilever box girder deck is illustrated in Figure 13.50. The cantilever tendons are

Figure 13.50 Prestress layout for balanced cantilever deck (reproduced courtesy of Hyder Consulting (UK) Ltd, copyright reserved)

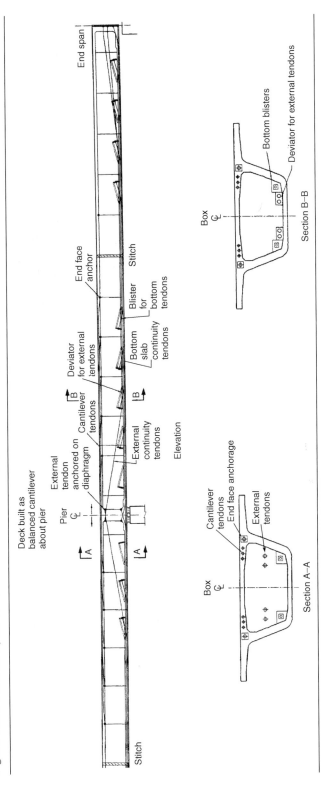

used to support the deck during segment erection, and the continuity tendons are installed after the cantilevers are joined together. Cantilever tendons are anchored on the end face of the segments or on top blisters inside the box. Continuity tendons are anchored on blisters to give access for installing and stressing. Where external tendons are used they are held in position by deviators and anchored on the diaphragms or on anchor blocks inside the box.

Simply supported decks require the tendons to be in the bottom of the section, and a simple draped profile, as illustrated in Figure 13.51, is used. In this example, external tendons are anchored on the diaphragms and at deviators along the box, with the deviators also holding the tendons in a draped profile.

Shear keys at joint
Shear keys are provided on the surface of the joint to help align the segments during erection and to transfer the shear forces between the segments. With epoxy joints the shear transfer through the keys is needed only during the erection and until the epoxy has set. Where dry joints are used the shear keys provide the shear transfer in both the erection and permanent conditions. Either a few large or multiple small shear keys are used, as illustrated in Figure 13.52. Both are designed to prevent a shear failure occurring through the joint. It is common practice to provide additional shear keys above the minimum number required by the design to allow for possible damage to some of the keys during storage or handling.

The design of match-cast shear keys is not covered by BS 5400 (BSI, 1990). Considering the shear friction concept as given in AASHTO (1998–2011), the capacity of the shear keys is estimated using a cohesion factor of 2.8 N/mm^2 and coefficient of friction of 1.4, giving:

$$V_k = 2.8A_{sk} + 1.4[A_{sk}f_{pk} + A_r f_y] \text{ with a maximum of } 10.3A_{sk}$$

with the values adjusted by the material factors or strength reduction factors as appropriate. Eurocode (BSI, 2004) gives a similar approach with the cohesion factor of 0.5f_{ctd} and coefficient of friction of 0.9.

Large shear keys are designed as a concrete corbel with appropriate reinforcement provided, as shown in Figure 13.53, with the reinforcement enhancing the shear capacity of the keys. Multiple small shear keys do not require reinforcement and normally they are unreinforced.

Design with epoxy or concrete joints
Segmental decks with epoxy joints are designed similar to in situ box girders. The main difference is that under serviceability conditions the longitudinal bending stresses along the deck are always kept in compression across the joint under all loading conditions to prevent the unreinforced concrete adjacent to the joints cracking.

The epoxy in the joint creates a bond of greater strength than the concrete itself, and the ultimate shear and moment design is carried out as if the deck is monolithic, although when designing to AASHTO (1998–2011) the strength reduction factors for ULS (Ultimate Limit State) moment are reduced for segmental construction with cast in place or epoxy joints and internal tendons, and for both ULS moment and shear with external tendons.

Where wide, cast in situ concrete joints are incorporated in the structure, BS 5400 Part 4 (BSI, 1990) requires the shear force under ultimate loads to be no greater than 0.7(tan α)(0.87P) to ensure against a

Figure 13.51 Prestress layout for simply supported deck (reproduced courtesy of Hyder Consulting (UK) Ltd, copyright reserved)

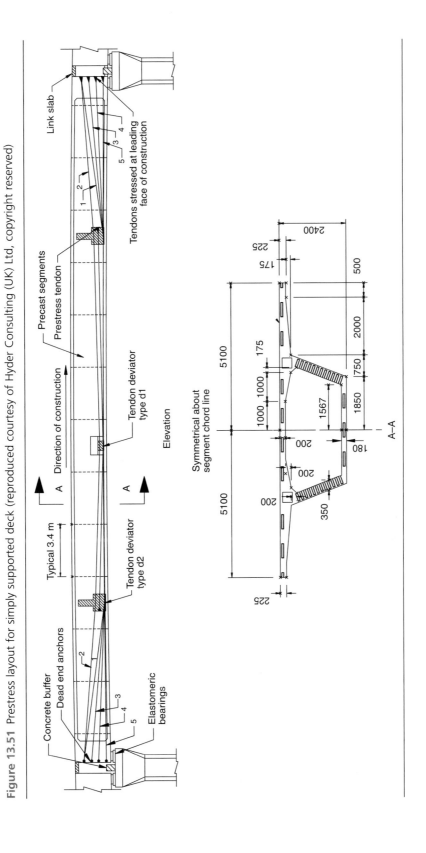

Figure 13.52 Shear key arrangements

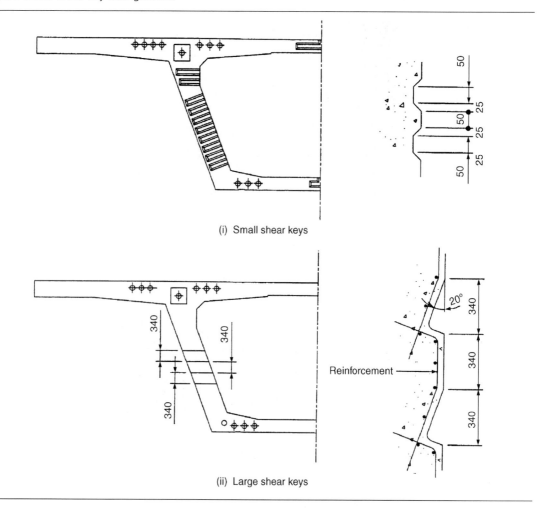

(i) Small shear keys

Reinforcement

(ii) Large shear keys

frictional failure across the joint. Tan α depends on the interface and for roughened faces is taken as 0.7 during erection and 1.4 after completion of the bridge. P is the horizontal component of the prestress force after all losses.

Design with dry joints

Tests carried out by MacGregor *et al.* (1989) and Rabbat and Sowlat (1987) demonstrated that a dry-jointed segmental deck behaves in a similar manner to an in situ or epoxy-jointed structure up to the point when longitudinal bending stresses give rise to decompression of the joints or shear stresses cause the joint to slip. With segmental construction the joints are always kept in compression under service conditions and the longitudinal stress check and other serviceability checks are carried out as if the deck is monolithic.

The shear keys at dry joints provide the mechanism for transferring shear forces across the joints at all stages of the structure's life. Tests carried out by Koseki and Breen (1983), and Buyukozturk *et al.*

Figure 13.53 Reinforcement for large shear keys

(1990) have shown that dry joints with multiple small shear keys spread out down the webs are more efficient than large individual keys. Joint slippage occurs after the initial frictional resistance of the mating concrete surfaces is exceeded. The tests indicated that slipping starts after mobilising a friction of at least 0.5.

The webs should be checked to ensure that the unfactored shear force at the joint is less than 40% of the compressive force on the web to provide an adequate factor of safety against joint slippage at the SLS condition. The shear force includes the vertical and torsional shears, less the vertical component of prestress. The compressive force includes the prestress modified by the bending stresses present. Suitable reduction factors should be applied to the prestress force estimated from the design. For this 'serviceability' check the shear key effects are not considered.

Ultimate shear behaviour of a dry-jointed deck relies on the friction capacity of the joint and shear keys. If the friction between the mating surfaces is exceeded slippage will occur until the shear keys are fully engaged. Increased shear is then resisted until either the bearing resistance capacity or the frictional resistance capacity of the keys is reached, at which stage the keys will either fail by crushing or by shearing off. The shear capacity of a dry joint is significantly less than for an epoxy joint or mono-lithic structure, with up to 40% reduction being observed in tests.

BS 5400 (BSI, 1990) does not cover dry jointed construction, while AASHTO (1998–2011) requires that joints between precast units shall be either cast-in-place closures or match-cast and epoxied joints. Eurocode (BSI, 2004) covers the design of dry jointed construction and gives guidance for the designs of bridge decks that incorporate dry joints.

Where dry joints occur it is necessary to take account of the effect of the discontinuity in determining the shear resistance globally along the deck. As the applied loading approaches the ultimate limit-state dry joints will open due to decompression, restricting the area available for the compression strut resisting the shear. This effect is not critical in simply supported decks with dry joints as the maximum shear is at the ends of the deck where the moment is small and decompression is unlikely to occur. In continuous decks, where the maximum moment may occur over the piers in a similar location to the maximum shear, the reduced depth needs to be taken into account when carrying out the ULS shear resistance design. Eurocode (2005) considers this effect and provides guidance on the adjustments to the compression struts angles and placement of the reinforcement in this case.

With dry-joints between the precast segments, the interface has to carry the full ULS shear force in the deck at the section being considered. The design shear resistance is made up of the sum of the shear resisted by the shear keys plus the shear resistance of the rest of the joint area. Both of these are calculated taking into account the variation in compressive stress across the joint due to the applied moments and prestress force, with more resistance being generated in the portion of the joints under greater compression.

Shear resistance of shear keys $\quad V_{Rdk} = (0.5f_{ctd} + 0.9f_{pk})A_{sk}$

Shear resistance from remaining area $\quad V_{Rdj} = (0.25f_{ctd} + 0.5f_{pk})A_{sj}$

To derive the compressive stress across the joint, the prestress tendon force is based on long-term after-losses, multiplied by a partial safety factor of 0.9. Where there are inclined external tendons passing the segment joint being considered, such that it counters the applied shear force, the vertical component of the prestress, multiplied by the partial safety factor of 0.9, is deducted from the applied ULS shear force before considering the design of the joint.

The transverse design is carried out as for other box arrangements, with an additional check done to ensure that there is sufficient frictional resistance on the joint surface of the top slab to distribute local wheel loads' effects longitudinally across the joint and to prevent joint slippage. To assist in the load transfer across the joint, small shear keys are placed across the top slab, as seen on the segment in Figure 13.46. Where an in situ concrete barrier or edge beam is built onto the top slab its joints should not coincide with the segment joints, so that they assist in providing additional continuity along the deck.

When dry-jointed decks approach their ultimate bending capacity they undergo high deflection with the curvature generated by rotations concentrated at individual segment joints. This is in contrast to monolithic or epoxy-jointed decks, where the rotations and tensile cracks are more evenly spread out. These concentrations of rotation cause a dry-jointed deck to fail at a lower moment than the equivalent monolithic or epoxy-jointed arrangement.

Figure 13.54 Ultimate behaviour of beam with dry joints and external tendons

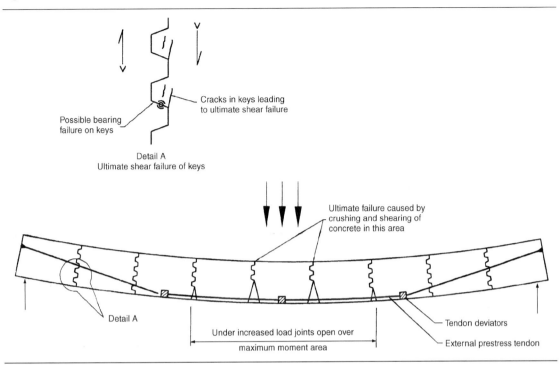

For the ultimate longitudinal moment design, when the deck is subjected to increased loading beyond the serviceability state the joints decompress and begin to open up, as illustrated in Figure 13.54. As the load increases more joints along the span open up and the openings extend over a greater part of the segment depth. As the ultimate moment capacity is approached the deck deflections rapidly increase and final failure occurs on the compression side by crushing of the concrete due to excessive strain. With external tendons the strain in the tendons is distributed out over the tendon length and the tendons do not reach their breaking load. The most appropriate method for estimating the ULS moment capacity of a deck with dry joints is to carry out a finite element analysis with the concrete and external tendons accurately modelled to provide a realistic behaviour up to the point of failure. This can be done using a full 3D FE analysis with solid elements and both geometric and material non-linear behaviour modelled. The model should allow slippage of the prestress tendons through the deviators under increasing loading, with failure being controlled by the strain limits of the concrete.

With the concrete faces in contact the shear keys are restrained from spalling and bursting. To prevent a bearing failure of the shear keys the total applied ultimate shear force in the webs should not exceed $0.6f_{cu} \times$ shear key bearing area, although this check is not included in Eurocode (2004, 2005).

Torsional shear stresses in the top and bottom slabs should not exceed the frictional resistance available at the joints, derived in a similar manner to that of the webs as described above.

Typical segment details
The features and details of precast concrete segments are discussed above and depend on the particular requirements of the project where they are being used. Figure 13.55 illustrates a typical segment section

Figure 13.55 Typical segment arrangement (reproduced courtesy of Hyder Consulting (UK) Ltd, copyright reserved)

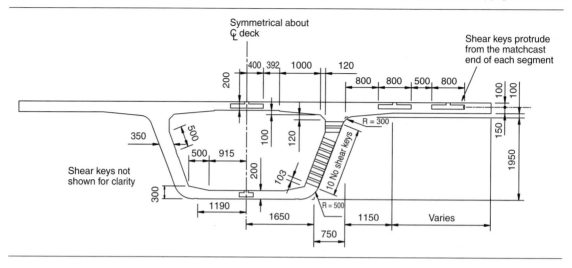

for an urban viaduct with spans of 45 m. The length of the segments should be standardised as much as possible although small variations in length are unavoidable in urban environments where at-grade limitations may govern the exact position of each pier. If possible the range of segment lengths should be kept to less than 400 mm. If segment lengths vary to a greater degree than this it affects the casting cell shutters set up, with adjustments disrupting the segment production cycle.

The webs and bottom slab are thicker near the piers than at mid-span to cater for the higher shear forces and bending compression respectively. The transition in the web and bottom slab thickness is either taken out uniformly over one segment length or is stepped up or down at one end of the segment, as illustrated in Figure 13.56. The stepped arrangement simplifies the internal shutter configuration

Figure 13.56 Web and bottom slab transition detail

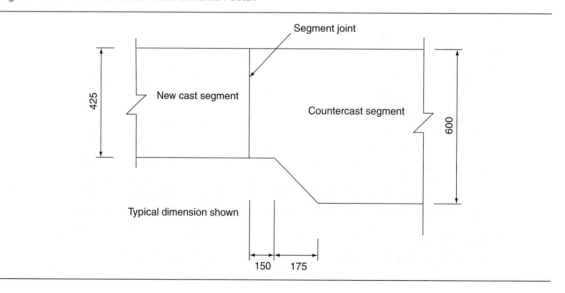

and is the most common way of changing the web or bottom slab thickness with precast segmental decks.

Reinforcement in the segments is standardised as much as possible with repetitive arrangements in the bottom slab, webs and top slab to facilitate the factory assembly of the units.

Deck erection

As with in situ box girders, the stresses, strength, and stability of the deck and substructure and the temporary works designs are checked at each stage of erection.

In addition to the loading noted in Chapter 12, precast segmental decks are checked during erection for the load applied from the specialist erection equipment used. These include the segment-lifting beam, where typically a load of between 30 kN and 50 kN is applied at the end of the cantilever. Where a lifting frame is fixed on the deck to lift the segments into place, the weight of this is typically between 200 kN and 500 kN depending on the size and weight of the segment being lifted. A stressing platform is often attached to the end of the cantilever to provide access for installing and stressing the temporary and permanent prestressing tendons, as shown in Figure 13.57 which shows a stressing platform fixed to a segment about to be erected. A stressing platform typically weighs about 10 kN.

Gantries used to erect the segments are usually supported off the substructure or completed deck, and they apply significant loads which may govern the design of some parts of the structure. For a deck erected by the balanced cantilever method the gantry supports only one segment at a time and it typically weighs between 200 tonnes and 400 tonnes for spans between 50 m and 120 m. For a deck erected by the span-by-span method the gantry supports a complete span of segments and typically weighs between 150 tonnes and 350 tonnes for spans between 35 m and 50 m.

Figure 13.57 Stressing platform on segment

The design and stability of a deck erected as a balanced cantilever is usually checked in the temporary condition with one segment out of balance, to match the construction sequence adopted. If there is a risk of a segment being dropped, resulting in a situation with two segments out of balance in the permanent structure, then this is also checked in the design. Several projects have developed erection schemes where segments are erected in pairs to keep the cantilevers balanced during the construction and to minimise the risk of any out-of-balance forces from occurring.

The design of a deck erected by the span-by-span method is generally less sensitive to temporary loading during construction as all the segments are supported from a truss or gantry until the permanent prestress is applied; however, each stage of their construction is checked to ensure that the design and stability of the permanent and temporary works is adequate. The heavy load of a span-by-span gantry results in high forces being imposed on the bearings supporting both the previous span and rear of the gantry, and also on the piers, where eccentric loading may produce high moments and deflections.

Decks erected using the progressive placing of segments supported by props are affected by settlement of the props, which generate additional stresses in the deck. A jacking system is incorporated into the prop arrangement and used to compensate for any expected settlements. Decks built span-by-span on gantries or trusses are checked during application of the prestressing and during the transfer of load from gantry to the deck structure to ensure that the stresses are within acceptable limits. As the prestress is applied the deck lifts up from the gantry; however, as the weight is relieved the gantry deflects upwards and continues to apply a force onto the deck. The gantries are usually more flexible than the deck structure and continue to apply significant load onto the deck even after the deck is fully prestressed. This upward force on the deck could overstress the concrete if not controlled by releasing the deck from the gantry in stages as the prestress is installed.

During the segment erection process the erection contractor frequently wants to apply loading onto the freshly formed epoxy joints. On some projects using balanced cantilever construction the erection contractors have erected three segments on a cantilever in a single day, with the moments and shears being applied to the epoxied joints before the epoxy has fully hardened. With span-by-span construction the erection contractor will want to lower the complete span of segments onto the bearings as soon as the last epoxy joint is formed and some of the prestress is installed. The capacity of the joints to resist this early loading must be checked at each stage based on the corresponding early-age strength of the epoxy. This early strength gain in the epoxy is dependent on the material being used and the ambient temperature, with typical values from one site shown below:

| | Typical strength of epoxy in N/mm^2 | |
Age after mixing	Compression	Shear
2 hrs	0.3	0.5
3 hrs	3.6	6.2
4 hrs	6.2	10.8
5 hrs	7.8	13.6
6 hrs	9.1	15.8

REFERENCES

AASHTO (1998–2011) *Load and resistance factor design (LRFD) bridge design specifications.* 4th edition. AASHTO, Washington, DC.

BSI (1990) BS 5400: Steel, concrete and composite bridges. Part 4. Code of practice for the design of concrete bridges. BSI, London.

BSI (2004) BS EN 1992–1-1:2004: Eurocode 2: Design of concrete structures. Part 1–1. General rules and rules for buildings. BSI, London.

BSI (2005) BS EN 1992–2:2005: Eurocode 2: Design of concrete structures. Part 2. Concrete bridges – design and detailing rules. BSI, London.

Buyukozturk O, Bakhoun MM and Beattie M (1990) Shear behavior of joints in precast concrete segmental bridges. *Journal of Structural Engineering* **116(12)**: doi: 10.1061/(ASCE)0733-9445(1990)116:12(3380).

Koseki K and Breen JE (1983) *Exploratory study of shear strength of joints for precast segmental bridges.* Centre for Transportation Research, The University of Texas at Austin, Research report 248–1, September.

MacGregor RJG, Kreger ME and Breen JE (1989) *Strength and ductility of a three-span externally post-tensioned segmental box girder bridge model.* Centre for Transportation Research, The University of Texas at Austin. Research report 365–3F, January.

Podolny W and Muller JM (1982) *Construction and Design of Prestressed Concrete Segmental Bridges.* Wiley, New York.

Rabbat BG and Sowlat K (1987) Testing of segmental concrete girders with external tendons. *PCI Journal* **March-April**: 86–107.

Prestressed Concrete Bridges, 2nd edition
ISBN: 978-0-7277-4113-4

ICE Publishing: All rights reserved
doi: 10.1680/pcb.41134.277

Chapter 14
Precast full-length box girders

Introduction

Several major bridge crossings and long urban viaducts have been constructed using concrete box girder decks precast in long sections and lifted into place. The investment needed to set up large casting facilities and to provide special heavy transporting and lifting equipment limits this form of construction to projects where long lengths of deck are being built which require a large number of units to make the investment viable.

There are several variations to the technique, but they all involve precasting a complete span or part span in a casting yard, transporting the units to site and placing them in position. The units may be over 100 m long and weigh thousands of tonnes, and usually they require special equipment to move and erect them.

The technique was adopted for the viaducts on the 25 km long Saudi Arabia–Bahrain Causeway shown under construction in Figure 14.1. The 50 m spans were formed using a combination of 66 m and 34 m long precast units. The construction was carried out over a four-year period and was completed in 1985.

The Central Viaduct in Figure 14.2, part of the Vasco da Gama Crossing in Portugal, used 77.6 m long precast units weighing 2200 tonnes each to form the deck. The sections were cast on-shore and transported on barges before being lifted into place. Initially positioned as simply supported on the piers, the units were connected to make the deck continuous.

In many modern cities, new road and rail systems are being constructed to relieve the congestion in the existing infrastructure with the requirement for rapid construction and minimum disruption. These systems often include urban viaducts extending over many kilometres and they provide the opportunity to use full-length precast units to construct the deck. Figure 14.3 shows the Singapore MRT, built in the 1980s and 1990s, which adopted a standard full-length precast deck design for the long lengths of viaduct. The units were cast adjacent to the viaduct under construction and transported along the completed deck before being lifted into place by a gantry.

The deck for the 13 km long Confederation Bridge, in Figure 14.4, was precast using two types of units. The main units were precast in lengths symmetrical about the pier and then lifted in as a balanced cantilever. Shorter precast units were placed over the mid-span, between the ends of the two adjacent cantilevers, to complete the arrangement.

Full-length precast box girders have the advantages of rapid deck erection and repetition in their construction with factory conditions for casting the units. They are an economical choice for very

Figure 14.1 Saudi Arabia–Bahrain Causeway (reproduced courtesy of Ballast Nedam BV, copyright reserved)

Figure 14.2 Vasco da Gama Crossing, Portugal (reproduced courtesy of Yee Associates and Hyder Consulting (UK) Ltd, copyright reserved)

Figure 14.3 Singapore MRT

Figure 14.4 Northumberland Strait Bridge (reproduced courtesy of Ballast Nedam BV, copyright reserved)

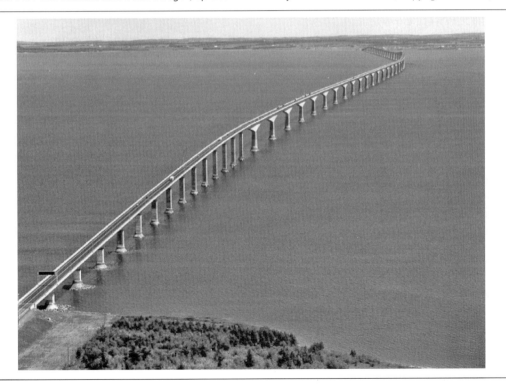

long viaducts with many units, where the cost of the casting facilities and special equipment is spread out over many uses. Their disadvantages include the high cost of the large lifting and transporting equipment and the extensive areas needed to cast and store the units. They are not suited for highly curved decks as these complicate the transport and erection of the units.

Smaller structures, such as footbridges, often have precast decks spanning the full length between piers and these could also be classified as precast full-length girders, but are more similar to the precast beam arrangement described in Chapter 10. This chapter covers the application of the technique to larger structures and longer spans.

General arrangement

Cross-sections for full-length precast decks are usually similar to the single-cell box girder arrangement used for in situ or precast segmental construction, as described in Chapters 12 and 13. A box section is efficient from the design aspects, and provides robust units for moving and erecting.

The transverse cross-section arrangement depends on the traffic requirements which dictate the top slab width. For light-rail systems, as shown in Figure 14.3, the top slab is usually only 5 m or 6 m wide if it carries a single track, or up to 10 m wide with a double track. This results in a narrow box arrangement with short side cantilevers. For highway bridges, with a single box carrying several lanes of traffic, the top slab width is usually greater than 12 m allowing a wider box arrangement to be used, similar to Figure 12.3.

The precast deck units have diaphragms at the pier positions, as in the other forms of box construction. The units are usually placed on bearings to simplify the construction process, but they may be made integral with the piers with an in situ concrete connection.

The units are often placed directly on the piers and left as simply supported spans, or they are joined together with an in situ concrete stitch and continuity prestress tendons to form a continuous structure.

The main precast units on the Confederation Bridge were 192 m long and weighed 8200 tonnes. With spans of 250 m the deck units were haunched, 14 m deep at the piers and 4.5 m deep over the mid-span. The maximum size of the precast units depends on the capacity of the lifting and transporting equipment available; however, on the large-scale projects where this type of construction is used it is common to design and build new equipment specifically for the length and weight of the units to be handled.

Either internal or external prestressing tendons, or a combination of both, are used to provide the prestress with the normal anchorage arrangements and blister and deviator details adopted. Tendon layouts depend on the deck arrangements and the erection techniques adopted, with layouts similar to the in situ box girder construction described in Chapter 12.

Casting and storage of the units

The casting yard is usually located near the bridge site to facilitate the transfer of the units. A large area is needed to provide sufficient space to build and store the precast units. The casting yard is arranged with areas for bending and assembling the reinforcement and for casting the segments. Separate areas are provided to store the units while they gain strength and wait their turn to be erected.

Figure 14.5 Casting cell shutters (reproduced courtesy of Hyder Consulting (UK) Ltd, copyright reserved)

(i) Outer shutters   (ii) Inner shutters

Good ground conditions are required at the casting yard to support the heavy units; otherwise piling must be used under the casting and storage areas to prevent differential settlements, which may cause the units to distort and become overstressed.

Where adequate concreting facilities are available, the units are often cast in one pour with full-length outer and inner shutters used, as shown for the Taiwan High-Speed Rail project in Figure 14.5. Casting the units in one pour reduces the construction time although the inner shutter arrangement is usually more complex, and placing the large volume of concrete requires careful planning and control.

The large units may be cast in several stages to simplify the formwork and concreting operation, using techniques similar to in situ concrete box girders, as described in 'Construction, span by span', Chapter 12. Long units are sometimes divided into a number of shorter segments, cast in stages similar to the long-line technique described in 'Casting of segments', Chapter 13.

The factory conditions for manufacturing the units allow maximum repetition within the different construction activities. To reduce the casting cycle-times the reinforcement is often pre-assembled and the complete cage moved into the casting cell. Figure 14.6 shows the reinforcement for a full-length span unit being assembled with the prestressing ducts laid out along the top of the webs.

In Figure 14.7 the reinforcement and shutters have been assembled and the concrete is being placed for the complete unit. Concreting usually starts down each web, until it flows into the bottom slab, at which time the bottom slab is poured directly from above. The webs are then filled up and finally the top slab is cast. The large quantities of concrete and long lengths of pour can cause problems in keeping the surface of the concrete fresh without cold joints forming, and retarders may be needed in the concrete mix in some circumstances. After casting each unit the concrete is cured until it reaches sufficient strength for the prestress tendons to be installed and stressed, and for the unit to be moved. The unit is then taken to a storage area to complete its curing, and left there until it is needed for erection. A completed unit supported on temporary plinths in the storage area is shown in Figure 14.8. The temporary bearings must be carefully set to provide an even support to the corners of the unit and to prevent the units from twisting or warping while the concrete is still maturing.

Figure 14.6 Reinforcement being assembled

Figure 14.7 Unit being concreted (reproduced courtesy of VSL Intrafor Asia, copyright reserved)

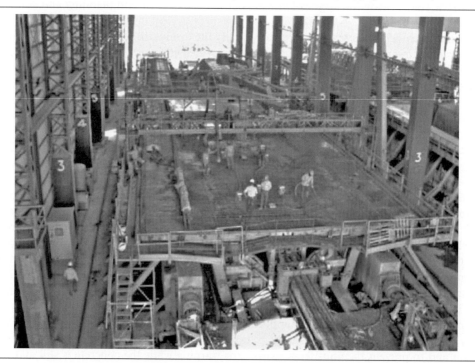

Figure 14.8 Completed precast unit stored

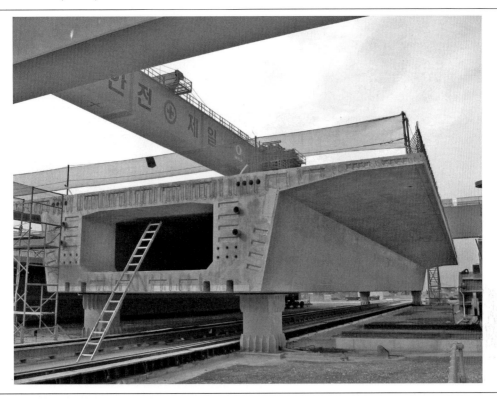

Transportation of the units

Moving the large units requires special equipment to carry the heavy loads. The units are moved about the casting and storage areas using gantries or lifting frames, as seen in Figure 14.9. The gantries or frames are either rail mounted or run on wheels depending on the casting and storage yard set up. Rail mounted arrangements are often simpler to assemble and operate, while wheels provide more flexibility when the units are being moved about the yard.

Special transporters, similar to the one shown in Figure 14.10, are used for moving the units longer distances over land. The multi-axial arrangement spreads out the large loads on the highway pavement. Problems often occur when units are moved over existing bridges or highways with utilities underneath. The existing bridges and highways may need strengthening or protection to allow the transporters to pass over.

The casting and storage areas for the Singapore MRT were next to the line of the viaduct and the units were lifted directly on the completed deck, as shown in Figure 14.11. The 165 tonne and 23 m long units were moved along the deck on a twin bogey arrangement to an erection gantry that lifted them into position on the piers.

Barges are a convenient way to transport the units where a bridge is being constructed over water. Often a combined barge and lifting facility is used to pick up the units from a jetty and carry them out to the erection site. The casting and storage yards are located next to the river or sea being crossed

Figure 14.9 Beam with lifting gantry (reproduced courtesy of Tony Gee and Partners LLP, copyright reserved)

to provide easy access to the barges. The units are moved around the yard by gantries or lifting frames and are lifted or slid onto the jetty, where the barges collect them. Figure 14.12 shows a precast unit for the approach viaduct of the Incheon Second Crossing being positioned by the floating crane and barge arrangement.

Figure 14.10 Beam transporter (reproduced courtesy of Tony Gee and Partners LLP, copyright reserved)

Figure 14.11 Beam being moved on to deck (reproduced courtesy of Tony Gee and Partners LLP, copyright reserved)

Figure 14.12 Beam being transported by barge (reproduced courtesy of Samsung C&T Corporation, copyright reserved)

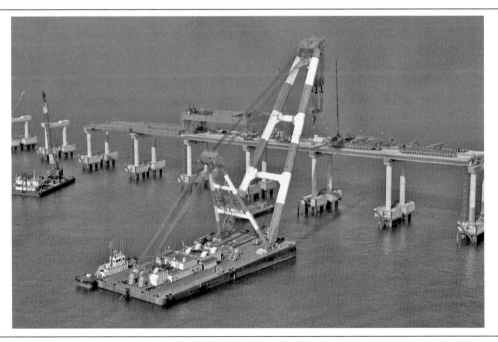

Erection of the units

Erection of the units requires special equipment to lift the heavy loads and to carefully position the beam. The large number of units being erected often justifies the cost of manufacturing special erection equipment for a particular project, with the arrangement chosen to suit the individual project requirements. Over the years the number of existing gantries, barges and other heavy-lifting equipment specially developed for bridges constructed with large precast units has built up around the world, so that they are available for any new project.

For viaducts over land, the units are usually transported along the completed deck and moved into position using a gantry. After placing the unit on its bearings, the gantry launches itself forward over the next span to repeat the process. Figure 14.13 shows a precast unit on the Taiwan High-Speed Rail project being lifted into place by an overhead gantry spanning between the completed deck and the next pier.

Where the bridge is over water and the units are transported by barge they are placed directly onto bearings on the piers. For the Central Viaduct, in Figure 14.2, the 77.6 m long units spanning between the approach viaduct piers were erected as simply-supported beams and subsequently connected with an in situ concrete stitch to make the deck continuous.

On the Saudi Arabia–Bahrain Causeway, in Figure 14.1, the deck for the 50 m spans was assembled from two different lengths of precast unit. The longer 66 m units were first placed in every other span. The units rested on the bearings at the piers with short cantilevers extending into the adjacent spans. Shorter 34 m long units were positioned to fill the gap, with a half-joint arrangement between the two.

The 250 m long spans on the Confederation Bridge in Figure 14.14 were constructed using 192 m long precast units placed as a 'balanced cantilever' about each pier with the remaining 58 m gap filled with a shorter precast unit dropped into place. The units placed as a balanced cantilever were supported at the piers to provide adequate stability in the temporary condition.

Figure 14.13 Placing beams by gantry (reproduced courtesy of Tony Gee and Partners LLP, copyright reserved)

Figure 14.14 Cantilevered deck being lifted into place (reproduced courtesy of Ballast Nedam BV, copyright reserved)

Where the units span between piers and are arranged as simply supported beams they are usually supported by two bearings at each end, with a typical arrangement shown in Figure 14.15. For spans less than 50 m the small deck movements and loads generated allow either mechanical or rubber bearings to be used. For spans longer than 50 m the higher loads and greater deck movements

Figure 14.15 Deck on bearings (reproduced courtesy of Tony Gee and Partners LLP, copyright reserved)

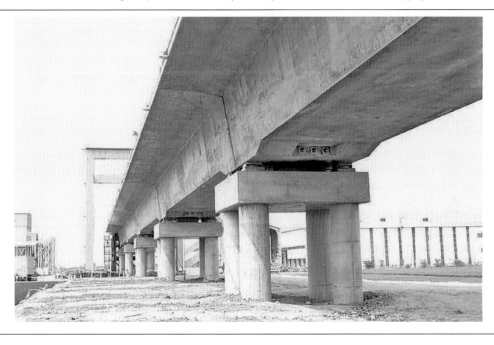

Figure 14.16 Stitch for deck continuity (reproduced courtesy of Hyder Consulting (UK) Ltd, copyright reserved)

usually require mechanical bearings. With rubber bearings, the precast units are placed directly onto them without any bedding mortar needed. If mechanical bearings are used the units are held above the top plate on temporary supports while the gap is filled with grout.

For decks made continuous at the piers the units are initially placed on temporary supports. After erecting the units in adjacent spans the gap between their ends is filled with in situ concrete with the permanent bearing installed beneath. Figure 14.16 shows two units in place at a pier with the reinforcement being fixed before placing the formwork and concreting the stitch. After the concrete has reached sufficient strength, permanent prestressing tendons are installed across the connection to complete the deck.

A rapid erection rate is achieved when using full-length precast units because of the minimal work needed on site. For smaller units and simpler arrangements, an erection rate of one unit per day is often achieved, although the very large units may take longer than this.

Design of full-length precast box girder decks

The design of full-length precast box girders is similar to the in situ box girders, covered in Chapter 12. The main difference is that with regard to the former it is necessary to check the forces and stresses during the transportation and lifting of the units.

The design of the units in the temporary condition is checked to ensure that sufficient reinforcement and prestress are present at each stage to cater for the forces and stresses set up. The weight from any construction equipment and other temporary loads is applied in the temporary condition as with the other types of construction. The stability of the units is checked during storage, transportation and erection, and temporary supports are provided as necessary.

As the units are moved, lifted and placed into position they are subjected to dynamic loading from the handling process. The magnitude of the dynamic loading depends on the equipment and procedures adopted, and every effort should be made to minimise any potential impact loading on the precast concrete. In normal circumstances it is sufficient to check the deck in the temporary condition when being moved for a 20% increase in the dead load to cater for the dynamic effects.

Prestressed Concrete Bridges, 2nd edition
ISBN: 978-0-7277-4113-4

ICE Publishing: All rights reserved
doi: 10.1680/pcb.41134.291

Chapter 15
Incrementally launched box girder bridges

Introduction
Where the bridge alignment is straight or on a constant radius curve, incrementally launching a concrete deck overcomes access problems or avoids obstructions at ground level. Incrementally launched decks usually have a box girder section, although beam-and-slab decks have also been successfully launched incrementally.

Many major bridges have utilised this technique for their construction, such as the deck for the Dornoch Firth Viaduct in Figure 15.1 which was launched into position with spans of 43.5 m and an overall length of 890 m.

The Hasdel Viaduct, shown in Figure 15.2, was a contractor's alternative design developed to simplify the construction over the deep valley. The 325 m long deck with spans up to 58 m was pushed out from behind the abutment in stages after each 29 m long section was cast. Using the incremental launch technique reduced the temporary works needed for the deck construction and avoided any associated works in the valley.

The technique involves casting the deck in sections behind one of the abutments. After each section is cast the deck is pushed or pulled out over the piers, as seen in Figure 15.3. The first segment is cast and moved forward on temporary bearings. The second segment is then cast against the first and both are moved forward by a further increment with subsequent segments cast and the deck moved until it reaches the opposite abutment and its final position.

Generally used for spans less than 60 m, the technique often utilises temporary supports which are placed to reduce the effective span length during launching. It is possible to launch longer spans, but the construction becomes less economical as the spans increase due to the deeper box and higher prestress required.

The advantages of incrementally launching a deck include the need for only limited temporary works, and the minimal disruption to the ground below. The disadvantages include higher concrete and prestress quantities in the deck, and the relatively slow construction rate.

General arrangement
The deck arrangement is similar to the in situ concrete box girders described in Chapter 12. The main difference is the requirement to provide sufficient strength and bearing capacity at the bottom of each web for the temporary bearings supporting the deck during the launch. This results in the area at the junction between the web and bottom slab being thickened up, as illustrated in Figure 15.4.

Figure 15.1 Dornoch Firth Viaduct, Scotland (reproduced courtesy of Tony Gee and Partners LLP, copyright reserved)

Figure 15.2 Hasdel Viaduct, Istanbul

Figure 15.3 Deck during launch over piers (reproduced courtesy of Freyssinet International, copyright reserved)

The outer face of the bottom of the webs is often made vertical to simplify the guidance arrangement during the launch; however, it is possible to guide the deck with the webs sloping.

Referring to the deck in its final position, the width of the web over the mid-span region is governed by the shear design requirements during the deck launching. As each section passes over the temporary

Figure 15.4 Typical launched deck section (reproduced courtesy of Tony Gee and Partners LLP, copyright reserved)

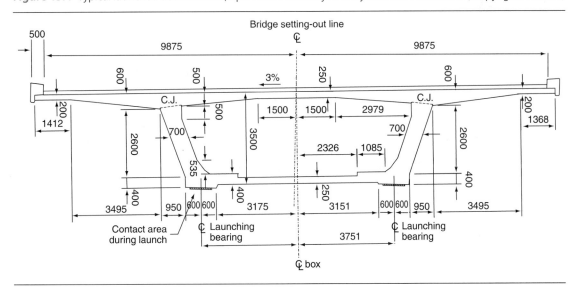

bearings it is subjected to significant shear forces. The width of the webs near the supports is governed by the shear requirements in the permanent condition. The webs are often kept a constant thickness throughout the length of the deck to simplify the construction and cater for the shear stresses during the launch.

Box girder decks usually have a single-cell arrangement, with the box being supported by a temporary bearing under each web during the launch. Using a twin- or multi-cell box with several webs causes difficulties with arranging the temporary bearings at each support. If two bearings are used, one under each outer web, the box structure may not be strong enough to transfer the load in the other webs out to the bearings during the launch stages. Using more than two bearings at each support makes it difficult to control the load distribution between them due to the construction tolerances. This may cause the bearings to be overloaded or the deck to be overstressed.

A temporary nosing arrangement, as described in 'Launching the deck' later in this chapter, is connected to the front of the deck during the launch. The launching nose is fixed against a diaphragm or wall across the end of the deck. It is not necessary to provide other diaphragms in the box during the launch. The permanent diaphragms may be cast after the deck launch is completed.

The launching requirements for a beam-and-slab deck dictate an arrangement with two main longitudinal girders and the deck slab spanning between or supported on secondary transverse beams. The main longitudinal girders are used for the launch with the temporary bearings positioned beneath.

To allow a deck to be launched its soffit surface must have a uniform alignment profile to slide along. Decks may be launched with significant gradients and crossfalls, and with tight horizontal curvature as long as these provide a constant launching geometry along the length of the bridge. When the alignment has both horizontal and vertical curvatures, these are arranged to give a constant 'arc' for the deck soffit to launch along.

The range of moments to be taken by each section requires a deeper deck than that used with other forms of structure for the same span length. The ratio of launched span-to-deck depth is usually 16:1 or less.

Segment lengths are standardised for an individual bridge deck and are usually between 15 m and 30 m long depending on the span arrangement and overall deck length. Longer segments reduce the total number required, giving a shorter construction period. Shorter segments require a smaller casting area with less formwork and labour needed.

The deck is usually continuous between abutments; however, it is possible to include expansion joints along the length if required. Temporary prestress is used to connect the deck across the expansion joint and make it continuous for launching. After the launch is completed the temporary prestress is removed, and the deck is separated to form the expansion joint.

The longitudinal prestress in the deck is divided into two groups: the launch prestress, installed as the deck is being built, and the supplementary prestress installed after the launch is completed. The launch prestress usually consists of straight tendons placed in the top and bottom slabs. These are anchored on the construction joint between segments and coupled or lapped to extend through each new length of deck cast. The supplementary prestress consists of additional tendons in the top and bottom slabs and draped tendons running through several spans anchored at blisters within the box. A typical

Figure 15.5 Tendon anchorage details (reproduced courtesy of Tony Gee and Partners LLP, copyright reserved)

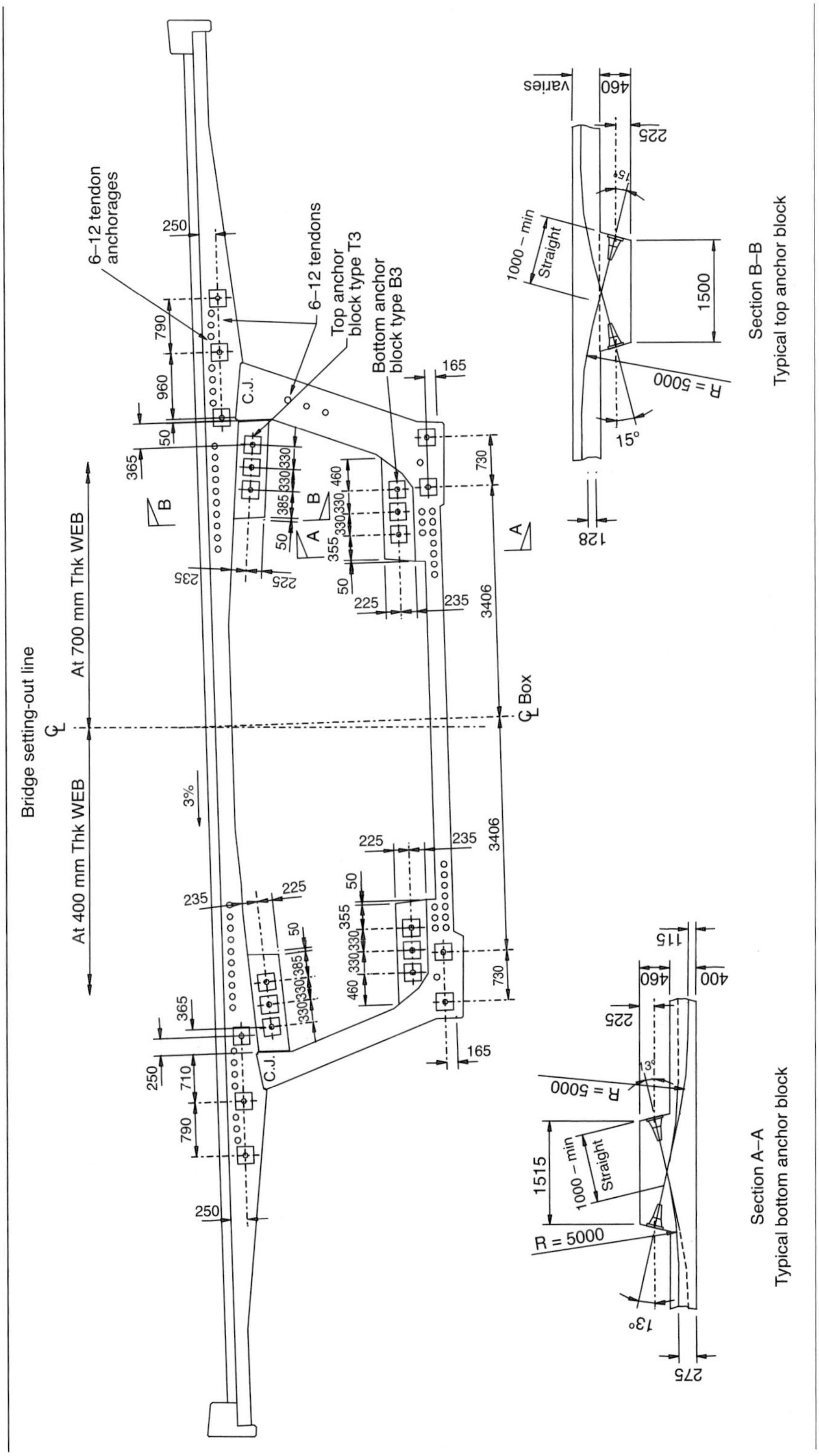

arrangement for the end face anchors on the construction joint and the blisters inside the box is illustrated in Figure 15.5.

Casting the deck

A specially prepared casting yard is located behind the abutment, as seen under construction in Figure 15.6. The yard is arranged with separate areas to assemble the reinforcement, to concrete and to launch the deck. The formwork is placed along the line of the deck and set back behind the abutment at a distance sufficient to provide stability to the deck in the initial stages of the launch. The shutters for casting are a mixture of in situ and precast technology, with facilities for striking and resetting the formwork in an efficient manner.

The launching nose is assembled in position at the front of the casting cell at the beginning of the deck construction process, as seen in Figure 15.6. The nose is placed on temporary supports, as illustrated in Figure 15.7, with the first deck section cast up against it. The number and location of the temporary supports depends on the position of the casting cell behind the abutment, the length and design of the segments, and the stability of the deck. After casting the first deck section up against the launching nose the two are stressed together to provide the connection, as seen in Figure 15.8. The connection between the launching nose and end of the deck is subjected to a range of sagging and hogging moments as the nose and deck pass over a support during launching, requiring a stressed connection at both the top and bottom of the nose–deck interface. It also is subjected to significant shear force and this can be catered for at the connection by the creation of a step, as seen at the bottom of Figure 15.8.

There are several different sequences for assembling the reinforcement, casting the deck and launching. The reinforcement is either pre-assembled and lifted into position or fixed behind the cast deck and pulled into position when the deck is launched forward. The top slab to the deck is often cast as a separate operation after the bottom slab and webs to simplify the internal formwork.

Figure 15.6 Casting bed behind abutment

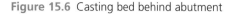

Figure 15.7 Typical casting and launching sequence (reproduced courtesy of Tony Gee and Partners LLP, copyright reserved)

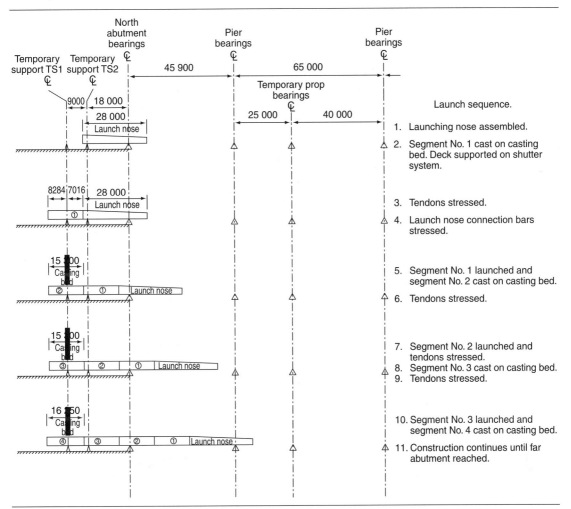

Four of the most common sequences are outlined below:

1. Casting bottom slab first
   (*a*) bottom slab reinforcement and tendon ducts assembled
   (*b*) bottom slab cast
   (*c*) first-stage launch prestress installed
   (*d*) deck launched forward with bottom slab moved into next section of casting cell
   (*e*) web and top slab reinforcement and tendon ducts fixed
   (*f*) web and top slab concrete cast at the same time as the next section of bottom slab in the bay behind
   (*g*) second-stage launch prestress installed
   (*h*) deck launched forward and next sections of webs, top slab, and bottom slab constructed.

Figure 15.8 Launch nose connection to the deck

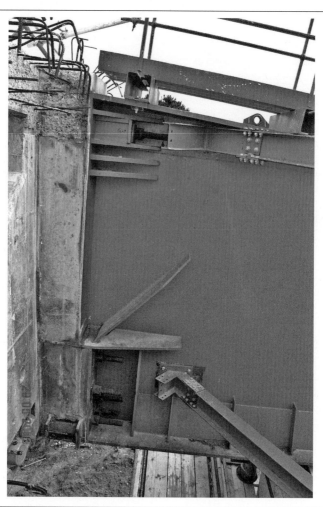

2. Casting bottom slab and webs first
   (a) bottom slab and web reinforcement, and tendon ducts assembled
   (b) bottom slab and webs cast
   (c) first-stage launch prestress installed, if needed
   (d) deck launched forward with bottom slab and webs moved into next section of casting cell
   (e) top slab reinforcement and tendon ducts fixed
   (f) top slab concrete cast at the same time as the next section of bottom slab and webs in the bay behind
   (g) second-stage launch prestress installed
   (h) deck launched forward and next sections of bottom slab, webs and top slab constructed.
3. Casting complete section in stages
   (a) bottom slab and web reinforcement, and tendon ducts assembled in casting bay
   (b) bottom slab and webs cast over full segment length
   (c) shutters positioned for top slab

Figure 15.9 Formwork arrangement (reproduced courtesy of VSL International, copyright reserved)

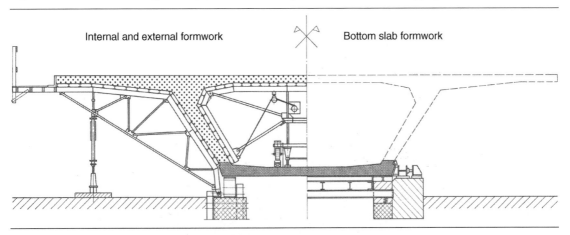

Internal and external formwork   Bottom slab formwork

(*d*) top slab reinforcement and tendon ducts fixed

(*e*) top slab concrete cast over full segment length

(*f*) launch prestress installed

(*g*) deck launched forward and next section constructed.

4. Casting complete section in single pour

(*a*) reinforcement and tendon ducts for complete section assembled in casting bay

(*b*) shutters positioned, including internal forms to inner web and the soffit of the top slab

(*c*) complete deck section cast over full segment length

(*d*) launch prestress installed

(*e*) deck launched forward and next section constructed.

A typical shutter arrangement for sequence (1) above is illustrated in Figure 15.9. After the bottom slab concrete has hardened it is able to support the internal form for the webs and top slab, which greatly simplifies the shutter design. The outer formwork system is designed to be lowered, leaving the deck on temporary supports ready for launching.

A typical casting area for sequence (3) above is shown in Figure 15.10, with the outer formwork in place and the reinforcement being fixed before casting the bottom slab and webs. In this example the formwork is supported on steel beams with jacks underneath to allow the entire outer shutter to be lowered in one piece. An alternative arrangement, often used, is to provide temporary concrete walls under each of the webs. These support the soffit shutter and provide a runway for the deck to slide along during the launch.

After setting up the outer formwork and fixing the reinforcement, as seen in Figure 15.11, the inner form for the webs is installed. Where form ties are allowed through the web the shutters consist of standard panels placed and supported in the traditional way. Where form ties are not permitted the inner form is supported using a frame to hold it in place and to resist the concrete pressures.

With the top slab cast after the bottom slabs and webs, a table-form is used to support the fresh concrete between the webs. The table-form is placed after the bottom slab concrete has hardened so that it can be easily supported. After casting the top slab the table-form is usually left in the

Figure 15.10 Outer formwork assembled

Figure 15.11 Casting cell looking towards deck

Figure 15.12 Concreting deck slab

deck during launching. It is then moved into position when the next section of top slab is ready for construction.

Casting of the deck is done using conventional techniques with the concrete skipped or pumped into position, as seen in Figure 15.12. Early high-strength concrete is used to allow the shutters to be stripped and the prestress to be installed as soon as possible after concreting. The deck is then launched forward freeing up the casting bed for the construction of the next segment to start.

By separating the reinforcement fixing and concreting into different areas of the casting cell, as for sequences (1) and (2) above, these operations are progressed at the same time to minimise the construction cycle time. Similarly, prefabricating the reinforcement and lifting the cages into the shutters reduces the cycle time. A casting and launching cycle typically takes between seven and ten days.

Launching the deck

As described above, the deck is cast in sections and incrementally launched into position, with a typical sequence being illustrated in Figure 15.7. As the deck slides out it is supported at the piers, abutments and other temporary support positions by temporary bearings, as seen in the foreground of Figure 15.13. These temporary bearings support the deck under the soffit as it passes above. Each bearing consists of a steel plate, surfaced with a stainless steel sheet, supported on a mechanical rubber bearing. A teflon pad is fed in between the bearing's stainless steel sheet and the concrete soffit as the deck slides over to provide a low-friction interface. When the launch is complete the deck is raised on temporary jacks to allow the temporary bearings and teflon pads to be removed and the permanent bearings installed.

Figure 15.13 Temporary bearings for launch

A combined permanent and launch bearing is shown in Figure 15.14. These bearings are set in their permanent position with the bottom plate placed horizontally and the normal slide mechanism locked into position. The bearing's top plate is tapered to match the longitudinal gradient of the deck, and a stainless steel sheet is fixed over the top. During the launch, a teflon pad is fed in between the stainless steel sheet and the concrete deck in the same way as with temporary bearings. After completing the launch the deck is jacked up and the stainless steel sheet and teflon pads removed. The gap between the deck and bearing is then filled with grout.

Figure 15.14 Combined permanent and launch bearings (reproduced courtesy of VSL/CTT-Stronghold, copyright reserved)

Stainless steel plate

Figure 15.15 Push launching jack

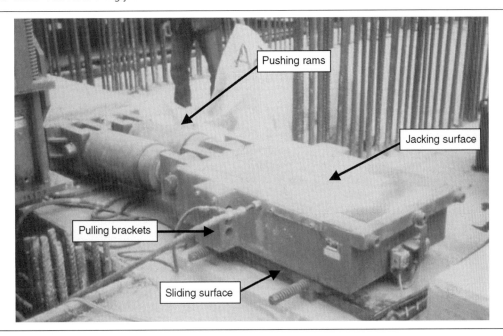

The deck is launched either by pushing or pulling out over the spans. The launching device is usually fixed to the abutment which provides the resistance against the horizontal thrust needed to move the deck. A typical launch rate is between 4 m and 6 m an hour.

Push-launching jacks, as seen in Figure 15.15, are positioned on the abutment in front of the bearing positions and under each web. The launching jacks raise the deck a small amount to generate sufficient friction to grip the structure. Then they push the deck forward a short distance before lowering it on to the temporary bearings. After releasing the structure the launching jacks move back ready to start another stroke.

The interface between the launching jack and the concrete deck is usually able to generate a frictional force of at least 40% to 60% of the vertical load carried. Launching jacks rely on being able to generate sufficient friction with the deck to generate the launching force needed. This is not always possible at the beginning of a launch when the vertical reaction on the jack is small. The launching nose may be sitting on the launching jack with very little load present. To overcome this, most launching jacks include a pulling bracket through which a bar is attached to the front of the deck and the jack used to pull the deck out until there is sufficient weight to generate the required friction.

Similarly, insufficient friction may be available when launching long lengths of deck. This becomes critical towards the end of the launch when the back of the deck approaches the launching jack reducing the vertical load. If this occurs, it may be more appropriate to use a pulling device throughout the launch.

An arrangement for pulling the deck into position by the use of strand jacks is illustrated in Figure 15.16. The jacks are located at the abutment which provides the anchor point and resistance against

Figure 15.16 Pulling arrangement (reproduced courtesy of VSL International, copyright reserved)

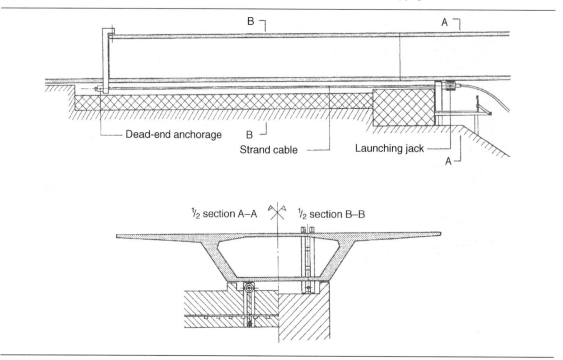

the force generated. The strands are fixed to a frame located at the back of the deck section just cast and are pulled forward by the jacks. After each section is launched the frame is removed and fixed to the back of the next section.

An alternative to pushing or pulling at the abutment in front of the casting area is to set up a series of pushing devices at each pier. The pushing devices jack up the deck at each pier and push it forward in a synchronised operation. This technique was successfully used for the launch of the approach viaduct deck to the Sungai Johor cable-stayed bridge in Malaysia, in 2010. By jacking up and pushing at each pier the launching forces were balanced at the pier tops and the large thrust forces on the abutments eliminated.

Lateral guides, as seen in Figure 15.17, are fixed to the abutment and piers to keep the deck correctly aligned during the launching sequence. The guides prevent the deck from moving sideways and resist any transverse loads that may be imposed on the deck. It is not necessary to provide lateral guides at every pier or temporary support, although there must always be a sufficient number to adequately restrain the deck. The lateral guides are removed after the launch of the deck is completed and the permanent bearings installed.

As the deck is pushed out and extends beyond the leading pier, large cantilever moments occur until the next pier is reached. To reduce these moments a temporary steel launching nose is fixed to the front of the deck, as shown in Figure 15.18. The launching nose allows the deck to be supported by the next pier at an earlier stage in its launching cycle and reduces the bending moments and shears that occur over the front part of the deck.

Figure 15.17 Guides fixed to piers and bearing plinths being prepared

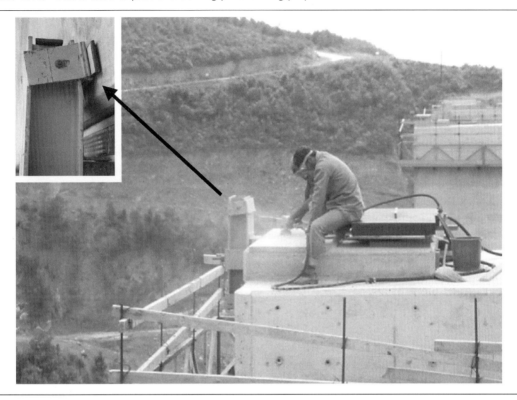

Figure 15.18 Launching nose on deck

The effectiveness of the launching nose is a function of its length and stiffness, and the optimum arrangement needs to be chosen to balance the cost of the nose against the cost of catering for higher moments and forces in the deck. The launching nose length is typically 60% of the launch span length, and usually has a stiffness (EI) of between 10% and 15% of the concrete deck.

Steel girders or trusses are used for the launching nose. A typical arrangement using twin girders separated with cross-bracing is seen in Figure 15.13. The nose may be of constant cross-section throughout to simplify its fabrication, but it is usually tapered, reducing in depth at the front end to minimise weight and save cost.

An alternative to using a launching nose is to utilise stay cables to reduce the bending moments and shears. The stays are supported on a temporary tower located back from the front of the deck. The force in the stays is adjusted to control the moments and forces imposed on the deck as it moves forward. This technique has been used on a number of projects in the past, but it is generally not favoured due to the complexity involved in controlling the temporary stay forces during the launch.

The forces in the temporary supports and the line and level of the deck are regularly monitored during the launching of the deck to ensure that the deck is behaving as expected. The pushing force and verticality of the piers and temporary supports are also monitored with any change from the expected situation being an indication of problems with the launch.

Design aspects associated with launched box girder decks

The design of launched decks is similar to in situ decks when they are in their final position, but during the launching process they are subjected to a different force and stress regime that has to be catered for. The design of the deck during the launch is described in the following section. A paper by Rowley (1993) and a book by Rosignoli (2002) discuss the different aspects of launching in more detail.

Longitudinal design during launch

The deck is checked at each stage during its launch, and prestress and reinforcement provided to suit. The longitudinal analysis is carried out using a line beam model which includes the launch nose structure. The model changes as each section of deck is cast and as the deck is launched out over the supports.

A typical moment distribution along the deck, just before the deck reaches a pier and just after, is illustrated in Figure 15.19. As the deck moves over the piers, the moments and shears at each section change. At some stage during the launch each section will be over a pier or temporary support and subjected to hogging moments or in a mid-span position and subjected to sagging moments. A typical envelope of moments at each deck section during the launch phase of a three-span bridge is given in Figure 15.20. The launch prestress is designed to cater for the range of bending stresses at each section. The largest hogging moments occur near the front of the deck, just before the launching nose reaches the next pier. The largest sagging moments also occur near the front after the launching nose has reached the pier but just before the front of the deck reaches it. Elsewhere the range of moments is more uniform, reducing at the back for the last section to be cast.

The launch prestress is usually arranged to give a uniform compressive stress across the section, typically about $5 \, \text{N/mm}^2$. Additional prestress is provided at the front of the deck to cater for the higher moments generated. After completion of the launch the supplementary prestress is installed to balance the bending moment profile generated by the deck in its final position.

Figure 15.19 Bending moments in deck during launching

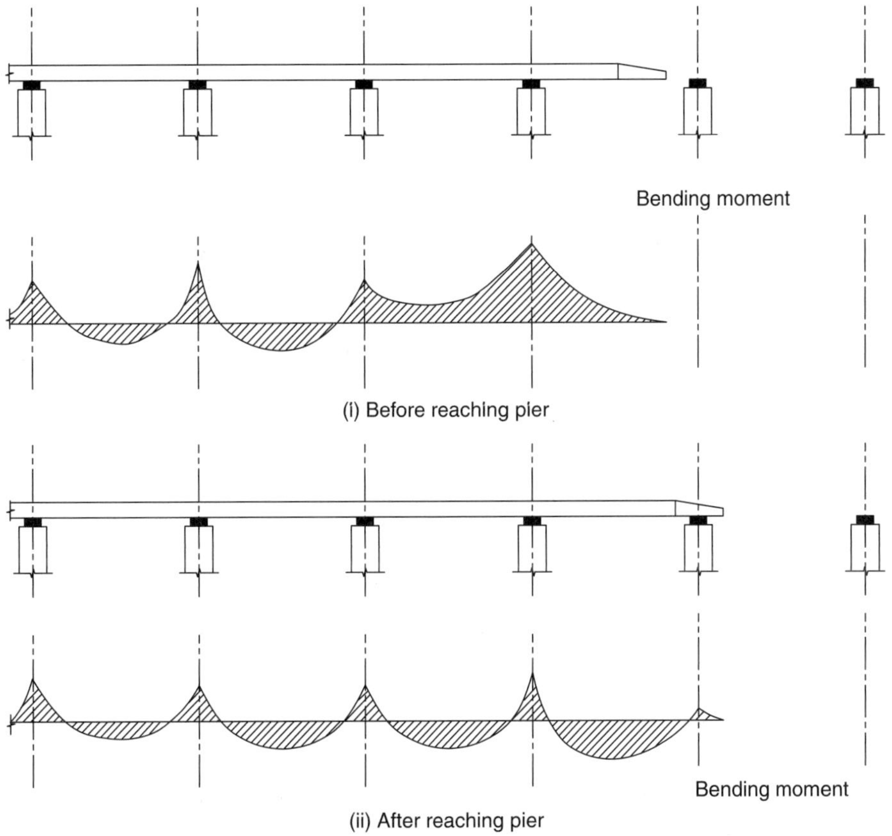

(i) Before reaching pier

(ii) After reaching pier

Transverse and local design during launch

Each section of deck has to be strong enough to resist the loads from the temporary bearings as it passes over them.

It is preferable for the temporary bearings during the launch to be located directly under the centre of the webs. If this is not possible, bending moments and shear forces are induced in the bottom slab and the bottom of the web in both the longitudinal and transverse directions. The structural behaviour in this region is analysed locally using a three-dimensional finite element model, as illustrated in Figure 15.21. The stresses derived from the analysis are used to determine the tensile and shear forces to be catered for by the reinforcement and concrete.

The analysis and design of the deck above the temporary bearings must allow for the increase in load and adjustment in support position due to the construction tolerances discussed below.

Construction tolerances

Unevenness of the concrete surface, misalignments of the bearings and differential settlement of the piers and temporary supports generate additional moments, shears and torsion in the deck during the launching operation.

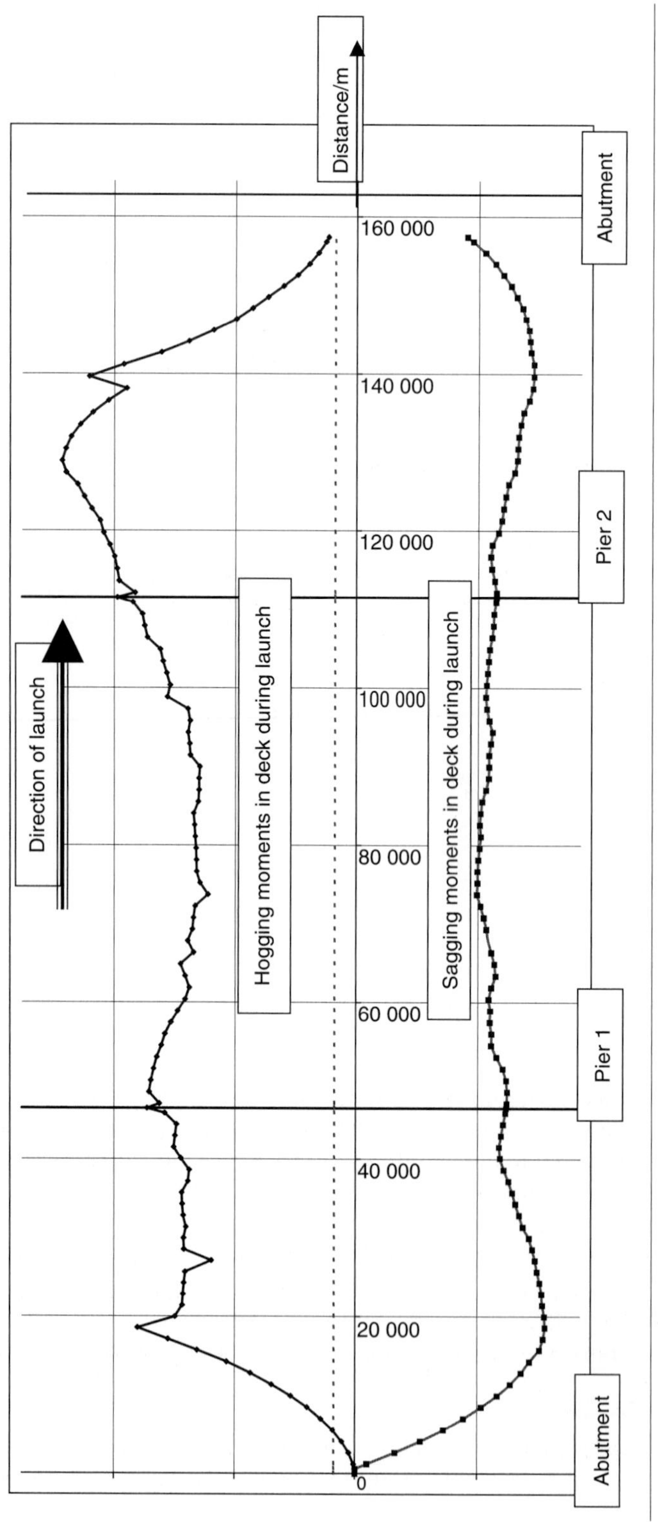

Figure 15.20 Bending moment range in deck during launching (reproduced courtesy of Tony Gee and Partners LLP, copyright reserved)

Figure 15.21 Analysis of bottom slab and web during launch (reproduced courtesy of Tony Gee and Partners LLP, copyright reserved)

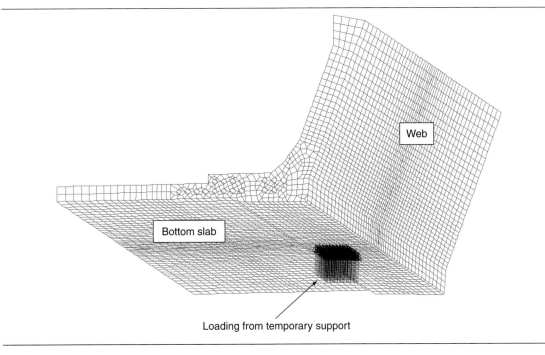

Loading from temporary support

Typical values assumed in the design are:

Soffit concrete level $= \pm 3$ mm from design level
Level of temporary bearings $= \pm 2$ mm from design level
Differential settlement between adjacent supports $= \pm 5$ mm

The differential settlement depends on the soil properties and the type of foundation used, while tight construction controls are needed to achieve the tolerances for the concrete and bearing level.

The above tolerances may combine to give a total displacement of ± 10 mm on the deck soffit, causing differential effects longitudinally between adjacent piers and transversely between the bearings. The change in levels redistributes the loads in the temporary bearings, causing bending, shear and torsional effects in the deck.

Construction tolerances for the lateral position of the deck and temporary bearings produce a misalignment between the two compared to their theoretical positions. This increases if there is any lateral movement of either the deck or the piers during the launch. To allow for this in the design the position of the temporary bearings under the deck is assumed to vary by up to 75 mm transversely. This allowance may be reduced if suitable measures are taken to limit the misalignment that can occur.

When setting the bearings after completing the launching the deck is jacked up and the bearings installed with a specified load built in. In this way the construction tolerances are compensated for in the final structure.

Construction loading

As the deck is being launched there is usually very little temporary load imposed on the deck, while the launching nose weight is already included in the structural analysis model. If storage of material or heavy equipment is not permitted on the deck the construction loading allowed for in the design is typically taken as:

(i) In front of abutment bearing
   Point load = 10 kN anywhere on the deck for miscellaneous equipment
   General working load = 2 kN/m length of deck
(ii) Behind abutment bearing and in casting area
   Point load = 20 kN anywhere on the deck slab for miscellaneous equipment
   General working load = 4 kN/m length of deck

Weight of formwork for deck slab between webs = 1 kN/m^2 of deck

Figure 15.22 Forces on substructure during launching

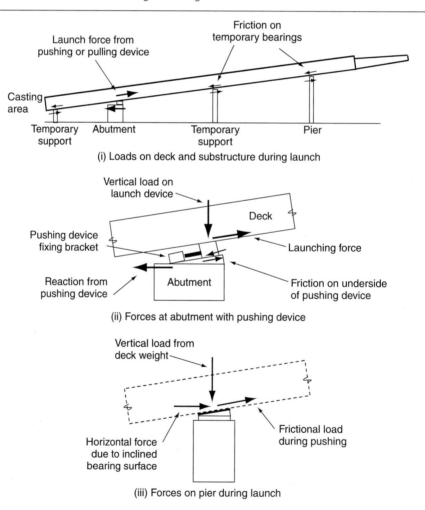

(i) Loads on deck and substructure during launch

(ii) Forces at abutment with pushing device

(iii) Forces on pier during launch

Loads on supports during launching

During the deck launch the friction in the temporary bearings creates a load at the top of the piers and temporary supports, as illustrated in Figure 15.22(i). When greased slide pads are used, the friction is typically between 2% and 6% of the vertical load.

The top surface of the temporary bearings is aligned parallel to the deck rather than horizontally, and this introduces an additional horizontal load on to the support, as illustrated in Figure 15.22(iii). This force is equal to the vertical load times the percentage gradient. The piers and temporary supports are designed to resist these horizontal loads in combination with the co-existent vertical loads. Providing stays or guys from the top of the support to an adjacent pier stiffens up the support and reduces the effects of the loads, although the relative flexibility of any stays reduces the effectiveness.

As the deck moves out and gets longer the launching force increases to overcome the increased frictional force on the temporary bearing. Greater pushing forces are required when a deck is being launched up a slope, while a braking device is needed when going down a slope.

For the pushing arrangement, shown in Figure 15.22(ii), the upper surface in contact with the deck is subjected to the launching force generated from the friction resistance at all the other supports. The lower surface of the pushing device runs over a stainless steel sheet and generates a frictional resistance as the device moves forward, proportional to the load on the pushing arrangement. The force needed to launch the deck is the sum of the frictional resistance at all the temporary supports and the underside of the pushing device.

Where the deck is pushed or pulled into position by a launch device located at the abutment, the abutment is designed to resist the horizontal forces generated during the launch and to prevent sliding or overturning. Additional resistance is mobilised by providing the casting area with a ground slab as a working platform and by connecting this slab to the abutment.

REFERENCES

Rosignoli M (2002) *Bridge Launching*. Thomas Telford, London.
Rowley F (1993) Incremental launch bridges: UK practice and some foreign comparisons. *The Structural Engineer* **71(7):** 111–116.

Prestressed Concrete Bridges, 2nd edition
ISBN: 978-0-7277-4113-4

ICE Publishing: All rights reserved
doi: 10.1680/pcb.41134.313

Chapter 16
Cable-stayed bridges

Introduction

Many recent major cable-stayed bridges have utilised post-tensioned concrete in the main deck elements. The use of stay cables to support the deck results in a slim section and makes long spans possible. Concrete is a versatile material that is able to combine efficient deck arrangements with an aesthetically pleasing appearance which is enhanced when combined within a cable-stayed bridge. The design and construction of concrete cable-stayed bridges uses many of the traditional techniques associated with the other types of concrete bridges, although their greater size and flexibility require special consideration.

As concrete is good in compression it is an ideal material to resist the high compression forces generated in the deck by the stays, especially adjacent to the pylons. Elsewhere, where the compression from the stays is less, the deck is prestressed to overcome the tension and bending stresses generated by the deck behaviour. The stiff concrete section assists in distributing the load along the deck and between the stays, while its high mass and damping characteristics reduce its susceptibility to vibrations or aerodynamic movements.

Sunshine Skyway Bridge, Florida, shown in Figure 16.1, uses a concrete box girder deck to give an elegant cable-stayed arrangement. It was built in 1987 after the previous bridge was damaged by a ship collision. The 366 m long main span, back spans and approach viaducts are all constructed from precast concrete box girder segments. A 29 m wide deck slab is part of a 4.5 m deep single-box deck arrangement, supported by a single plane of stays anchored down the centre of the box.

The Main Bridge of the Vasco da Gama crossing, Portugal, shown in Figure 16.2, is part of a much longer estuarine crossing. The 420 m long main span of the cable-stayed bridge provides the required clearance to the shipping lanes below. The slim beam-and-slab deck is only 2.6 m deep and is supported by two planes of stays, one down either side of the deck. Development of the bridge design was part of a design-and-build tender, with the concrete arrangement giving the most cost-effective solution.

The 'A' shaped pylon and asymmetric arrangement of the Flintshire Bridge in Figure 16.3 provides a distinctive landmark crossing over the River Dee in Wales. The main span, with a length of 194 m, has a beam-and-slab concrete deck arrangement and an edge beam depth of 1.7 m. The back span was cast on falsework built up from the ground, while the main span over the river used a form traveller to cast the deck in situ as a cantilever.

For long cable-stayed spans, the deck is often a combination of steel and concrete elements. Normandie Bridge in France, with a main span of 856 m, uses a steel box over the centre part of the main span and a concrete box adjacent to the pylons in the main span and over the back spans.

313

Figure 16.1 Sunshine Skyway Bridge, Florida (reproduced courtesy of VSL International, copyright reserved)

Figure 16.2 Vasco da Gama Bridge, Portugal (reproduced courtesy of Yee Associates, copyright reserved)

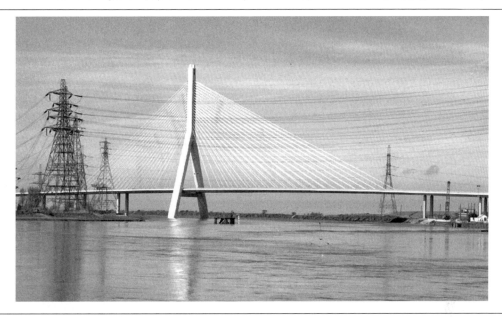

The light steel box reduces the dead load over the long main span. The concrete box section is better able to resist the high compressive forces in the deck near the pylons, and it provides a heavier weight in the shorter back spans to counterbalance the longer main span.

The design and construction of cable-stayed bridges is complex and requires many different aspects to be considered. This chapter looks at the general aspects involved with prestressed concrete decks when used on cable-stayed bridges. For a more detailed description of cable-stayed bridges, reference should be made to specialist books and other literature, noted at the end of this chapter and in Appendix C.

Cable-stayed bridge arrangements

Of all the types of prestressed concrete bridges, the cable-stayed form gives the most diverse range of structural arrangement with each bridge usually being unique. Concrete cable-stayed bridges are built with one, two or more pylons; stays are placed in a single plane or twin plane; span arrangements are symmetrical or asymmetrical; and piers may be used in the back spans. The choice of the final arrangement is often due only to the designer's preference and the client's vision.

Concrete cable-stayed decks are either a box or a beam-and-slab arrangement, with typical layouts illustrated in Figure 16.4. The beam-and-slab arrangement produces the lightest decks, but it needs two planes of stays for support, while the box girder arrangement provides a stiff deck and is used with either one or two planes of stays. Box cross-sections are often precast to give a fast and simple erection process.

The box girder arrangement illustrated in Figure 16.4(i) has inherent torsional strength, which makes it suitable for use with a single plane of stays. This deck shape is used on the Brotonne Bridge in France, shown in Figure 1.18, which has a main span of 320 m and a box depth of 3.8 m. It is also used on the Sunshine Skyway Bridge, shown in Figure 16.1.

Figure 16.4 Concrete deck arrangements for cable-stayed bridges

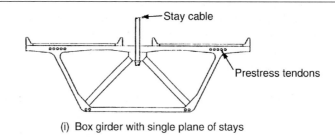

(i) Box girder with single plane of stays

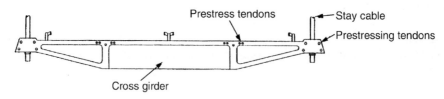

(ii) Semi-box girder with twin plane of stays

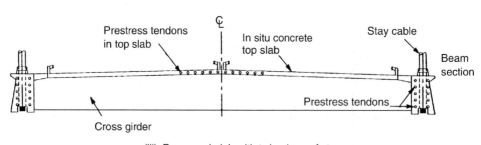

(iii) Beam-and-slab with twin plane of stays

Twin planes of stays provide some torsional stiffness to the overall structural behaviour, and less-stiff deck arrangements such as a 'semi-box', shown in Figure 16.4(ii), or a beam-and-slab, as shown in Figure 16.4(iii), are usually adopted.

A semi-box arrangement is used on the Pasco-Kennewick Bridge, USA, with a main span of 299 m and a deck depth of 2.13 m. Typical examples of beam-and-slab decks are the Vasco da Gama Bridge, in Figure 16.2, and the River Dee crossing, in Figure 16.3.

The length of the main span and height of the deck above the ground are dictated by the obstruction being crossed or the minimum clearance requirements of any traffic passing underneath. The width of the deck is chosen to suit the users' requirements. Footbridges may be only a few metres wide, yet a road bridge, carrying a dual carriageway, may be over 30 m wide.

Decks are typically between 1.4 m and 3 m deep for a beam-and-slab arrangement and between 2.5 m and 4.5 m deep for a box arrangement, depending on the spacing of stays, the deck width and the live load being carried.

The shape of the pylons is usually based on 'H', 'A' or 'I' layouts, although there are many variations of these. The pylon shape has the biggest influence on the appearance of a cable-stayed bridge and the final arrangement is often developed on the basis of aesthetic considerations. To obtain an efficient design for the cable-supported structure the height of the pylon above the deck is normally between 20% and 25% of the main span length where two pylons are used, or between 40% and 50% of the main span length where a single pylon is used.

At the pylon the deck is either built into the legs, supported on bearings, or left hanging freely from the stays. Building the deck into the pylon provides a fixed support, but it causes a 'hard spot' along the deck which attracts load and may overstress the concrete. Placing the deck on bearings on a crossbeam softens the support, while hanging the deck freely on the stays results in a more uniform distribution of bending moment along the deck.

Early cable-stayed bridges used only a few stays widely spaced along the deck, whereas, more recently, most projects use multi-stay systems with closely spaced stays. Fan, harp or semi-fan layouts of stays are illustrated in Figure 16.5.

A fan arrangement is the most efficient in terms of generating the support to the deck, resulting in least weight of stay material, but the anchorage on the pylon is congested. A harp arrangement is often

Figure 16.5 Multi-stay arrangements

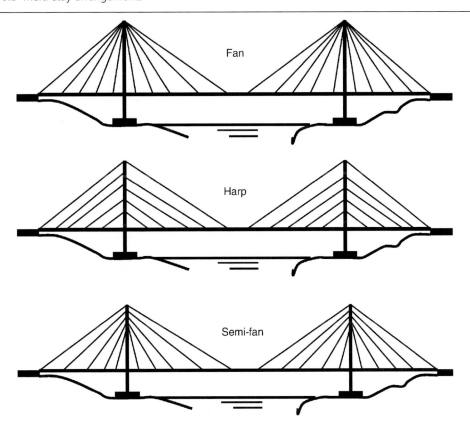

considered aesthetically better, but the stays work less efficiently. The semi-fan arrangement is a compromise between the two, with the stay anchors on the pylon spread out to ease congestion and the stay angle maximised to give reasonable efficiency. The semi-fan arrangement is the most commonly adopted arrangement.

With multi-stay arrangements the stays' spacing along the deck is typically between 8 m and 12 m. This provides a good support to the deck and allows thin, flexible decks to be used.

Several different types of stay are used with concrete cable-stayed bridges, the most common being based on 7-wire prestressing strand. Multi-strand stays, with up to 127 strands, capable of carrying forces of 1600 tonnes are available, and special anchorage and sheathing systems have been developed to ensure the integrity and durability of the stay. The strands are usually galvanised and individually sheathed inside a grease-filled high-density polyethylene duct to provide a multiple protection system.

Lock-coil, stressbars and parallel wire stays are also used where preferred by the designer. All stays adopt a multiple protection system, with galvanised wires or bars surrounded by additional protection such as ducting or painting.

Construction of concrete cable-stayed bridges

Concrete deck sections are usually constructed in situ as segments cast inside a form traveller or with precast elements transferred to site and lifted into position. Both methods involve cantilevering the

Figure 16.6 Balanced cantilever construction (reproduced courtesy of VINCI Construction Grands Projets and Hyder Consulting (UK) Ltd, copyright reserved)

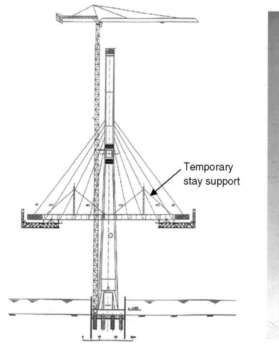

deck out from the pylons in a balanced manner and installing the stay cables as the cantilever extends, as illustrated in Figure 16.6.

As the deck is cantilevering during construction, high bending moments are generated over the end section due to the weight of the construction equipment, form traveller and wet concrete of the new segment. Temporary stay systems, such as the arrangement indicated in Figure 16.6, are used to distribute the high loads on the end of the deck back along the structure and into the permanent stays.

Where there is good access under the deck the concrete may be cast in situ on full-height falsework, as shown in Figure 16.7, before the stays and prestressing tendons are installed. The advantage of casting the deck on full-height falsework is the simpler concreting procedure and stay installation, which results in faster construction. However, with difficult access, high structures or long decks, the cost of extensive falsework is prohibitive.

Several projects have combined the use of full-height falsework for the back spans, when the access is good, with a form traveller to cantilever the main span out over the obstruction being crossed. This gives the advantages from both systems.

Stressing of the stays is normally done from the deck anchorage, with the jacks located on the form traveller or access platform. Some projects have stressed the stays from the pylon anchorage end,

Figure 16.7 Deck cast on falsework (reproduced courtesy of Gifford, part of Ramboll, copyright reserved)

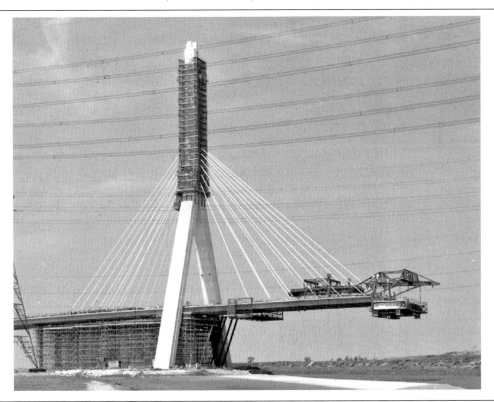

although this is usually where the pylons are large enough to provide sufficient room to arrange the lifting equipment and access platforms to position the jacks.

The deflections, forces and stresses in a cable-stayed bridge continually change as the deck is being built, and comprehensive monitoring is carried out to ensure that the structure is behaving as expected. As well as geometry surveys, the stresses, temperatures and stay forces are monitored, as described in the paper by Al-Qarra (1999).

Setting out and surveying the position of the deck requires careful control due to the many factors that influence the structural behaviour. As the temporary loads and construction equipment move along the deck the pylon and deck deflect. The natural variations in the properties of the concrete used affect the achieved alignment at each stage of the construction.

The temperature of the deck, stays and pylon affects the vertical, longitudinal and transverse alignment of the deck. Temperature differences between the concrete and stays cause bending in the deck and a change to the deck levels. Differential temperature gradients through the concrete also change the deck profile. It is difficult to determine accurately the temperature differentials within the structure at any particular time of day and hence to predict the theoretical position of the structure at that time. It is usual to do all of the setting out and surveying activities during the early hours of the morning, when any temperature differentials are at a minimum.

Box girder decks

Concrete box girders are either cast in situ, precast, or built in a combination of the two. The techniques used in their construction are similar to those for other types of box girder decks, described in Chapters 12 and 13, with stay cables installed to support the deck during the construction and in the final arrangement.

In situ box girders are cast in short lengths in a form traveller, as described in Chapter 12. The length of the segments is usually between 3 m and 5 m, to suit the stay spacing, with stays being anchored at every second or third segment. During the construction the deck is subjected to high bending moments at the end of the cantilever and either reinforcement or prestress is used to cater for this.

Precast box girders use match-cast segments, formed and transported as described in Chapter 13. Their erection follows a similar sequence as for a balanced cantilever deck except that they have stays attached as the cantilever is extended out. Temporary prestressing is used to join the segments together and support the deck before the stays and permanent prestressing tendons are installed.

Precast segments are usually positioned by a lifting frame or crane located on the completed deck. The precast match-cast segments on the Sungai Prai Bridge in Malaysia were raised into place using a lifting frame, as seen in Figure 16.8. Using precast, match-cast segments and lifting them into place minimises the work done on site and greatly reduces the construction time. For the Sungai Prai Bridge, the central part of the deck was cast and erected first, creating a spine beam supported by the stays. Additional precast side panels were subsequently lifted and stressed to the central box with in situ stitches and transverse prestress to give the full width required for the dual two-lane carriageway above.

Several projects have combined precasting and in situ construction. The box of the Brotonne Bridge in France, seen in Figure 1.18, used thin precast web elements lifted into the form traveller with the slabs cast in situ to achieve a construction cycle of three days for a 3 m long segment.

Figure 16.8 Liifting frame on deck for segment erection

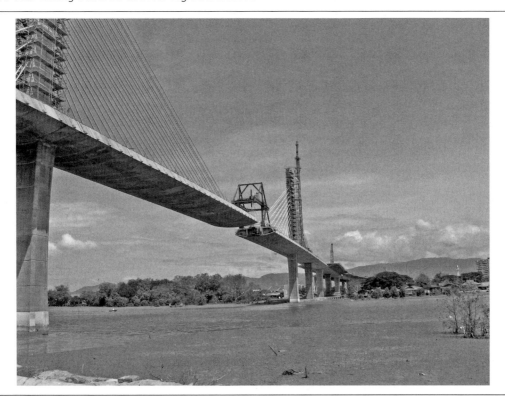

Beam-and-slab decks

Decks consisting of prestressed concrete edge beams, transverse girders and concrete deck slabs have become popular in recent years because they produce light and slender decks, thus combining both economy and aesthetics. Two planes of stays are used, one down either side of the deck, to provide transverse and torsional stiffness to the structure.

Cable-stayed decks utilising concrete beam-and-slab arrangements have been built with main spans of up to 500 m. The 830 m long Vasco da Gama crossing's cable-stayed bridge, shown in Figure 16.9, illustrates a typical layout for this type of deck arrangement. The additional piers in the back spans provided stability during the construction and stiffened up the cable structure in the permanent arrangement.

Another example, the Yamuna cable-stayed bridge in India, is illustrated in Figure 16.10. The cable-stayed deck is continuous with the side spans which stiffen up the cable-supported structure and provide a stiff support for the outer stays to be anchored on. By placing the footpaths outside the plane of stays the distance between the edge beams was reduced and the deck slab arrangement more balanced.

The concrete edge beams provide the anchorage for the stays. They stiffen the deck longitudinally and support cross-girders spanning transversely. The position of the cross-girders usually matches the stay locations, with additional cross-girders between to give a typical spacing of between 4 m and 5 m. The deck slab thickness is usually between 250 mm and 300 mm.

Figure 16.9 Vasco da Gama Bridge layout (reproduced courtesy of VINCI Construction Grands Projets, copyright reserved)

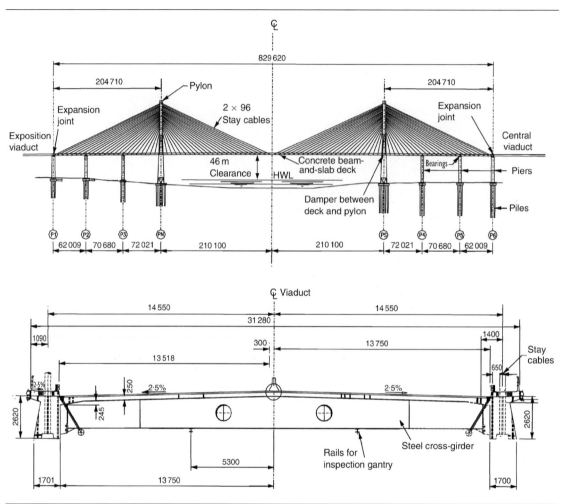

Using steel beams for the cross-girders minimises the deck weight and simplifies the form traveller construction and operation. The Vasco da Gama crossing's cable-stayed bridge, with a distance of 27.5 m between the edge beams, uses steel cross-girders spaced at 4.5 m centres supporting a 250 mm thick deck slab.

Concrete cross-girders usually have a simple 'I' or 'T' arrangement and are either precast or cast in situ. The concrete cross-girders on the Yamuna cable-stayed bridge in India are spaced at 5 m centres with a 250 mm deck slab, spanning 16.8 m between the edge beams. They were cast in situ and post-tensioned with multi-strand tendons. Concrete cross-girders may be simply reinforced instead of prestressed where the deck width and design allow.

Edge beams and deck slabs are usually cast in situ inside a form traveller, although precast sections have been used on several projects. Figure 16.11 shows the form traveller of the Yamuna Bridge in

Figure **16.10** Yamuna Cable-Stayed Bridge layout (reproduced courtesy of COWI)

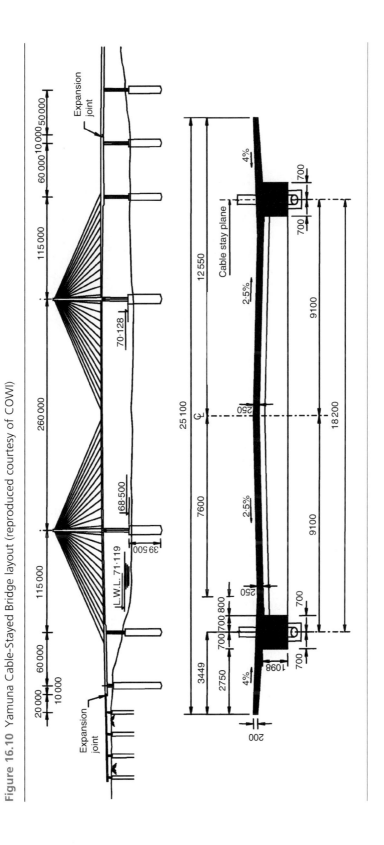

Figure 16.11 Form traveller for casting deck

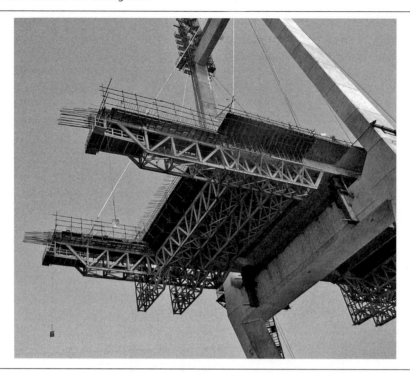

India, set up ready to cast the next 10 m long segment of deck. Segment lengths usually match the stay spacing. The construction time for each segment is typically between seven and 12 days.

A special formwork arrangement is required for the first section of deck at the pylon. The pylon table is usually cast using full-height scaffolding from the ground or with the standard form travellers adapted and lifted into place, as seen in Figure 16.12. After positioning of the deck formwork the reinforcement and tendon ducts are installed and the concrete cast.

When the deck concrete has reached sufficient strength the first stays are installed to support the deck, enabling the falsework to be removed or the form travellers released and moved forward. The form travellers are then set up for the construction of the next section of cantilevered deck, as shown in Figure 16.13. On the Yamuna Bridge, the form traveller was arranged to allow the deck to be cast in two phases, with the edge beams cast first, followed by installation of the stays to allow the form traveller to be moved forward and the deck slab cast. Figure 16.13 shows the preparations for casting the deck slab after installing the stays and moving the traveller forward. This sequence reduced the overall construction cycle time and allowed a lighter gantry to be used.

Prefabrication of parts of the reinforcement cage, as seen in Figure 16.14, simplifies the overall construction process and reduces the cycle time for the construction of each deck section. Prefabricating is of particular benefit in the area where the stay is fixed to the deck. The stay anchor, adjacent cross-girder and edge beam requirements result in a complex reinforcement and tendon arrangement which is easier to pre-assemble in a 'factory' environment before being lifted into place.

Figure 16.12 Form travellers being lifted into position (reproduced courtesy of VINCI Construction Grands Projets, copyright reserved)

On the Vasco da Gama crossing the multi-strand stays were installed before the deck section was concreted, as seen in Figure 16.15. The stays were initially anchored onto the form traveller with only a small load applied, and they supported the traveller under the weight of the fresh concrete. Temporary stays, as seen in Figures 16.6 and 16.11, were used to help distribute the load at the end of the cantilever back into the deck and other stays. After the edge beam concrete had gained sufficient strength, the stay anchors were transferred to their permanent position and fully stressed.

With in situ beam-and-slab construction the deck is usually designed as a reinforced concrete section as it cantilevers out. Sufficient reinforcement is provided to cater for the moments and shears generated. The longitudinal prestress tendons are installed after the closure stitches in the main span and back spans are completed.

Design aspects associated with concrete cable-stayed bridges
The design of cable-stayed bridges is complex and must take into account many different factors to build up the overall structural behaviour. The influence of the deck, pylon and stay arrangements

Figure 16.13 Beam-and-slab deck under construction

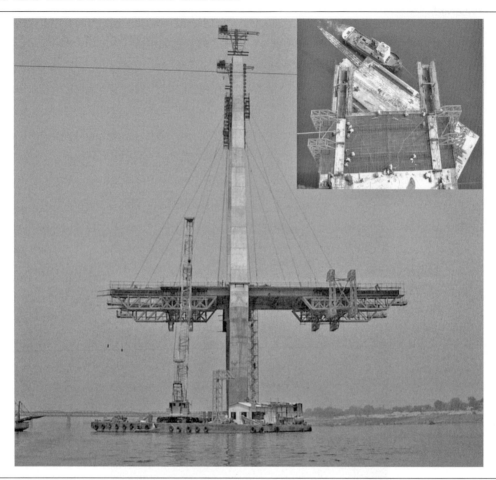

and the behaviour of the stayed structure under dead, superimposed and live loads are described in the books by Troitsky (1988) and Walther *et al.* (1988). The effects of wind loading and the associated dynamic behaviour are described in the TRL State-of-the-art review 5 by Hay (1992).

Typical sections for concrete cable-stayed decks are illustrated in Figure 16.4, although many variations of these basic arrangements can be used. The optimum arrangement for any particular project will depend on specific requirements such as deck width, traffic arrangements and minimum span length. The choice of stay arrangement and pylon shape also influences the deck layout, while aesthetics and architectural requirements play a key role in the final arrangement chosen.

The main difference between the design of concrete and other cable-stayed bridges is that the creep and shrinkage in the concrete of the deck and pylon changes the forces in the stays, which alters the forces, stresses and deflections in the prestressed concrete deck over the life of the structure. For longer spans the aerodynamic behaviour of the cable-supported structure may become critical and full wind studies are needed to ensure that the structure is stable under the full range of wind design speeds.

Figure 16.14 Prefabricated reinforcement cage (reproduced courtesy of VINCI Construction Grands Projets, copyright reserved)

Figure 16.15 Stays being installed (reproduced courtesy of VINCI Construction Grands Projets, copyright reserved)

Analysis of cable-stayed bridges

In simplistic terms, the deck behaves like a beam-on-elastic foundation with each stay providing a spring support to the deck. Initial sizing of the deck and stays during the conceptual stage of the design may be carried out using a line-beam model, as illustrated in Figure 16.16(i). The equivalent stiffness of the support at each stay location is given by:

$$K_{s} = (E_{i}A_{\text{stay}} \sin^{2} \alpha)/L_{\text{stay}}$$

The modulus of elasticity of the stay is modified due to the sag in the cable which is dependent on the force in the stay as follows:

$$E_{i} = E_{s}/[1 + \gamma_{s}^{2} H_{\text{stay}}^{2} E_{s}/(12 T_{\text{stay}}/A_{\text{stay}})^{3}]$$

As an initial estimate, the stress in the stays, $T_{\text{stay}}/A_{\text{stay}}$, may be taken as between 35% and 40% of the UTS, to represent the typical situation with all the dead load and some of the live load present.

A simplistic model is adequate to develop the structural arrangement and element sizing during the concept design, but a more sophisticated model is needed to take into account all the aspects involved with the detailed analysis of cable-stayed bridges.

The deck behaviour interacts with the stay and pylon behaviour and the design of the deck itself is an integral part of the overall design of the cable structure. To carry out the detailed design of the

Figure 16.16 Simplistic analysis model

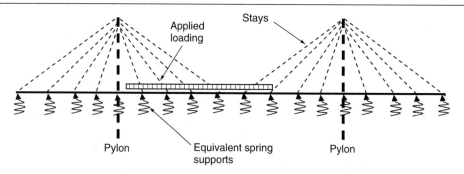

(i) Beam on elastic foundation model

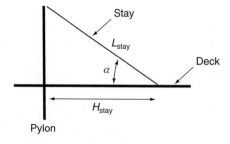

(ii) Stay geometry

Figure 16.17 Three-dimensional analysis model, showing deflections under load

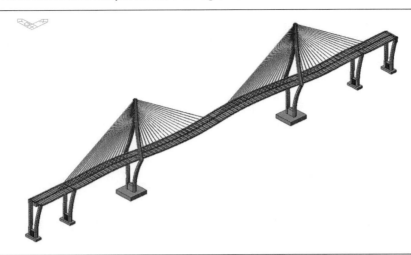

structure a full model of the superstructure and substructure is established to derive the moments, shears and axial forces in the deck under the different loading conditions. Figure 16.17 is an example of a 3D finite-element model established using the Midas software to derive the overall behaviour and dynamic parameters of a cable-stayed bridge, and shows the deflected shape under asymmetrical loading in the main span.

Cable-stayed bridges often have wide slender decks, and shear lag effects are significant close to the supports. An effective width approach, as described in Chapter 5, may be used when designing the structure with a two-dimensional beam analysis model. With a three-dimensional model using plate elements for the deck the shear lag is automatically taken into account in the overall structural analysis.

When applying the effective width approach the stays are usually considered as soft supports and do not give the shear lag effect associated with a rigid support. The effective width of the deck slab is the full width of the slab over most of the cable-supported section of deck. A reduced width is taken adjacent to the piers and pylons if the deck is rigidly supported at these points.

Specialist software, such as Midas described in Chapter 8, is required to combine the long-term effects, such as creep and shrinkage of the concrete and relaxation of the prestress and stays, with the overall behaviour of the structure.

The analysis of cable-stayed bridges must incorporate the stage-by-stage construction of the deck and include for any built-in stresses that occur, as described in Chapters 5 and 8.

The analysis output shown in Figure 16.18 illustrates the deflected shape of a concrete deck for a cable-stayed bridge after all the long-term effects have occurred. The exaggerated deflections shown assume the deck was built without any precamber imposed on the alignment. In practice the deck is pre-cambered during construction to give the correct deck profile on completion.

During the construction, the forces in the stays are set to optimise the deck design in the temporary stages. At the end of the deck construction the stays are re-stressed to give the required deck profile

Figure 16.18 Deflections of cable-stayed deck

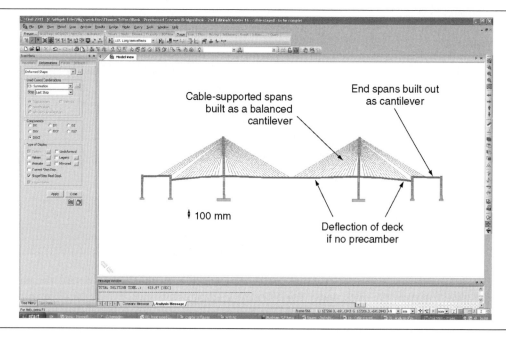

and to produce a uniform distribution of bending moments and shears along the length of the deck. Figure 16.19 illustrates a typical bending moment distribution in the deck elements towards the end of the construction.

Figure 16.19 Dead load bending moment profile

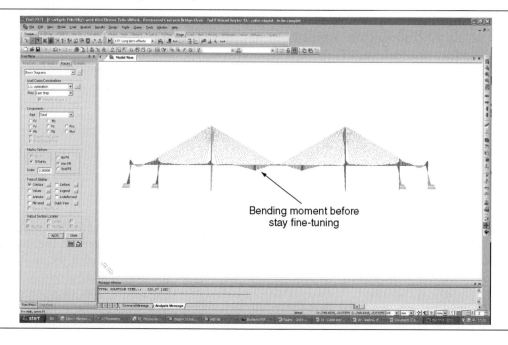

The final dead-load bending moments in the deck are set by adjusting the force in the stays during the final re-stressing. The analysis model must incorporate the adjustments to the stay forces during the construction stages and interact with the design process to derive the optimum installation and final force for the stays.

Deck design and behaviour

The concrete deck resists the moments and forces from the imposed loading and from the deck and stay interaction. Adjacent to the pylons the longitudinal compression generated by the stays is usually sufficient to counter the tensile stresses from the bending and shear in the permanent condition. At mid-span and near the ends of the deck there is little compression from the stays, and prestressing tendons are installed to keep the stresses within the acceptable limits. Figure 16.20 illustrates a typical axial force profile along the length of a concrete cable-stayed deck, assuming prestressing tendons are installed over the mid-span and ends of the deck as indicated.

The prestress tendon profiles are usually 'straight', with the number and extent of the tendons determined by the stresses generated in the concrete from the loading and overall structural behaviour. With a box girder deck the tendons run through the top and bottom slabs and are anchored on blisters inside the box, as illustrated in Figure 6.6. With a beam-and-slab deck the tendons are positioned in the edge beams and deck slab as illustrated in Figure 16.4(iii). Tendons in the slab are usually anchored in recesses on the concrete face, while the edge beam tendons are anchored on blisters formed on the beam.

The deck design is often governed by the forces and stresses generated during the construction. Each stage of the erection sequence is checked for both the strength and serviceability requirements and for overall stability.

Figure 16.20 Axial forces in deck

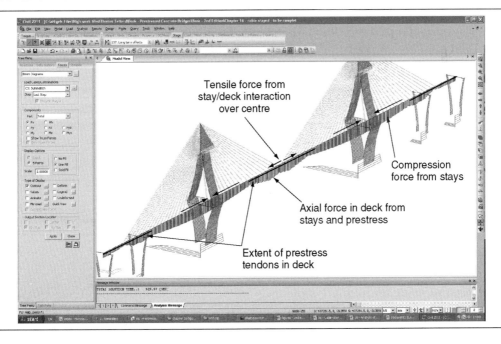

Figure 16.21 Bending moments during deck cantilevering

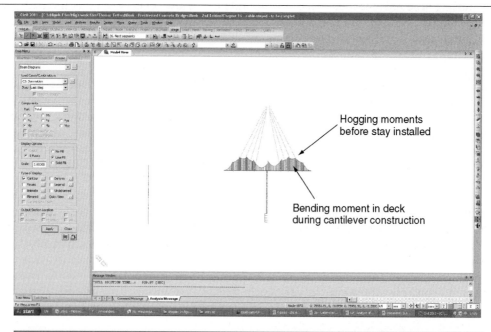

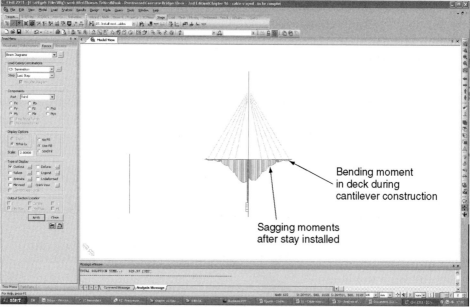

To check the design at each erection stage, the partly built structure is modelled with all the construction loads considered to derive the moments, forces, stresses and deflections. Figure 16.21 shows the range of bending moments from a typical analysis during deck cantilevering. These moments are combined with the co-existent axial and shear forces to check the section capacity and serviceability stresses or crack widths.

During casting of each deck section the additional weight at the end of the cantilever causes hogging in the previously completed deck, as illustrated in Figure 16.21(i). The hogging is increased as the construction equipment and form traveller are moved forward. When the next stays are installed they pull the deck up and create a sagging moment along the completed deck, as illustrated in Figure 16.21(ii). The cycle of hogging and sagging moments continues as the deck is built out.

The compression in the concrete deck generated by the stays during the construction stages is normally insufficient to keep the concrete stresses within acceptable limits. Reinforcement is provided in in situ decks to give the required strength and control crack widths in the concrete. With precast decks additional prestressing is installed to keep the joints in compression.

The deflection of the deck at each construction stage, as illustrated in Figure 16.22, is extracted from the analysis for use in monitoring the deck behaviour and to estimate the precamber required. When a section of deck is cast the additional concrete weight causes the deck to deflect downward, as indicated in Figure 16.22(i). On installing the stay on one side of the pylon, the deck is pulled up on that side while the profile on the other side is also affected. Installing the next stay on the other side of the pylon balances the forces and deflections with a general upward deformation, as indicated in Figure 16.22(ii).

The changes in forces and stresses in the deck and the deflections that occur during the construction are much greater in cable-stayed bridges than in other bridge types. Although this does not cause any particular problem in their design, they are more sensitive to variations in the properties of the concrete and to secondary effects in the structural behaviour.

Large deflections in the pylon and deck of a cable-stayed bridge cause the p–δ effect to be significant, requiring the use of structural analysis software that takes into account geometric non-linear behaviour.

Concrete modulus of elasticity, E_c, values stated in design codes and books are typical values that can vary by up to $\pm 20\%$ compared to a particular mix. The creep and shrinkage of concrete depends on many factors and can vary greatly from the typical values stated in the design codes. Whereas this may not be significant for other bridge types, for cable-stayed bridges it may be necessary to carry out laboratory testing on the actual concrete mix being used to determine accurately the concrete parameters assumed when monitoring the behaviour on site.

The deck often acts as a reinforced concrete section during construction and the concrete is designed to crack. Where the concrete is heavily reinforced or significant cracks occur, this changes the stiffness of the section, which affects the expected deflections and again must be taken into account.

Deck dynamic behaviour

It is necessary to check a cable-stayed bridge for its dynamic behaviour under both live and wind loading. The natural frequencies of the cable-supported structure and stays are determined using one of the proprietary three-dimensional finite-element software programs available, such as Midas, Lucas or Staad-Pro. To prevent potential problems, the natural frequency must be different from the vibration frequency of the traffic using the bridge. The natural frequencies of the structure are also used to develop the models for the wind engineering studies.

The theory and analysis of dynamic behaviour are explained in the guide by Maguire and Wyatt (1999). The different aspects of wind engineering are described in the book by Hay (1992). Dynamic behaviour

Figure 16.22 Deck deflections during cantilevering

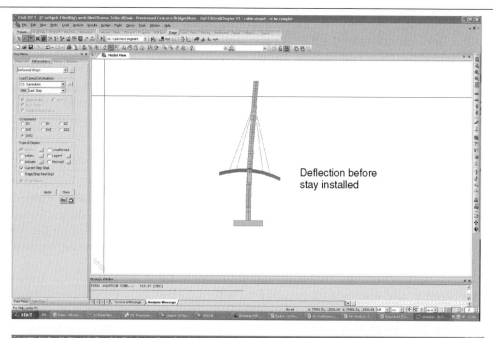

Deflection before stay installed

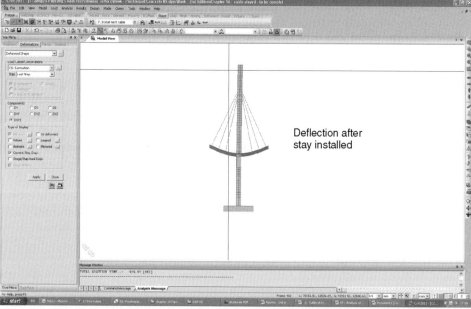

Deflection after stay installed

of the complete structure under wind is usually determined from wind-tunnel tests and involves confirming that wind effects are not critical over the design wind speeds expected. If wind-tunnel tests indicate that vibrations or instability might occur, the usual solution with concrete decks is to make small adjustments to the deck arrangement. Changing the shape of the edge of the deck or altering the flow path of the wind above or below the deck is usually sufficient to overcome the problem.

Wind effects are checked at critical stages during the construction as well as on the final structure. The construction stages with the deck cantilevering out are often more susceptible to dynamic behaviour than the completed structure. In the temporary situation the deck is stabilised by fixing it to the pylons or providing tie-down stays between the deck and foundations. Alternatively, temporary piers or damping devices are used to provide stability.

The heavier weight and higher damping of concrete decks compared to steel decks makes them less susceptible to vibrations and dynamic effects. The damping, expressed as the logarithmic decrement, may be 3% for a prestressed concrete deck and 5% for a reinforced concrete deck.

Stays

The most common type of stay is made up of 7-wire prestressing strands. Lock-coil cables, multi-bar and parallel wire systems are also used, but they are less popular. Stays are designed to have a maximum stress of less than 45% UTS under normal loading to minimise the risk of fatigue failures. This is usually increased to 55% UTS for exceptional loading conditions.

The stays impart a large force locally onto the deck, and the transfer of this force into the concrete section influences the deck layout and requires careful consideration. The possible stay locations are restricted by the traffic layout on the deck slab. The stays are set back out of the way of the traffic and fully protected from any possible accidental damage.

With beam-and-slab arrangements the stiff edge beams provide the ideal location to place the stay anchors. On box girders the webs do not always coincide with the possible locations for the stays and the anchors are located under the deck slab with local stiffening or additional transverse structural members to distribute the load across the section, as illustrated in Figure 16.4(i).

Where the stays are fixed to the concrete deck the arrangement is similar to an external prestressing anchorage, although with extra protective systems incorporated and always detailed to be re-stressable. The concrete around the stay anchor is designed in the same way as for a prestressing anchor, described in Chapter 6. A typical edge beam anchor arrangement is illustrated in Figure 16.23.

Stays are vulnerable to vibrations from wind effects and traffic loading, and damping devices are usually incorporated into the stay arrangement. The anchorage detail often includes a circular rubber bearing, positioned at the end of the stay pipe, that acts to reduce any local bending in the ends of the stay and to dampen out any vibrations. Other damping methods adopted include hydraulic pistons fixed between the stay and the deck, and damping ropes fixed across several stays to connect them together. In recent years, a phenomenon known as wind-rain-induced vibration has been observed under a combination of rain and low wind speeds between 7 m/s and 20 m/s. To prevent this, a ribbed or roughened sheathing is used to encase the stay and to control the water and wind interaction.

Temporary loading

For in situ construction the temporary loading on the deck usually includes

- form traveller at the end of the deck cantilever, typically between 1000 kN and 2500 kN depending on deck size and construction sequence
- stressing jacks for stays, typically 30 kN, located on the form traveller or on the pylon as appropriate

Figure 16.23 Stay anchorage arrangement

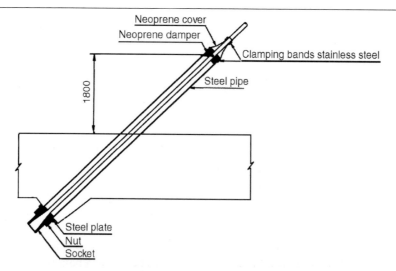

- winch for installing stay, typically 30 kN, placed on the deck, either at the pylon or near the stay anchorage
- deviator to guide stay into anchorage, typically 10 kN, placed near the deck anchorage
- uncoiler for stay, typically 30 kN, placed near the deck anchorage or at the pylon
- construction equipment, typically taken between 30 kN and 40 kN, placed on the tip of the deck cantilever
- construction live load, typically taken as 0.5 kN/m^2, applied to the deck and form traveller to give the most adverse effect

- wind loading on the deck, including 'upward' wind under one of the cantilevers and dynamic wind on the partially completed structure.

For decks with precast elements, the form traveller is replaced by lifting equipment, which is often of a similar weight.

In addition, it is usual to consider an out-of-balance dead load, with the concrete 2.5% heavier on one side of the balanced cantilever.

REFERENCES

Al-Qarra H (1999) The Dee Estuary Bridge – control of geometry during construction. *Proceedings of the Institution of Civil Engineers, Civil Engineering* **132**: 39.

Hay J (1992) TRL State-of-the-art review 5. HMSO, Norwich.

Maguire JR and Wyatt TA (1999) *Dynamics, An Introduction for Civil and Structural Engineers.* ICE Design and Practice Guides. Thomas Telford, the Wind Engineering Society and SECED, London.

Troitsky MS (1988) *Cable Stayed Bridges, 2nd* edition. Blackwell, Oxford.

Walther R, Houriet B, Isler W and Moia P (1988) *Cable Stayed Bridges.* Thomas Telford, London.

Prestressed Concrete Bridges, 2nd edition
ISBN: 978-0-7277-4113-4

ICE Publishing: All rights reserved
doi: 10.1680/pcb.41134.339

Chapter 17
Other prestressed concrete bridge types

Introduction
As well as the more common types of construction, presented in Chapters 9 to 16, prestressed concrete is used with several other forms of bridge deck arrangement as discussed in the following sections.

Extradosed bridges
The extradosed bridge is a variation on the cable-stayed arrangement, and it combines a stiff concrete deck with shallow cable stays anchored to a reduced height pylon. A typical arrangement is illustrated in Figure 17.1(i). The pylon height above the deck is approximately 10% of the span length, compared to between 20% and 25% for a similar cable-stayed bridge. Extra-dosed bridges are constructed with either a concrete box or a beam-and-slab deck, and the result is an elegant and aesthetically pleasing structure. They are considered an economical choice in the 100 m to 200 m span range.

The relative stiffness of the deck and shallow angle of the stays gives a structural behaviour bearing closer resemblance to an externally prestressed deck than to a cable-stayed structure, and the design is often carried out on that basis. A typical example of an extra-dosed bridge is the Tsukuhara Bridge in Japan, shown in Figure 17.2. The twin-deck structure has a single-cell box girder on each side with the stays anchored along the edge of each box. The 323 m long three-span bridge has a main span length of 180 m and short pylons rising to 16 m above the deck level.

The Sunniberg Bridge, Switzerland, shown under construction in Figure 17.3, utilised a slender beam-and-slab deck with spans of 140 m and a pylon height of 15 m above the deck. The concrete deck was constructed as a balanced cantilever built out from each of the pylons using form travellers, with the stays installed as the construction progressed.

The design and construction of extradosed bridges combine many of the features described in Chapters 12, 13 and 16.

Fin-back bridges
These are similar to the extradosed bridge, but with the stays replaced by tendons encased in a concrete wall or 'fin', as illustrated in Figure 17.1(ii). The fin is an extension of the deck section and stiffens the deck over the pier while also providing protection to the tendons.

There are only a few examples where this type of structure has been used and they usually have a concrete box girder deck with the fin extending up above a central web. The Barton Creek Bridge, shown in Figure 17.4, has a main span of 104 m and an overall length of 210 m. It was built using the in situ balanced cantilever technique with a constant deck cross-section throughout. A detailed description of the Barton Creek Bridge is given in a paper by Gee (1991).

Figure 17.1 Extra-dosed and fin-back bridge arrangements

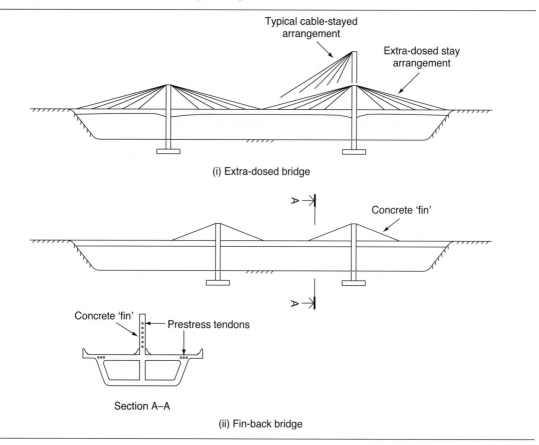

Typical cable-stayed
arrangement

Extra-dosed stay
arrangement

(i) Extra-dosed bridge

A

Concrete 'fin'

A

Concrete 'fin'
Prestress tendons

Section A–A

(ii) Fin-back bridge

The design and construction of fin-back bridges are similar to the in situ or precast concrete box girders described in Chapters 12 and 13.

Truss bridges

Trusses are often used with steel bridges, and they produce light and efficient structures. However, they are less common on concrete structures because of the complex design and construction of the joint between the diagonals of the truss and the deck slabs. Aside from that, several long-span precast segmental concrete viaducts and cable-stayed decks have utilised trusses for the webs to reduce dead weight, resulting in savings in quantities of prestress, stays and concrete.

Bubiyan Bridge, Kuwait, shown in Figure 17.5, was constructed using precast segments with the top and bottom slabs and the diagonal members all in concrete. The 2.5 km long crossing has typical spans of 40 m with the deck continuous in 200 m or 240 m sections. Segments were match cast using the long-line method described in Chapter 13. The diagonal members were cast first and lifted into the bed for the top and bottom slabs to be cast. Segment erection was span-by-span using an overhead gantry to position the segments. After placing a complete span of segments the deck was prestressed with external post-tensioned tendons.

Figure 17.2 Tsukuhara Bridge, Japan (reproduced courtesy of Freyssinet International, copyright reserved)

In France, the Boulonnais Bridge, shown in Figure 17.6, uses steel sections for the diagonal truss members. The deck was constructed using precast match-cast segments lifted into place by an overhead gantry and erected by the balanced cantilever technique. External tendons provide the longitudinal prestress in the deck.

On early projects, where the diagonal truss members were plain reinforced concrete sections, cracking occurred in the concrete. The diagonals are subjected to high tensile forces and applying prestressing along the diagonals helps to prevent the concrete cracking from occurring.

Figure 17.3 Sunniberg Bridge, Switzerland (reproduced courtesy of BBR Systems Ltd, copyright reserved)

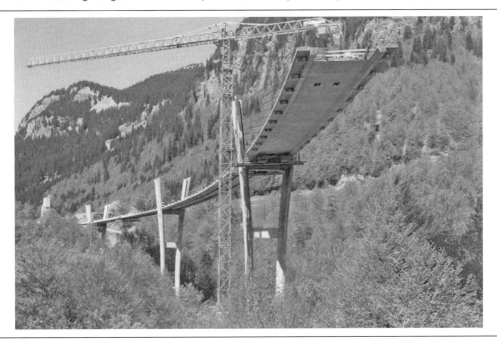

Figure 17.4 Barton Creek Bridge, USA (reproduced courtesy of Tony Gee and Partners LLP, copyright reserved)

The Vecchio Bridge in Corsica, shown in Figure 17.7, is an innovative structure combining balanced cantilevering construction with an unusual truss deck arrangement. The triangular sections of web are efficient in transferring the shear forces and provide a striking appearance. It was designed by Berds Mikaelion, Lanslas Paulik, and Razel Technic & Methods and constructed by Razel Technic & Methods using DYWIDAG prestressing systems.

Arch bridges

Arches are used to create long spans, with the arch either above or below the deck level. Arches extending above the deck provide support through hangers, and lighter steel or composite decks are usually preferred over heavier concrete arrangements. When the arch is below the deck level, concrete decks provide a suitable solution.

The arch section is naturally in compression and seldom needs to be prestressed, while the deck often utilises prestressed concrete elements. The deck is usually supported on columns resting on the arch and is similar in design and construction to a standard bridge deck with the same span lengths. In

situ concrete slabs, in situ boxes, precast prestressed beams and launched concrete box girders are all used with an arch supporting arrangement.

The Gladesville Bridge, Australia, described by Baxter *et al.* (1965), was built in 1964 and is shown in Figure 17.8. It has a 305 m span reinforced concrete arch that supports a concrete deck formed using prestressed, precast concrete 'T' beams with spans of 30 m. The arch was formed from precast hollow

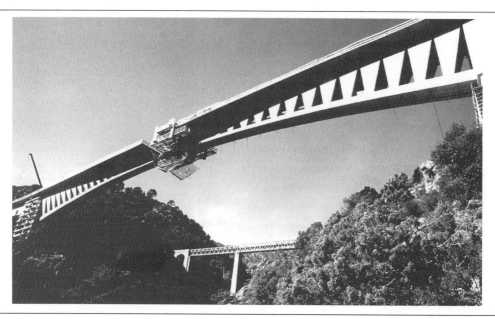

box sections erected on falsework. The columns were precast and lifted into place, while an overhead gantry placed the precast beams.

Figure 17.9 shows the Barelang Bridge under construction in Indonesia. The reinforced concrete arch has a span of 245 m and is a hollow box section. Vertical columns extend up from the arch at 35 m centres to support the prestressed concrete box girder deck. The 385 m long deck was incrementally launched over the columns and into position using the techniques described in Chapter 15.

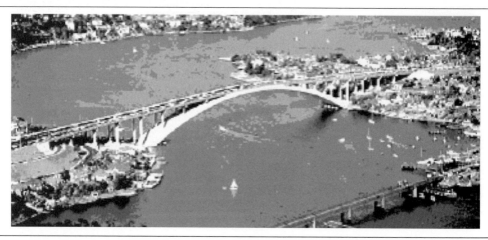

Figure 17.9 Barelang Bridge, Indonesia (reproduced courtesy of Freyssinet International, copyright reserved)

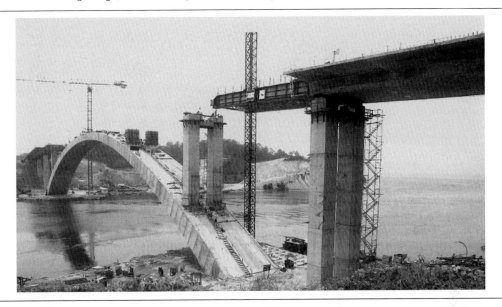

Footbridges

Footbridges are usually considered under a different category from the longer and heavier road and rail bridges, although the techniques used for their design and construction are similar. The lighter loading on footbridges plus the narrow widths result in lighter and more slender decks. The aesthetic

Figure 17.10 Footbridge in Singapore

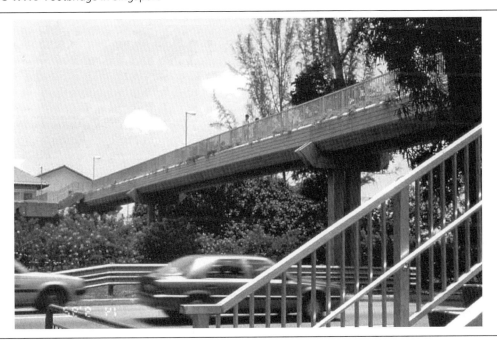

Figure 17.11 Footbridge in Hong Kong

Figure 17.12 Stressed ribbon footbridge, Ireland (reproduced courtesy of Roughan & O'Donovan, Dublin, copyright reserved)

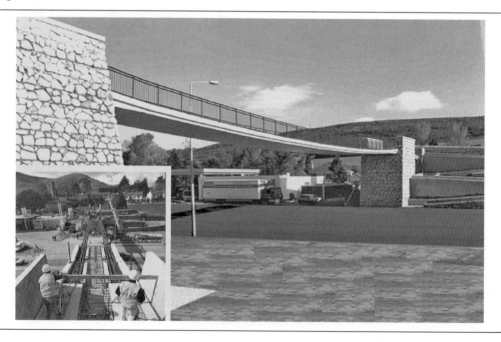

and performance requirements are more to the forefront in footbridge design and result in some innovative and unusual arrangements.

A standard footbridge arrangement used throughout Singapore is shown in Figure 17.10. The deck is either precast or cast in situ with the span length and arrangement varied to suit the location.

The footbridge in Figure 17.11 is typical of the arrangement found in Hong Kong, with the walkway and canopy cast in situ. With spans up to 30 m, the 'U' shaped lower section of walkway is prestressed with post-tensioned tendons running the full length in each edge beam.

The Kilmacanoque stressed-ribbon footbridge, shown under construction and when completed in Figure 17.12, uses a very thin concrete deck supported by cables anchored on the abutments at either end. The shallow deck section follows the draped profile of the cables. Precast concrete sections were first hung from the cables, which were encased by an in situ concrete topping.

The lighter decks of footbridges make them well suited to precasting, either in full length or in sections, ready to be rapidly assembled on site.

REFERENCES

Baxter JW, Gee AF and James HB (1965) Gladesville Bridge. *Proceedings of the Institution of Civil Engineers*, **30**: 489–530, doi: 10.1680/iicep.1965.9523.

Gee A (1991) Concrete fin-back bridge in USA. *Proceedings of the Institution of Civil Engineers, Part 1*, **90**: 91–122.

Prestressed Concrete Bridges, 2nd edition
ISBN: 978-0-7277-4113-4

ICE Publishing: All rights reserved
doi: 10.1680/pcb.41134.349

Chapter 18
Problems and failures

Introduction

The design and construction of bridges present some of the greatest challenges for structural engineers and occasionally this results in problems, ranging from minor cracking in the concrete to the collapse of a structure.

The more spectacular bridge failures, such as the Tay Bridge collapse in 1897, the Tacoma Narrows suspension bridge twisting itself apart in 1940, or the steel box collapses at Milford Haven, Yarra and Cologne in the early 1970s, are all etched in the memories of bridge engineers. Prestressed concrete bridges have also made headline news with collapses such as the Ynys-y-Gwas Bridge, Wales, in 1985 and the Injaka Bridge, South Africa, in 1998.

Other less dramatic problems occur more regularly and prestressed concrete bridges have suffered their own particular troubles in the past, some of which are discussed in the following sections. From these problems and failures has come a better understanding of the design and the construction requirements of concrete bridges.

Prestressing components

A critical stage for prestressing tendons and their components is during the stressing process when they are subjected to the largest forces in their design life. During lock-off the force in the tendon at the stressing anchorage reduces with subsequent reductions occurring over time due to the creep and shrinkage in the concrete, and relaxation in the tendon.

Wire, strand and tendon failures

Wires in a multi-strand tendon occasionally break during the stressing operation. This is normally identified by a sharp 'bang' heard during tensioning of the tendon. Individual wires break for a number of reasons such as damage occurring during installation or kinks in the duct causing stress concentrations. When the tendons are anchored the jaws of the wedges bite into the wires causing indents and, if the tendon is re-stressed, failures sometimes occur where the wires have been weakened.

Breakage of a single wire in a 7-wire strand, which is itself one of a number of strands making up a larger tendon, may not be significant, in which case no remedial action is required. By stressing the tendon to its specified jacking force the correct prestress is still applied to the structure with the other wires sharing the full load. If several wires in a tendon break then the remaining wires may not be capable of carrying the full load. In this case the tendon is removed and the damaged strands replaced.

Complete strands have occasionally broken during stressing. As well as a loud 'bang' being heard, the strand may 'jump' back out of the stressing jack when the breakage is close to the end of the tendon.

Figure 18.1 Failed transverse tendon anchor

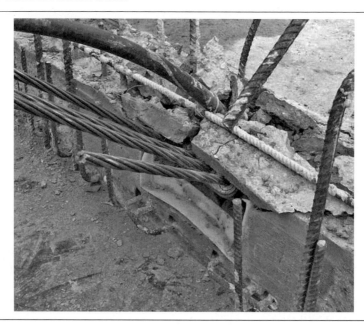

This is often accompanied by a drop in pressure in the stressing jack as the overall force in the tendon reduces. If there is insufficient spare capacity in the remaining strands, or in the overall design, the tendon is de-stressed and the broken strand replaced.

Strands have slipped through the wedges, either at the jack, at the dead-end anchorage during stressing, or at the stressing end anchorage during lock-off of the tendon. This may be caused by incorrect seating of the wedges around the strand or by dirt or grease getting between the jaws and the wires.

Couplers with prestressing bars have failed when the threaded connection has not been fully engaged. This is avoided by carefully measuring the length of thread engaged and by ensuring that the connections are complete.

There have been cases of a prestressing anchor breaking or moving into the concrete during the jacking operation. This is often caused by honeycombing or deficient concrete behind the cast-in bearing plate or trumpet. Figure 18.1 shows a failed transverse tendon anchorage of the flat-slab type. The anchor block rotated and moved into the concrete during stressing with the top concrete bursting off. Close inspection of the anchorage revealed that the bursting reinforcement around the embedded anchor block had been placed too far from the back of the anchorage to be effective.

Tendon extensions
With multi-strand post-tensioned tendons, the extensions of the strands during stressing inevitably varies from that expected, with a range of $\pm 5\%$ usually being considered acceptable. If the variation is outside this range it may indicate a problem along the tendon. Higher extensions may be due to lower friction than expected or it may be more serious if a strand has broken or slipped through the wedges at the dead-end.

Lower extensions usually indicate that the friction between the tendon and the duct is greater than that allowed for in the design, or wobbles and kinks in the duct have increased the resistance as the tendon is pulled. The first option is usually to increase the stressing force to compensate for the extra losses, although this is normally limited to a maximum force in the tendon of 80% UTS. Alternatively, the design is re-assessed for the lower prestress forces associated with the reduced extensions to determine if it is acceptable. If the design is not acceptable the tendon is usually de-stressed and re-stressed again, if necessary with soluble oil used to reduce the friction.

There is also a possibility that the tendon is snagged along its length, or concrete has got into the duct and prevented the complete length of tendon from being stressed, which would adversely affect the design of the deck. An indication of this is that the wedges at the dead-end anchorage do not fully pull in during the stressing. In this case the tendon is de-stressed, removed and re-installed to see if the blockage clears itself. If that fails to solve the problem, the blockage is located and cleared before the tendon is re-stressed.

Grouting and ducts

Grouting is another activity that has caused problems in the past, with voids being left in the ducts. When many existing post-tensioned concrete bridges have been inspected a significant percentage have been found to have voids present, although only a small proportion of these showed signs of tendon corrosion. Figure 18.2 is a negative from a radiograph survey clearly showing up a void in a duct.

Voids are caused due to a number of reasons, including blockages in the duct, entrapped air or leaking of the grout. Guidelines on grouting are given in TR 47 (Concrete Society, 2002) which should go some way to help eliminate this problem in the future.

On precast segmental decks there have been occasions when the pre-stressing ducts have 'crossed over' within a segment. When the tendon is threaded in at the anchorage of one duct it emerges at the

Figure 18.2 Incomplete grouting

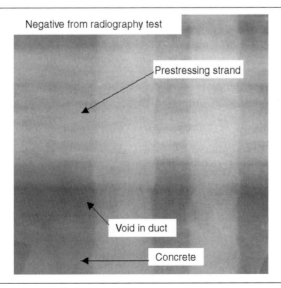

Negative from radiography test

Prestressing strand

Void in duct

Concrete

anchorage of the other one. This does not normally cause too much of a problem as the number of tendons at each section is still maintained; however, it can be of concern if any one tendon becomes too long, resulting in excessive friction losses along its length.

Corrosion

A number of bridges around the world have suffered from corrosion to the prestressing tendons or anchors well before the end of their design life. A survey carried out in the UK by the Highways Agency during the late 1990s revealed that 4% of the bridges inspected had heavy to severe corrosion of the tendons with another 8% having moderate corrosion present. The reasons for the corrosion range from poor detailing and specifications to deficiencies in construction and workmanship.

The need for robust protection to the tendons is now established and, as the industry becomes more aware of past problems, incidents of corrosion should become less frequent.

Concrete and reinforcement

Problems relating to the concrete and reinforcement are commonplace although most are no more than relatively minor inconveniences that are simply dealt with on site.

Concrete cracks

Concrete cracks for many reasons, both design and construction related, and it is common for cracks to appear at some time or other in most concrete bridges. Reinforced concrete design assumes that cracks occur in order for the reinforcement to fully work, with crack widths of up to 0.3 mm accepted in most design codes.

Prestressed concrete bridges are usually designed with tensile stresses limited to a level where cracks should not occur. However, this applies only to the concrete that is prestressed, and other areas designed as reinforced concrete sections, such as the transverse members, diaphragms and anchorage areas, are liable to exhibit cracks. Despite this, it always seems to surprise engineers when cracks appear and steps are usually taken to seal them, such as injecting low viscosity resin into the cracks.

Cracks can occur due to early thermal effects within the fresh concrete and differential shrinkage between concrete cast at different times. These cracks are controlled by adopting a suitable concrete-mix design, good curing of the concrete and providing sufficient distribution reinforcement based on the approach given in BD 28/87 (Highways Agency, 1987) and BA 24/87 (Highways Agency, 1989).

Plastic settlement after placing of the concrete causes reflective cracking above the reinforcement, which is prevented by using a suitable concrete mix design and by good compaction of the concrete during placing.

With post-tensioning, partially stressing some of the tendons early, when the concrete is still young, helps to prevent cracks from opening up. Most prestressing systems have anchorages designed to be loaded when the concrete has reached $25 \, \text{N/mm}^2$ or $30 \, \text{N/mm}^2$; however, this is based on the full load being applied. If the tendons are initially stressed to only 50% or less of their design force it should be possible to stress them when the concrete is at a lower strength.

High stresses are generated around the tendon anchor block as the force spreads out into the concrete section. The local tensile stresses generated can cause cracking in the concrete. These cracks are

Figure 18.3 Spalling at deviator

controlled by reinforcement arranged to counter the tensile stresses. It is difficult to determine the precise stress pattern for arrangements with complex concrete or anchorage layouts, and large cracks can occur. Spalling of the concrete on the end-face of an anchorage looks unsightly, but it does not adversely affect the strength of the anchor block and needs repairing only to reinstate the protection to the reinforcement and anchor arrangement. Small cracks in the bursting region of an end block are acceptable provided their width is within the acceptable design limit, while excessive crack widths may indicate that remedial measures are needed. Cracks that occur in the adjacent concrete, to the side and behind the anchorage area are assessed in a similar way.

External tendons impose a large concentrated force onto their anchor blocks and deviators. Cracking of the concrete is controlled by closely spaced reinforcement placed to counter the tensile stresses generated. Where the tendons or holes are misaligned this changes the distribution of the force along the deviator and, in the worst cases, causes kinks in the tendons as they emerge at the end of the deviator. This can give rise to cracks in the concrete or spalling of the edges, as seen in Figure 18.3.

Several concrete box girders with external tendons have experienced failures at their anchor blocks and deviators. Built in the 1970s, anchor blocks on the A3/A31 bridge near Guildford in the UK exhibited severe cracking during the construction and were strengthened to enable the bridge to be completed. Elsewhere there have been deviator failures with the tendon pulling out of the concrete, as seen in Figure 18.4, requiring the deviators to be repaired and strengthened. This emphasises the importance of ensuring that the anchor blocks and deviators fully secure the external tendons with the applied force evenly spread out.

Curved tendons have burst out of the concrete when running near the concrete surface. Adequate cover must always be provided to the prestressing ducts. Extra reinforcement must be placed, where the tendons are highly curved, to tie the tendon into the main body of the concrete.

Figure 18.4 Deviator failure

Honeycombing

Prestressed concrete bridges are usually highly stressed, and require high-strength, dense concrete. Achieving well-compacted concrete is not always easy, as construction sites present many practical problems to be dealt with. Additionally, the tendons and reinforcement in post-tensioned concrete decks are often closely spaced, hindering the concreting operation. These factors can lead to honeycombing in the concrete, as seen in Figure 18.5.

The anchor-block reinforcement, shown in Figure 18.6, was made more congested by the designer doubling up on the bursting reinforcement, with both spirals and links detailed. When the rest of the web, diaphragm and top slab reinforcement were placed it left little room for the concrete.

Although these regions are always heavily reinforced, careful selection of the concrete shape and the reinforcement detailing can ease construction and improve the concreting.

In some areas it is necessary to use smaller aggregate in the concrete, with super-plasticisers added to the mix to help the concrete flow around the reinforcement, ducts and anchors. Temporary windows should always be left in the shutters so that the concrete is seen to fill up the space and to allow vibrators to be inserted where necessary. It also helps to adjust the reinforcement locally to leave gaps for the concrete and vibrators to be inserted.

With proper planning and procedures it should always be possible to achieve a good dense concrete, although site conditions sometimes work against this and honeycombing and other concreting problems do occasionally occur. However, it was difficult to envisage why the honeycombing and missing concrete occurred in the web of the deck in Figure 18.7.

In most cases, the honeycombing is simply cleaned out and the voids are filled up with fresh concrete; although there have been several bridge decks demolished and rebuilt because of severe defects.

Figure 18.5 Honeycombing around anchors

Concrete cover

A common failing in the past has been inadequate concrete cover to the reinforcement and prestressing ducts. Construction tolerances in the formwork and in the bending and fixing of the reinforcement can result in the design cover not being achieved. This reduces the protection to the reinforcement and tendons and adversely affects the durability of the structure. The need for remedial works must be considered on a case-by-case basis. In the past these have included breaking out and recasting the concrete, spraying on additional concrete to increase the cover, applying a protective coating to the concrete or doing nothing if this is acceptable.

Problems during construction

Bridges are often at their most vulnerable during their construction, and prestressed concrete decks have experienced their share of failures and mishaps. Most failures are due to mistakes during the construction process or deficiencies in the design of the temporary or permanent works under the temporary loading. Accidents can happen at any time.

Figure 18.6 Congested reinforcement

Figure 18.7 Honeycombing and voids in concrete

Failures due to design

A few concrete box girder bridges have experienced problems during the launching process. In 1988, the Mainbrücke Stockstadt viaduct in Germany collapsed while being pushed out over the piers. Temporary stay cables had been used to support the deck during the launch, a technique that had been used successfully before. As the deck moves forward it is difficult to control the forces in the stays and the deck, and the use of stays in this manner is seldom seen today.

The deck of the Injaka Bridge, South Africa, collapsed during its launch in 1998. Failure of the bottom slab above the temporary bearings and subsequent overstress of the launching nose and deck is thought to have resulted in the collapse.

Full-height falsework used with in situ concrete construction suffered a number of notable failures in the UK in the 1960s resulting in the publishing of the Bragg Report (Bragg *et al.*, 1975). Deficiencies in design, lack of understanding of some imposed loading conditions, poor construction practices and inadequate founding of the falsework all contributed to the problems encountered. The investigations carried out during the drafting of the report led to more stringent design codes and construction guidance being developed.

These failures demonstrate the need to clearly understand the behaviour of concrete bridges and their interaction with the temporary works. The importance of the detailing, especially in highly stressed regions, and of the design during each stage of the construction as well as in the permanent condition, must be recognised by the designer.

Construction procedures

Concrete bridges often require sophisticated construction techniques with the design and construction closely linked. Good communication between the design and construction teams is essential to minimise the risk of problems. The designer must understand the construction requirements and the contractor must appreciate the design requirements and limitations.

The construction must also be carefully controlled to prevent problems. In the early 1990s an in situ post-tensioned concrete multi-cell box girder deck in Hong Kong came near to collapse when the falsework supporting it was released by the scaffolders before the prestress had been installed. Fortunately, there was sufficient reinforcement in the deck to hold it up and the cracks closed up again when the prestress was subsequently applied.

Less fortunate were the workmen and engineers on the Vasco da Gama Bridge, in Portugal, when the form traveller fell from the deck during construction of the concrete deck of the cable-stayed bridge. The traveller was being moved forward and adjusted to pass the side pier when the accident happened.

Precast concrete elements require lifting, transporting, and positioning, during which they run the risk of being dropped or damaged. Figure 18.8 shows an incident when a precast beam undergoing a two-crane lift was dropped, due to one of the crane jibs failing. Two-crane lifts are always difficult and require careful control to ensure each crane properly carries the load.

Precast beams need supporting when placed on their bearings and before the deck slab is constructed. Although the bearings are usually horizontal it has been known for beams to slide off or topple over.

Figure 18.8 Precast beam dropped

Precast box girder segments have been dropped during handling and when being transferred to the deck. Although not a common occurrence, several projects have suffered from the dropping of a segment, such as the A13 viaduct in Dagenham and the Second Severn Crossing approach viaducts.

Gantries used for erecting precast box girder segments or precast beams have encountered problems, although collapses are rare. On the Second Severn Crossing approach viaducts a segment was dropped when it moved onto the wrong section of the gantry and caused the girder to fail. Gantries are complex structures with numerous mechanical and moving parts that need careful operation and regular inspections and maintenance.

Storing precast segments can give rise to problems, as seen in Figure 18.9 where the segments toppled over in the storage area. As well as instability of the segments, they can also be damaged in their handling and by being hit from passing traffic.

Structural behaviour problems

Bridge decks do not always behave as expected and this is often first noticed during the construction phase. Deflections and movements of the deck are often different from the estimated design value, due to the many different factors that affect the construction and concrete parameters. Usually the differences are relatively small and do not adversely affect the design, although several bridges have suffered from significantly higher movements than expected which have caused problems and necessitated remedial works.

The Long Keys Bridge in Florida was reported to have experienced higher longitudinal movements than expected which may have contributed to problems with the expansion joints and substructure. In the 1970s and 1980s several long-span balanced cantilever decks developed significant vertical sags over their spans, due to higher concrete creep than expected.

Figure 18.9 Segments toppling in storage yard

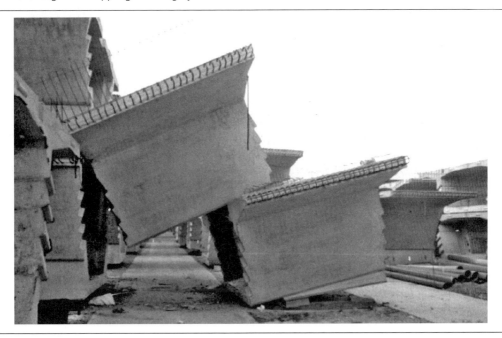

Under-estimating the prestress effects on curved bridges has resulted in several box girder decks lifting off one or more of their bearings. Post-tensioned tendons placed symmetrically in a continuous box girder deck, with a tight horizontal curvature, generate significant torsional effects that redistribute the bearing loads.

Problems after opening
Most prestressed concrete bridges perform well after opening, provided inspections and maintenance are carried out regularly.

Durability
The problem of corrosion and other durability issues are described in Chapter 3, with the collapse of the Ynys-y-Gwas Bridge, Wales, due to corrosion of the tendons, prompting improvements in design and construction procedures.

Another reported incidence of corrosion was on the Mid-Bay Bridge in Florida, where two of the external tendons were replaced in 2000. The 6 km long multi-span viaduct was constructed in 1993 with draped external tendons used to prestress the 41 m long box girder spans. The external tendons were encased in cement grout inside HDPE ducts. One tendon was completely de-stressed and hanging slack while the other tendon had ruptured strands protruding from the duct. It is thought that the existence of voids and grout bleed water had established an environment where corrosion occurred.

Rehabilitation and modifications
The Koror-Babelthaup Bridge, on the Pacific Islands, collapsed soon after strengthening in 1996. Originally built in the late 1970s, the in situ concrete box girder deck had a 241 m long main span with a hinge joint at mid span. Designed as a pair of cantilevers, the main span experienced excessive

deflections and in 1996 the deck was made continuous with additional external tendons installed under the top slab along the complete length of deck. The modifications were successfully completed, but after several months the deck collapsed at a time when very little live load was present. The reasons for the collapse have not been published, but it would appear that the change to the structural system, plus the additional tendons, caused an overstress in the deck adjacent to the piers.

REFERENCES

Bragg SL, Ahm P, Bowen FM, Champion S, Kemp LC, Mott JCS *et al.* (1975) *Final Report of the Advisory Committee on Falsework*. HMSO. London.

Concrete Society (2002) *Technical Report no. 47. Durable Bonded Post-tensioned Concrete Bridges*. 2nd Edition. Concrete Society, Slough.

Highways Agency (1987) (MDRB) Departmental Standard, BD 28/87. Early Thermal Cracking of Concrete. HMSO. Norwich.

Highways Agency (1989) (MDRB) Departmental Standard, BA 24/87. Early Thermal Cracking of Concrete (including amendment No. 1 (1989)). HMSO, Norwich.

Prestressed Concrete Bridges, 2nd edition
ISBN: 978-0-7277-4113-4

ICE Publishing: All rights reserved
doi: 10.1680/pcb.41134.361

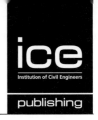

Appendix A
Definitions

Anchor	A mechanical element placed at the end of post-tensioned tendons to hold the tendon and transfer the load from the tendon to the concrete
Bar	Round drawn out steel element, typically 15 mm to 75 mm diameter, for prestressing
Blister	A concrete block cast on the side of the deck to anchor the tendons against
Bonded tendons	Tendons which are bonded to the concrete member by being cast-in, as for pre-tensioning, or through concrete grout for post-tensioning
Box girder	A deck arrangement with a box cross-section, comprising of a top slab, webs and a bottom slab
Cable	A number of strands placed together to form a stressed element
Cable-stayed	An arrangement where the deck is supported along its length by stays fixed to a pylon
Concordant profile	The profile and layout of the prestressing tendons giving no secondary moments
Coupler	A mechanical device for connecting tendons together to make a longer element
Creep	The change in strain within the concrete over time under a constant applied stress
Deviator	A concrete or steel support for external tendons, fixed to the deck
Diaphragm	A concrete wall placed across the deck beneath the top slab to provide rigidity to the structure and/or transfer the loads in the webs to the bearings
Duct	A steel, plastic, or high-density polyethylene (HDPE) tube into which tendons are placed
External tendons	Tendons which are placed outside the concrete section
Extradosed	Similar to a cable-stayed bridge with a shallow angle of stays and a short pylon
Fin-back	Similar to an extradosed arrangement, but with the stays encased in a concrete wall extending up from the deck
Grout	A cement–water mixture or grease compound used to fill the void around the tendon within a duct
Incrementally launched	A technique of casting the deck in short sections at one end of the structure and launching the deck into position in stages
In situ concrete	Concrete cast directly into its final position
Internal tendons	Tendons placed within the concrete section
Post-tensioning	Where the tendon is stressed after the concrete has hardened, and with load transfer during the stressing operation

Precast concrete	Concrete cast away from its permanent position, cured and moved into place
Prestressing	Applying a stress to the concrete by pre-tensioning or post-tensioning
Pre-tensioning	Where the tendon is stressed before placing the concrete, with load transfer occurring after the concrete has set
Primary moment	The direct moment applied at a section due to the prestress tendon, equal to the prestress force multiplied by the offset from the N-A
Pseudo-box	A deck arrangement achieved by placing precast 'I' or 'M' beams side-by-side and connecting their top and bottom flanges with an in situ slab to form a multi-box structure
Relaxation	The reduction in stress in the tendon over time under a constant strain
Secondary moments	The moments set up by the applied prestress in a statically indeterminate structure in addition to the primary moments
Segmental	A technique of precasting the deck in short segments and assembling on site
Shrinkage	The change in strain of the concrete over time as the concrete matures
Strand	Normally 7-wire strand, typically 13 mm or 15 mm diameter, for prestressing; made up of six wires twisted around a central straight wire
Tendon	A wire, strand, a bar or bundle of strands used to prestress the concrete
Thixotropic grout	Special grouting material which is fluid when shaken but becomes gelled when standing
Unbonded tendons	Tendons which are not connected to the concrete other than at the anchor positions, allowing the tendons and concrete to act independently along the tendon length
Wires	Round drawn out elements of steel, typically 5 mm to 7 mm diameter, for prestressing

Prestressed Concrete Bridges, 2nd edition
ISBN: 978-0-7277-4113-4

doi: 10.1680/pcb.41134.363

Appendix B
Symbols and notation used

AASHTO	American Association of State and Highway Transportation
A or A^1	Area of reinforcement to be provided
A_c	Area of concrete section
A_L	Area of concrete considered for longitudinal shear check
A_o or A_k	Area enclosed by median wall lines around the box
A_p	Area of non-prestressed reinforcement
A_r	Area of reinforcement across failure plane
A_s	Area of slab
A_{sj}	Area of web other than the shear keys
A_{sk}	Area of shear key
A_{sl}	Area of longitudinal reinforcement
A_{st}	Area of leg of link around section
A_{stay}	Area of stay cable
A_{sv} or A_{sw}	Area of shear reinforcement
A_t	Cross-sectional area of tendon
b or b_w	Breadth of member or web
b_e	Minimum effective width of element resisting applied torsion
b_s	Width of slab
C	Compressive force generated in concrete at ultimate moment capacity
d	Distance from tendons to compression face, or effective depth of section
d_c	Depth of compression in concrete at ultimate moment capacity
d_t	Distance from reinforcement to compression face
d_1	Larger dimension from line of action of anchor force to the boundary on non-symmetrical prism
d_2	Smaller dimension from line of action of anchor force to the boundary on non-symmetrical prism
D_t	Nominal diameter of tendon
E_c	28-day secant modulus of elasticity of concrete
E_i	Equivalent modulus of elasticity of stay cable
E_s	Modulus of elasticity of stay cable
E_t	Modulus of elasticity of tendon
f_{bpt}	Bond stress between tendon and concrete
f_c	Stress in the concrete at point considered
f_c' or f_{ck}	28-day cylinder strength of concrete
f_{cd}	Design compressive strength of the concrete
f_{ci}'	Cylinder strength of concrete at age being considered
$f_{ck(t)}$	Cylinder strength of concrete at time t

f_{ci}	Cube strength of concrete at age being considered
f_{cp}	Compressive stress at centroid of section due to prestress
f_{ctd} or $f_{ctd(t)}$	Concrete design tensile value of strength based on 28 days strength or at given time
$f_{ctk,0.05}$	5% fractile of mean axial tensile strength of concrete (taken as 70% of tensile strength)
f_{cu}	28-day cube strength of concrete
f_d	Stress due to unfactored dead load at tensile face of section subject to M_{cr}
f_p	Initial stress in tendon
f_{pt}	Stress in tendon after time, t (hours) or at time being considered
f_{pc}	Stress due to prestress only at the centroid of the tendons
f_{pe}	Effective prestress after all losses
f_{pi}	Stress increase in tendon
f_{pk}	Average compressive stress over shear keys or characteristic strength of tendon in Eurocodes
$f_{p0.1k}$	Characteristic 0.1% proof-stress of prestressing steel
f_{ps}	Average stress in prestressing steel at the time which the resistance of the member is required
f_{pt}	Stress due to prestress only at the tensile face or stress in prestress at transfer
f_{pu} or f_s'	Characteristic strength of tendon
f_{py}	Yield strength of prestressing steel
f_s	Initial stress in non-prestressed reinforcement
f_{si}	Increase in stress in non-prestressed reinforcement
f_{su}^*	Average stress in tendon at ultimate load
f_t	Tensile strength of concrete taken as $0.24\sqrt{f_{cu}}$
f_y or f_{yk}	Characteristic strength of reinforcement
f_y^*	Yield strength of tendon taken as:
	$0.9\,f_s'$ for low-relaxation strand
	$0.85\,f_s'$ for stress-relieved strand
	$0.85\,f_s'$ for Type I (smooth) bar
	$0.80\,f_s'$ for Type II (deformed) bar
f_{yl} or f_{yd}	Characteristic yield strength of longitudinal reinforcement
f_{yv} or f_{ywd}	Characteristic yield strength of link reinforcement
F	Force in tendon at point being considered
F_{bst}	Anchor bursting force
F_o	Force applied by the jack at the anchor
F_s	Force generated in top slab due to differential shrinkage
h	Overall depth of member
h_{max}	Larger dimension of the section
h_{min}	Smaller dimension of the section
h_{wo}	Web or slab thickness
HDPE	High density polyethylene
H_{stay}	Horizontal length of stay cable
I	Second moment of area of section
k	Wobble coefficient
k_i	Concrete bond coefficient
k_f	Coefficient for deck frequency
k_t	Coefficient dependent on type of tendon
K_L	Longitudinal shear coefficient
K_s	Equivalent spring stiffness for stay support

l	Lever arm at ultimate moment
l_c	Losses of stress in tendon due to creep of concrete
l_E	Losses of stress in tendon due to elastic shortening of concrete
l_f	Stress in tendon at point being considered expressed as amount below the 70% UTS level
l_r	Losses of stress in tendon due to relaxation
l_s	Losses of stress in tendon due to shrinkage of concrete
L_{stay}	Inclined length of stay
l_t or L_d	Transmission length for anchorage of pre-tensioned strand
L	Length of deck
L_e	Effective flange width
L_s	Width of longitudinal shear failure plane
L_{sp}	Length of span of deck
L_T	Free length of tendon
M	Moment at section due to ultimate loads
M_a	Moment generated by the change applied 'instantaneously'
$M_{as\text{-}built}$	Moment as constructed
M_{cr}	Cracking moment of section
M_{final}	Final moment after creep effects
$M_{inst.}$	Moment if the structure is built instantaneously
M_o	Moment necessary to produce zero stress in the concrete at the tensile face
M_p	Primary moment from prestress on section
M_r	Ultimate moment of resistance at a section
M_s	Prestress secondary moment
m_t	Mass per metre run of tendon
m_d	Mass per metre run of deck
m_1	Moment due to unit restraint moment applied at pier 1
m_2	Moment due to unit restraint moment applied at pier 2
N–A	Neutral axis of section
P	Total, unfactored prestress force acting on section
P_h	Horizontal force from prestress tendon
P_v	Vertical force from prestress tendon
p	Ratio of reinforcement, A_r/bh
S	First moment of area above and about the centroidal axis
S_v or s	Spacing of link reinforcement
S_L	Spacing of longitudinal reinforcement
S_t	Spacing of transverse reinforcement
$t_{ef,1}$	Wall thickness
T	Tensile force generated at ultimate moment
T_c	Torsional cracking moment
T_{Ed}	Design torsional moment
$T_{Rd,max}$	Design torsional resistance moment
T_s	Torsional moment due to serviceability loads
T_{stay}	Force in stay
T_u	Torsional moment due to ultimate loads
UTS	Ultimate tensile strength
v	Shear stress in the concrete due to ultimate loads
v_c	Ultimate shear stress allowed in concrete

v_L	Ultimate longitudinal shear stress in the concrete
v_t	Torsional shear stress
V	Shear force due to ultimate loads
V_c	Ultimate shear resistance of concrete at section
V_{co} or V_{cw}	Ultimate shear resistance of concrete uncracked in flexure
V_{cr} or V_{ci}	Ultimate shear resistance of concrete cracked in flexure
V_d	Shear force at section due to unfactored dead load
V_{Ed}	Design value of the applied shear force
V_k	Ultimate shear resistance of shear key
V_L	Longitudinal shear force per unit length
V_{nh}	Nominal horizontal shear strength
V_p	Vertical component of prestress
$V_{Rd,c}$	Design shear resistance of the member without shear reinforcement
$V_{Rd,s}$	Design value of the shear force which can be sustained by the yielding shear reinforcement
$V_{Rd,max}$	Design value of the maximum shear force which can be sustained by the member, limited by crushing of the compression struts
V_{Rdj}	Shear resistance of the rest of the joint area at match-cast segmental joint
V_{Rdk}	Shear resisted by the shear keys at match-cast segmental joint
V_s	Ultimate shear resistance provided by the reinforcement
x	Distance of point being considered from the tendon anchor
x_i	Smaller centreline dimension of torsion link
y or y'	Distance in section from N–A to point being considered
y_i	Larger centreline dimension of torsion link
y_o	Half length of side of anchor block
y_{po}	Half length of side of loaded area
y_t	Distance in section from N–A to tensile face
z_b	Elastic sectional modulus referred to bottom face (I/y)
z_i	Side length of wall
z_t	Elastic sectional modulus referred to top face (I/y)
γ_c	Partial factor for strength of concrete, given as 1.5 for persistent and transient situations
γ_L	Load factor
γ_m	Partial safety factor for strength
γ_{f3}	Analysis factor
γ_s	Density of stay
θ	Total angle change in the tendon over distance x in radians
μ	Friction co-efficient
$\sigma_{as-built}$	Stress in section due to construction sequence
σ_c	Stress in concrete adjacent to prestress tendon
σ_{cd}	Stress in concrete adjacent to prestress tendon due to change in dead load since tendon installed
σ_{ci}	Stress in concrete adjacent to prestress tendon at time of transfer
σ_{cp}	Stress in concrete at neutral axis due to prestress
σ_{final}	Final stresses in section
$\sigma_{inst.}$	Stress in section if built instantaneously
σ_{pm0} or σ_{pi}	Stress in tendon just after time of release
ϕ	Creep factor or nominal diameter of strand/tendon
Δ_l	Change in deck length due to Δ_t

$\Delta\sigma_{pr}$	Relaxation loss in prestress
Δ_t	Change in effective temperature
Δ_{cs}	Shrinkage strain deformation of the concrete
Δf_{pR1}	Relaxation loss in prestress up to time of casting slab
Δf_{pR2}	Relaxation loss in prestress after time of casting slab
ΔF_{td}	Additional tension force in longitudinal reinforcement due to shear force
Δ_s	Differential shrinkage stress between in situ slab and precast beam
ρ_{1000}	Value of relaxation in the prestress strand after 1000 hrs as a percentage
ξ_s	Depth factor for shear, given in Table 9 of BS 5400 Part 4
$\propto$	Coefficient of thermal expansion of the concrete per °C
γ_m	Partial safety factor for strength
Φ	Aashto strength reduction factor
ε_{cb}	Strain in concrete at bottom fibre
ε_{ct}	Strain in concrete at top fibre
ε_p	Strain in prestress tendon
ε_s	Strain in non-prestressed reinforcement
$\tau_{t,i}$	Torsional shear stress in wall

Prestressed Concrete Bridges, 2nd edition
ISBN: 978-0-7277-4113-4

ICE Publishing: All rights reserved
doi: 10.1680/pcb.41134.369

Appendix C
Bibliography, further reading and useful references

C.1. Introduction

The following references are intended as a guide for further reading as well as references for the preceding chapters. Also included is a list of website addresses which can provide further information and details of the prestressing systems and other equipment used, and of prestressed concrete bridges.

C.2. Books

Abeles PW (1949) *The Principles and Practice of Prestressed Concrete*. Crosby Lockwood, London.

Beeby AW and Narayanan RS (2009) *Designers Guide to Eurocode 2: Design of Concrete Structures*. Thomas Telford, London.

Calgaro JA, Tschumi M and Gulvanessian H (2010) *Designer's Guide to Eurocode 1: Action on Bridges*. Thomas Telford, London.

Clark LA (1983) *Concrete Bridge Design to BS5400*. Construction Press, Harlow.

England GL, Tsang NCM and Bush DL (2000) *Integral Bridges*. Thomas Telford, London.

Gulvanessian H, Calgary JA and Honicky M (2002) *Designer's Guide to EN 1990, Eurocode: Basis of Structural Design*. Thomas Telford, London.

Guyun Y (1951) *Prestress Concrete*. Edition Eyrolles, Paris.

Hambly EC (1991) *Bridge Deck Analysis*. 2nd Edition. Chapman and Hall, London. Highways Agency (1996) *The Appearance of Bridges and Other Highway Structures*. HMSO, London.

Kong FK and Evans RH (1987) *Reinforced and Prestressed Concrete*. 3rd Edition. E & FN Spon, Wokingham.

Lee DJ (1994) *Bridge Bearings and Expansion Joints*. 2nd Edition. E & FN Spon, London.

Magnel G (1950) *Prestressed Concrete*. Concrete Publications, London.

Neville AM (1995) *Properties of Concrete*. 4th edition. Longman, London.

Parke GAR and Hewson NR (eds) (2008) *ICE Manual of Bridge Engineering*. 2nd Edition. Thomas Telford, London.

Pennells E (1978) *Concrete Bridge Designers Manual*. 1st Edition. Cement and Concrete Association, London.

Podolny W and Muller JM (1982) *Construction and Design of Prestressed Concrete Segmental Bridges*. Wiley, New York.

Prichard B (1992) *Bridge Design for Economy and Durability*. Thomas Telford, London.

Prichard B (ed.) (1994) *Continuous and Integral Bridges*. 1st Edition. E & FN Spon, London.

Pucher A (1976) *Influence Surfaces of Elastic Plates*. 5th Edition. Springer-Verlag, Wien.

Ramberger G (2002) *Structural Bearings and Expansion Joints for Bridges*. International Association for Bridge and Structural Engineering, Zurich.

Rosignoli M (2002) *Bridge Launching.* Thomas Telford, London.

Schlaich J and Scheef H (1982) *Concrete Box-Girder Bridges.* International Association for Bridge and Structural Engineering, Zurich.

Sutherland J, Humm D and Chrimes M (eds) (2001) *Historic Concrete – The Background to Appraisal.* London, Thomas Telford.

Troitsky MS (1988) *Cable-Stayed Bridges.* 2nd Edition. Blackwell, Oxford.

Walther R, Houriet B, Isler W and Moia P (1988) *Cable-Stayed Bridges.* Thomas Telford, London.

C.3. Design guides and technical reports

Bragg SL, Ahm P, Bowen FM, Champion S, Kemp LC, Mott JCS *et al.* (1975) *Final Report of the Advisory Committee on Falsework.* HMSO. London.

Comité Euro-International Du Béton (1989) *Durable Concrete Structures.* 2nd Edition. Thomas Telford.

Concrete Society (1996) *Technical Report No. 47. Durable Bonded Post-Tensioned Concrete Bridges.* Concrete Society, Slough.

Concrete Society (2002) *Technical Report no. 47. Durable bonded post-tensioned concrete bridges.* 2nd Edition. Concrete Society, Slough.

Concrete Society and IStructE. (1971) *Technical Report TRCS 4, Falsework.* Concrete Society, London.

Ciria (1976) *Guide No. 1. A guide to the design of anchor blocks for post-tensioned prestressed concrete members.* Ciria, London.

Ciria (1977) *Guide No. 2. The design of deep beams in reinforced concrete.* Ciria, London.

Ciria (2001) *Bridge detailing guide.* Ciria, London.

Ciria (1985) *Report 106. Post-tensioning systems for concrete in the UK: 1940–1985.* Ciria, London.

Ciria (1996) *Report 155. Bridges – design for improved buildability.* Ciria, London.

Emerson M (1977) TRRL Laboratory report 765: Temperature difference in bridges: Basis of design requirements. Transport and Road Research Laboratory, Crowthorne.

Fédération International de la Préconstraint (FIP) (1990) *Guide to Good Practice, Grouting of Tendons in Prestressed Concrete.* Thomas Telford, London.

Hay J (1992) *Response of Bridges to Wind. TRL State-Of-The-Art Review 5.* HMSO, Norwich.

Highways Agency and TRL (1999) *Post-tensioned concrete bridges Anglo-French liaison report.* Thomas Telford, London.

Koseki K and Breen JE (1983) *Exploratory study of shear strength of joints for precast segmental bridges.* Centre for Transportation Research, The University of Texas at Austin, Research report 248–1, September.

MacGregor RJG, Kreger ME and Breen JE (1989) *Strength and ductility of a three-span externally post-tensioned segmental box girder bridge model.* Centre for Transportation Research, The University of Texas at Austin. Research report 365–3F, January.

Maguire JR and Wyatt TA *Dynamics, An Introduction for Civil and Structural Engineers.* ICE design and practice guides, Thomas Telford, the Wind Engineering Society, and SECED.

Maisel BI and Roll F (1974) *Methods of Analysis and Design of Concrete Box Beams with Side Cantilevers.* C&CA, London.

Nicholson BA (1997) *Simple Bridge Design Using Prestressed Beams.* Prestressed Concrete Association, Leicester.

Powell LC, Breen JE and Kreger ME (1988) *State of the art externally post-tensioned bridges with deviators.* Centre for Transportation Research, The University of Texas at Austin, Research report 365–1, June.

TRRL (1977) *Laboratory Report 765, Temperature difference in bridges: basis of design requirements*. Transport and Road Research Laboratory, Crowthorne.

West R (1973) *Recommendations on the use of grillage analysis for slab and pseudo-slab bridge decks*. C&CA/Ciria, London.

C.4. Articles

Aalami BO (1990) Load Balancing: A comprehensive solution to post-tensioning. *ACI Structural Journal* **87(6)**: 162–170.

Al-Qarra CH (1999) The Dee Estuary Bridge – control of geometry during construction. *Proceedings of the Institution of Civil Engineers, Civil Engineering* **132**: 31–39, doi: 10.1680/icien.1999.31228.

Baxter JW, Gee AF and James HB (1965) Gladesville Bridge. *Proceedings of the Institution of Civil Engineers* **30**: 489–530, doi: 10.1680/iicep.1965.9523.

Beaney NJ and Martin JM (1993) Design and construction of the Dornock Firth Bridge: construction. *Proceedings of the Institution of Civil Engineers* **100(3)**: 145–156, doi: 10.1680/itran.1993.24296.

Brockman C and Rogenhofer H (2000) Bang Na Expressway, Bangkok, Thailand – world's longest bridge and largest precasting operation. *PCI Journal* **45(1)**: 26–38.

Burgoyne CJ and Stratford TJ (2001) Lateral instability of long-span prestressed concrete beams on flexible bearings. *The Structural Engineer* **79(6)**: 23–26.

Buyukozturk O, Bakhoun MM and Beattie M (1990) Shear Behavior of Joints in Precast Concrete Segmental Bridges. *Journal of Structural Engineering* **116(12)**: doi: 10.1061/(ASCE)0733-9445(1990)116:12(3380).

Catchick BK (1978) Prestress analysis for continuous beams: some developments in the equivalent load method. *The Structural Engineer* **2(56B)**: 29–36.

Clark LA (1984) Longitudinal shear reinforcement in beams. *Concrete* **18(2)**: 22–23.

Garrett RJ and Cochrane RA (1970) The analysis of prestressed concrete beams curved in plan with torsional restraints at the supports. *The Structural Engineer* **3**: 128–32.

Gallaway TM (1980) Design features and prestressing aspects of long key bridge. *PCI Journal* **25(6)**: 84–96.

Gee A (1991) Concrete fin-back bridge in USA. *Proceedings of the Institution of Civil Engineers, Part 1* **90**: 91–122.

Hewson NR (1992) The use of dry joints between precast segments for bridge decks. *Proceedings of the Institution of Civil Engineers, Civil Engineering* **92**: 177–184, doi: 10.1680/icien.1992.21499.

Hewson NR (1993) The use of external tendons for the Bangkok Second Stage Expressway. *The Structural Engineer* **71**: 412–415.

Inversen N, Faulds JR and Rowley F (1993) Design and construction of the Dornock Firth Bridge: design. *Proceedings of the Institution of Civil Engineers*, Transp., **100**: 133–144.

Lewis CD, Robertson AI and Fletcher MS (1983) Orwell Bridge – design. *ICE Proceedings, Part 1* 19831, **74(4)**: 765–778, doi: 10.1680/iicep.1983.1363.

Lopes SMR and Do Carmo RNF (2002) Bond of prestressed strands to concrete: transfer rate and relationship between transmission length and tendon draw-in. *Structural Concrete* **3(3)**: 117–126.

Magura DD, Sozen MA and Siess CP (1964) A Study of Stress Relaxation in Prestressing Reinforcement. *PCI Journal* **9(2)**.

Moreton AJ (1990) Segmental bridge construction in Florida: a review and perspective. *ICE Proceedings*, Part 1, **88(3)**: 381–419, doi: 10.1680/iicep.1990.6842.

Muller J (1980) Construction of Long Key Bridge. *PCI Journal* **25(6)**: 97–111.

Podolny W and Mireles AA (1983) Kuwait Bubiyan Bridge – a 3-D precast segmental space frame. *PCI Journal* **Jan–Feb**: 68–107.

Rabbat BG and Sowlat K (1987) Testing of segmental concrete girders with external tendons. *PCI Journal* **March–April**: 86–107.

Rawlinson J and Stott PF (1962) The Hammersmith Flyover. *Proceedings of the Institution of Civil Engineers* **23**: 565–624, doi: 10.1680/iicep.1962.10813.

Rowley F (1993) Incremental launch bridges: UK practice and some foreign comparisons. *The Structural Engineer* **71(7)**: 111–116.

Smith LJ and Wood R (2001) Grouting of external tendons – a practical perspective. *Proceedings of the Institution of Civil Engineers* **1(146)**: 93–100.

Smith WJR, Benaim R and Hancock CJ (1980) Tyne and Wear Metro: Byker Viaduct. *Proceedings of the Institution of Civil Engineers, Part 1* **68**: 701–718.

Sriskandan K (1989) Prestressed concrete road bridges in Great Britain: a historical survey. *Proceedings of the Institution of Civil Engineers, Part 1* **86(4)**: 269–303.

Taylor HPJ (1998) The precast concrete bridge beam: the first 50 years. *The Structural Engineer* **76(21)**: 407–414.

Van Leonen J and Telford S (1983) Orwell Bridge – construction. *Proceedings of the Institution of Civil Engineers, Part 1* **(74)**: 779–804, doi: 10.1680/iicep.1983.1364.

Witecki AA (1969) Simplified method for the analysis of torsional moment as an effect of a horizontally curved multispan continuous bridge. *American Concrete Institute, First International Symposium on Concrete Bridge Design:* 193–204.

Wood RH (1968) The reinforcement of slabs in accordance with a pre-determined field of moments. *Concrete* **February**: 69–76.

Woodward RJ and Williams FW (1998) Collapse of Ynys-y-Gwas Bridge, West Glamorgan. *Proceedings of the Institution of Civil Engineers, Part 1,* **85**: 635–669.

C.5. Standards, Codes of Practice and Specifications
Eurocodes

BSI (1991) BS EN 1991: Eurocode 1: Actions on structures. BSI, London.

BSI (2002a) EN 1990:2002: Eurocode 0: Basis of structural design, incorporating amendment no. 1. BSI, London.

BSI (2002b) NA to BS EN 1990:2002: UK national annex for Eurocode 0: Basis of structural design. BSI, London.

BSI (2003) BS EN 1991–1–5:2003: Eurocode 1 incorporating corrigendum no. 1: Actions on structure. Part 1–5. General actions – thermal actions. Incorporating Corrigendum No. 1. BSI, London.

BSI (2004a) BS EN 1992–1–1:2004: Eurocode 2: Design of concrete structures. Part 1–1. General rules and rules for buildings. BSI, London.

BSI (2004b) NA to BS EN 1992–1–1:2004: UK national annex to Eurocode 2: Design of concrete structures. Part 1–1. General rules and rules for buildings. BSI, London.

BSI (2004–2005) BS EN 1992: Eurocode 2: Design of concrete structures. BSI, London.

BSI (2005a) BS EN 1992–2:2005: Eurocode 2: Design of concrete structures. Part 2. Concrete bridges – design and detailing rules. BSI, London.

BSI (2005b) NA to BS EN 1992–2:2005: UK national annex to Eurocode 2: Design of concrete structure. Part 2. Concrete bridges – Design and detailing rules. BSI, London.

BSI (2007a) BS EN 445:2007: Grout for Prestressing Tendons – Test methods. BSI, London.

BSI (2007b) BS EN 446:2007: Grout for Prestressing Tendons – Grouting procedures. BSI, London.

BSI (2007c) BS EN 447:2007: Grout for Prestressing Tendons – Basic requirements. BSI, London.

BSI (2000) BS EN 197–1:2000: Cement. Composition, specifications and conformity criteria for low heat common cements. BSI, London.

BSI (2007) BS EN 447: 2007: Grout for prestressing tendons – specification for common grout. BSI, London.

BSI (2003) BS EN 523:2003: Steel Sheaths for tendons-terminology, requirements, quality control. BSI, London.

The Highways Agency. Departmental Standards (DMRB). HMSO, Norwich

Highways Agency (2009) *Manual of Contract Documents for Highway Works, Volume 1: Specification for Highway Works.* HMSO, Norwich.

Highways Agency (1992) (DMRB) Departmental Standard, BD 20/92. Bridge bearings. Use of BD 5400: Part 9:1983. HMSO, Norwich.

Highways Agency (1992) (DMRB) Departmental Standard, BD 24/92. Design of concrete bridges. Use of BS 5400: Part 4:1990. HMSO, Norwich.

Highways Agency (1987) (DMRB) Departmental Standard, BD 28/87. Early thermal cracking of concrete. HMSO, Norwich.

Highways Agency (1994) (DMRB) Departmental Standard, BD 33/94. Expansion joints for use in highway bridge decks. HMSO, Norwich.

Highways Agency (2001) (DMRB) Departmental Standard, BD 37/01. Loads for highway bridges. HMSO, Norwich.

Highways Agency (1999) (DMRB) Departmental Standard, BD 47/99. Waterproofing and surfacing of concrete bridge decks. HMSO, Norwich.

Highways Agency (1993) (DMRB) Departmental Standard, BD 52/93. The design of highway bridge parapets. HMSO, Norwich.

Highways Agency (2001) (DMRB) Departmental Standard, BD 57/01. Design for durability. HMSO, Norwich.

Highways Agency (1994) (DMRB) Departmental Standard, BD 58/94. The design of concrete highway bridges and structures with external and unbonded prestressing. HMSO, Norwich.

The Highways Agency. Advice Notes. HMSO, Norwich

BA 24/87 Early Thermal Cracking of Concrete (including amendment No. 1 (1989))

BA 26/94 Expansion Joints for Use in Highway Bridge Decks

BA 36/90 The Use of Permanent Formwork

BA 41/98 The Design and Appearance of Bridges

BA 42/96 The Design of Integral Bridges

BA 48/99 Waterproofing and Surfacing of Concrete Bridge Decks

BA 57/01 Design for Durability

BA 58/94 Design of Bridges and Concrete Structures with External Unbonded Prestressing

British Standards

BSI (1959) CP 115: 1959. The structural use of prestressed concrete in building. BSI, London.

BSI (1973) BS 4447:1973. Specification for the performance of prestressing anchorages for post-tensioned construction. BSI, London.

BSI (1978) BS 5400:1978. Steel, concrete and composite bridges, part 4. Code of practice for the design of concrete bridges. BSI, London.

BSI (1980a) BS 4486:1980. Hot rolled and hot rolled and processed high tensile alloy steel bars for the prestressing of concrete. BSI, London.

BSI (1980b) BS 5896: 1980. High tensile steel wire and strand for the prestressing of concrete.
BSI (1990) BS 5400:1990. Steel, concrete and composite bridges.
  part 1. 1988. General statement.
  part 2. 1990. Specification for loads.
  part 4. 1990. Code of practice for the design of concrete bridges.
  part 9. 1983. Bridge Bearings.
BSI (1996) BS12:1991: Specification for Portland cement. BSI, London
BSI (2006) BS 5400: Steel, concrete and composite bridges, part 2. Specification for loads. BSI, London.

American Standards

AASHTO (1931–1999) Standard Specification for Highway Bridges. AASHTO, Washington, DC.
AASHTO (1994) LRFD Specification for Design of Highway Bridges. AASHTO, Washington, DC.
AASHTO (1996–2002) Standard Specification for Highway Bridges, 16th Edition. AASHTO, Washington, DC.
AASHTO (1999) Guide specification for design and construction of segmental concrete bridges. 2nd edition. AASHTO, Washington, DC.
AASHTO (2011) Load and resistance factor design (LRFD) bridge design specifications. 4th edition. AASHTO, Washington, DC.
ASTM (1999) ASTM A416/A416M-10 Standard Specification for Steel Strand, Uncoated Seven-Wire for Prestressed Concrete. ASTM, Pennsylvania.
ASTM (2003) ASTM A722/A722M-98 (2003) Standard Specification for Uncoated High Strength Steel Bar for Prestressing Concrete. ASTM, Pennsylvania.
ASTM (1998) ASTM A722M-98. Standard specification for uncoated steel bar for prestressing concrete. ASTM, Pennsylvania11.

Others

CEB-FIP (1993) Model code 1990. Comité Euro-International du Béton. Thomas Telford, London.

C.6. Websites

Prestressing systems

BBR, www.bbrnetwork.com – Details of prestressing systems. General descriptions of bridge construction and project data.
DYWIDAG, www.dywidag-systems.com – General website with comprehensive details of the DYWIDAG prestressing system.
Freyssinet, www.freyssinet.com – General website with description of products as well as project data.
Macalloy Bar Systems, www.macalloy.com – General website with details of Macalloy bars.
MK4, www.mekano4.com – Description of different forms of bridge construction. Brochures for prestressing system and bearings can be downloaded.
VSL, www.vsl.com – General website with details of different construction activities. Details of projects and services.

Material and products

Prestressing wire and strand

Carrington Wire, www.carringtonwire.com – Descriptions of wires available
Trefileurope, www.arcelormittalna.com – Details of wires and strands
Tycsa, www.tycsa.com – Comprehensive data on wire and strands available for bridgeworks
Bridon Wire, www.bridonltd.com – Details of wires for bridges

Precast Beams

Tarmac Building Products Ltd, www.tarmacbuildingproducts.co.uk – General website with reference to precast beams

Redlands, www.redlandprecast.com.hk – General website with bridge beams and precast parapets

Bearings, expansion joints and drainage

(See also companies and websites under prestressing systems above)

Bridge Joint Association, www.bridgejoints.org.uk – Details and drawings of different expansion joints used on bridges

Britflex, www.usluk.com – Details of bridge expansion joints and waterproofing systems

Bowman, www.bowmanconstructionsupply.com – Information sheets on a range of bearings and expansion joints

D.S. Brown, www.dsbrown.com – Downloadable details of expansion joints and bearings

Ekspan, www.ekspan.co.uk – Details of bearings, expansion joints and drainage

Mageba SA, www.mageba.ch – General website with examples of projects

Maurer Söhne, www.maurer-soehne.de – General description of services provided

Maclellan Rubber, www.maclellanrubber.com – Details of rubber bearings

Prismo, www.prismo.co.uk – Description of Thorma joints

Proceq SA, www.proceq.com – Details of bridge bearings and expansion joints

RJ Watson, www.rjwatson.com – Details of Disktron bearings

Parapets

Bridge Parapets Ltd, www.bridgeparapets.com – General website with reference to parapets

Varley and Gulliver Ltd, www.v-and-g.co.uk – General website with reference to parapets

HBS Ltd, www.hbsonline.co.uk – General website with reference to parapets

Deck waterproofing

ASL Contracts, www.aslcontracts.co.uk – Details of Servideck and Bridgeguard systems

Pitchmastic PmB, www.pitchmasticpmb.co.uk – Specialist services to the civil engineering and construction industries

Grace, www.grace-construction.com – Details of Bituthen waterproofing membrane

Stirling Lloyd, www.stirlinglloyd.com – General website with details of Eliminator(K) waterproofing system

Universal Sealants, www.usluk.com – Details of Britflex waterproofing system

Formwork and falsework

Peri Ltd, www.peri.ltd.uk – Descriptions of formwork and falsework systems with examples of projects

SGB Formwork, www.sgb.co.uk – General website listing services offered

Thyssen Hünnebeck, www.thyssen-huennebeck.com – Description of formwork and falsework systems

RMD, www.rmdformwork.com – General website listing services offered

Doka, www.doka.com – General website with descriptions of formwork and falsework systems

BCM GRC Ltd, www.bmcgrc.com – General website with some information on permanent formwork panels

Gleitbau Salzburg, www.gleitbau.com – General website with slipping forming and jump form examples

Erection gantries and construction equipment

Deal, www.deal.it – Description of erection equipment, moulds and formwork, with examples of projects

NRS, www.nrsas.com – Description of gantries, form travellers and movable scaffolding with comprehensive library of pictures

Paolo de Nicola S.p.A, www.paolodenicola.com – Examples of erection equipment

Cimolai S.p.A, www.cimolai.com – Examples of erection equipment

Bridge access systems

MOOG, www.moog-online.de – Description and examples of bridge inspection equipment

Barin S.p.A, www.barin.it – Technical details and sketches of bridge inspection equipment

Institutions, Societies and Research Bodies

American Concrete Institute (ACI), www. aci-int.org – General website listing activities and publications

American Segmental Bridge Institute, www.asbi-assoc.org – Details of publications and activities. Drawings of standard bridges

British Cement Association, www.bca.org.uk – General website of activities, publications and services

Concrete Bridge development Group, www.cbdg.org.uk – General website of activities

The Concrete Society, www.concrete.org.uk – General website of activities and publications

fédération international du béton, http://fib.epfl.ch – Details of activities, publications and industry news

IABSE, www.iabse.ethz.ch – Information and references on bridges and associated topics

Institution of Civil Engineers, www.ice.org.uk – Extensive list of library references

Institution of Structural Engineers, www.istructe.org.uk – Extensive list of library references

Post-Tensioning Institute, www.post-tensioning.org – General website of the Institutes activities

Precast/Prestressed Concrete Institute, www.pci.org – Details of publications, news and general information on precast and prestressed concrete

Portland Cement Association, http://www.cement.org/ – www.cement.org/bridges/br_resources.asp contains an extensive list of USA references on concrete bridges

Prestressed Concrete Bridges, 2nd edition
ISBN: 978-0-7277-4113-4

ICE Publishing: All rights reserved
doi: 10.1680/pcb.41134.377

Appendix D
Proprietary systems

D.1. Introduction

This appendix gives details of several of the major prestressing systems that are available and can be used as a guide when designing and detailing the prestressing arrangements. All the details published here are with the kind permission of the companies noted.

As well as these systems there are other similar products available in the different local markets around the world.

For any particular project, the details of the systems available locally should be confirmed before commencing the detailed design.

D.2. Multi-strand systems

Most of the major prestressing suppliers produce a multi-strand prestressing system. VSL, Freyssinet, DYWIDAG, BBR and Mekano4 all produce systems that have similar characteristics.

- Use of 13 mm (0.5′) or 15 mm (0.6′) diameter strands.
- Standardised tendon anchorages and couplers for use with up to 55 strands.
- Selection of anchorage types.
- Steel or polyethylene ducts.
- Grouting with cement mortar or other materials.
- All strands in a tendon stressed simultaneously, but with each strand individually anchored.
- Stressing in steps.

Details of the VSL system are given in Figures D1 to D7. The typical tendon units used and the associated steel duct sizes for internal tendons are given in Figure D1.

For enhanced corrosion protection and improved fatigue resistance of the tendons the VSL PT-PLUS™ system, or similar, with corrugated polyethylene ducts is adopted. Details of the VSL PT-PLUS™ system are given in Figure D2.

VSL use a variety of anchorage arrangements with a typical detail for their live end anchorage type EC given in Figure D3 and the dead-end anchorage type H given in Figure D4. Details of two coupler arrangements are given in Figure D5. Details of VSL's jacks, anchor pockets and clearances for the multi-strand tendons are given in Figure D6 and D7.

Figure D1 VSL tendon and steel duct sizes

13 mm (0.5″) strand				
			Min. breaking load	
Tendon unit	Number of strands	Duct diameter internal/external (mm)	Strand type Euronorm 138-79 or BS 5896:1980, Super (kN)	Strand type ASTM A416-85 Grade 270 (kN)
5-1	1	25/30	186	184
5-2	2	40/45	372	367
5-3	3	40/45	558	551
5-4	4	45/50	744	735
5-6	6	50/55	1116	1102
5-7	7	55/60	1302	1286
5-12	12	65/72	2232	2204
5-18	18	80/87	3348	3307
5-19	19	80/87	3534	3490
5-22	22	85/92	4092	4041
5-31	31	100/107	5786	5695
5-37	37	120/127	6882	6797
5-43	43	130/137	7998	7899
5-55	55	140/150	10 230	10 104

15 mm (0.6″) strand				
			Min. breaking load	
Tendon unit	Number of strands	Duct diameter internal/external (mm)	Strand type Euronorm 138-79 or BS 5896:1980, Super (kN)	Strand type ASTM A416-85 Grade 270 (kN)
6-1	1	30/35	265	261
6-2	2	45/50	530	521
6-3	3	45/50	795	782
6-4	4	50/55	1060	1043
6-6	6	60/67	1590	1564
6-7	7	60/67	1855	1825
6-12	12	80/87	3180	3128
6-18	18	95/102	4770	4693
6-19	19	95/102	5035	4953
6-22	22	110/117	5830	5735
6-31	31	130/137	8215	8082
6-37	37	140/150	9805	9646
6-43	43	150/160	11 395	11 210
6-55	55	170/180	14 575	14 339

Figure D2 VSL PT-PLUS^{TM} duct sizes

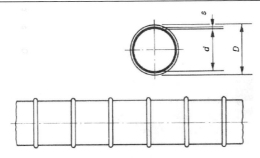

Strand type 13 mm (0.5″)	Strand type 15 mm (0.6″)	Dimensions of duct (mm)		
Tendon unit	Tendon unit	d	D	s
5-12	6-7	59	73	2
5-19	6-12	76	91	2.5
5–31	6-19/6-22	100	116	3
5-55	6-37	130	146	3

Figure D3 VSL anchorage type EC

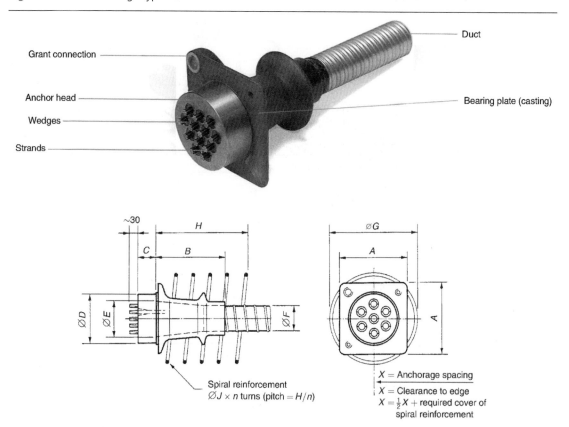

Tendon unit	A	B	C	ØD	ØE	ØF internal/external	ØG	H	ØJ	n	X
Strand type 13 mm (0.5″)											
5-3	120	130	50	90	50	40/45	130	150	10	3	155
5-4	135	125	50	95	55	45/50	160	150	10	3	180
5-7	165	155	55	110	74	55/60	205	200	12	4	235
5-12	215	215	60	150	104	65/72	285	250	14	5	305
5-19	270	285	75	180	135	80/87	365	300	16	6	385
5-22	290	335	85	190	150	85/92	395	360	18	6	415
5-31	340	365	95	230	172	100/107	470	400	18	8	490
5-37	370	360	105	240	188	120/127	510	420	20	7	535
5-55	430	460	130	290	230	140/150	620	540	22	9	655
Strand type 15 mm (0.6″)											
6-3	135	125	50	95	55	45/50	160	150	10	3	185
6-4	150	155	55	110	65	50/55	190	200	12	4	210
6-7	190	170	60	135	84	60/67	260	250	14	5	280
6-12	250	245	75	170	118	80/87	345	300	16	6	365
6-19	310	305	95	200	150	95/102	440	350	18	7	460
6-22	340	365	100	220	172	110/117	470	400	18	8	495
6-31	390	350	120	260	192	130/137	560	480	20	8	590
6-37	430	450	135	280	215	140/150	610	540	22	9	640
6-55	520	530	160	340	255	170/180	740	630	26	9	780

Figure D4 VSL Dead End anchorage type H

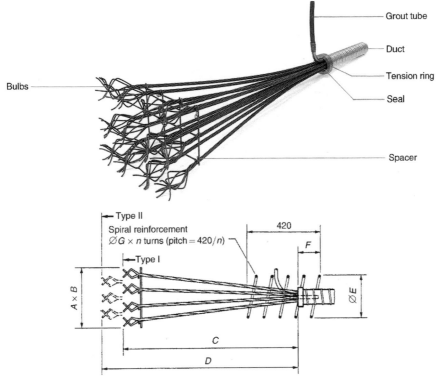

The force in the tendon is transferred by bond of the strands in the concrete and by the bulbs at the end of the strands

Strand type 13 mm (0.5″)										
Tendon unit	Alter-native	Type	A	B	C	D	ØE	F	ØG	n
5-3	1	I	230	70	930	–	–	–	–	–
5-4	1	I	310	70	930	–	–	–	–	–
	2	I	150	170	930	–	–	–	–	–
5-7	1	II	370	70	1130	1280	180	155	12	7
	2	II	170	190	1130	1280	180	155	12	7
5-12	1	II	350	190	1130	1280	200	155	14	7
	2	I	310	270	1130	–	200	155	14	7
5-19	1	II	470	190	1130	1280	230	155	14	7
	2	II	310	390	1130	1280	230	155	14	7
5-22	1	II	570	190	1130	1280	300	155	16	7
	2	II	390	390	1130	1280	300	155	16	7
5-31	1	II	670	310	1330	1480	350	155	16	7
	2	II	470	430	1330	1480	350	155	16	7
5-37	1	II	770	310	1530	1680	350	165	18	7
	2	II	470	550	1530	1680	350	165	18	7
5-43	1	II	870	350	1530	1680	400	165	18	7
	2	II	670	430	1530	1680	400	165	18	7
	3	II	570	550	1530	1680	400	165	18	7
5-55	1	II	1170	350	1830	1980	400	175	20	7
	2	II	870	430	1830	1980	400	175	20	7
	3	II	570	670	1830	1980	400	175	20	7

Strand type 15 mm (0.6″)										
Tendon unit	Alter-native	Type	A	B	C	D	ØE	F	ØG	n
6-3	1	I	290	90	950	–	–	–	–	–
6-4	1	I	390	90	950	–	–	–	–	–
	2	I	190	210	950	–	–	–	–	–
6-7	1	II	450	90	1150	1300	200	155	14	7
	2	II	210	230	1150	1300	200	155	14	7
6-12	1	II	430	230	1150	1300	230	155	14	7
	2	I	390	330	1150	–	230	155	14	7
6-19	1	II	570	230	1150	1300	300	155	16	7
	2	II	390	470	1150	1300	300	155	16	7
6-22	1	II	690	230	1150	1300	350	155	16	7
	2	II	470	490	1150	1300	350	155	16	7
6-31	1	II	810	260	1550	1700	400	165	18	7
	2	II	570	510	1550	1700	400	165	18	7
6-37	1	II	1050	370	1850	2000	400	175	20	7
	2	II	690	510	1850	2000	400	175	20	7
6-43	1	II	1050	370	1850	2000	400	175	20	7
	2	II	810	510	1850	2000	400	175	20	7
	3	II	690	650	1850	2000	400	175	20	7
6-55	1	II	1410	370	2150	2300	460	185	22	8
	2	II	1050	510	2150	2300	460	185	22	8
	3	II	690	790	2150	2300	460	185	22	8

Figure D5 VSL couplers types K and V

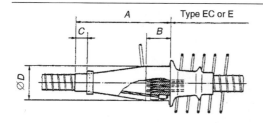

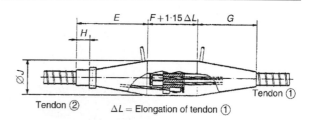

ΔL = Elongation of tendon ①

Coupler type K				
Tendon unit	A	B	C	⌀D
Strand type 13 mm (0.5″)				
5-3	430	140	40	130
5-7	550	140	60	170
5-12	650	140	65	200
5-19	740	140	80	240
5-22	830	140	90	260
5-31	1140	140	95	350
5-37	1320	180	125	390
5-55	1370	200	150	420
Strand type 15 mm (0.6″)				
6-2	380	150	30	130
6-3	490	160	60	150
6-4	520	160	60	160
6-7	630	160	75	190
6-12	730	160	80	240
6-19	860	160	95	280
6-22	930	160	95	310
6-31	1090	180	135	360
6-37	1390	200	135	430

Coupler type V				
Tendon unit	E	F	G	H	⌀J
Strand type 13 mm (0.5″)					
5-3	290	175	230	40	130
5-7	410	175	330	60	170
5-12	510	175	420	65	200
5-19	600	175	490	80	240
5-22	690	175	570	90	260
5-31	1000	175	870	95	350
Strand type 15 mm (0.6″)					
6-2	230	200	185	30	130
6-3	330	205	250	60	150
6-4	360	205	280	60	160
6-7	470	205	370	75	190
6-12	570	205	460	80	240
6-19	700	205	570	95	280

Figure D6 VSL Jack details

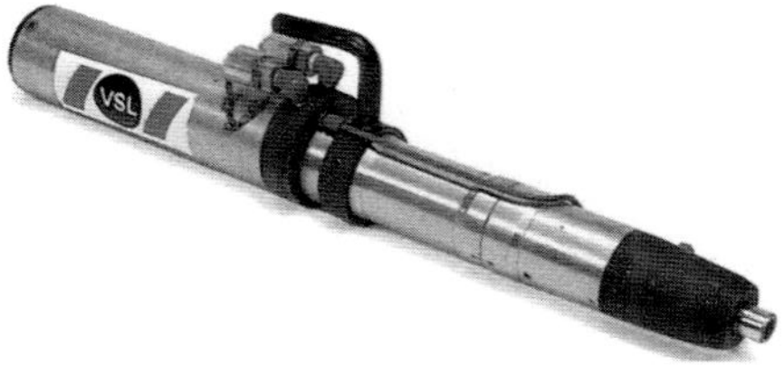

	Type I (ZPE-23 FJ)				Type II (ZPE-19)					Type III (ZPE-500J)			
Designation	ZPE-23FJ	ZPE-30	ZPE-3	ZPE-60	ZPE-7/A	ZPE-12/St2	ZPE-200	ZPE-19	ZPE-460/31	ZPE-500	ZPE-750	ZPE-1000	ZPE-1250
Type	I	III	III	III	III	II	III	II	II	III	II	III	II
Length (mm)	790	720	475	615	690	550	960	750	580	1000	1185	1200	1290
Diameter (mm)	116	140	200	180	280	310	315	390	485	550	520	790	620
Stroke (mm)	200	250	160	250	160	100	300	100	100	200	150	200	150
Piston area (cm^2)	47.10	58.32	103.6	126.4	203.6	309.4	325.7	500.3	804.0	894.6	1247.0	1809.5	2168.0
Capacity (kN)	230	320	500	632	1064	1850	2000	2900	4660	5000	7500	10000	12500
(bar)	488	549	483	500	523	598	614	580	580	559	601	553	577
Weight (kg)	23	28	47	74	115	151	305	294	435	1064	1100	2290	1730
Used for 13 mm/0.5″ tendon types	5-1	5-1	5-2, 5-3	5-2 to 5-4	5-6, 5-7	5-12	5-12	5-18, 5-19	5-22, 5-31	5-22, 5-31	5-31, 5-37	5-37 to 5-55	5-37 to 5-55
Used for 15 mm/0.6″ tendon types	6-1	6-1	6-2	6-2, 6-3	6-4	6-6, 6-7	6-6, 6-7	6-12	6-18, 6-19	6-18 to 6-22	6-31	6-31 to 6-43	6-31 to 6-55

Figure D7 VSL anchorage block out and jack clearances

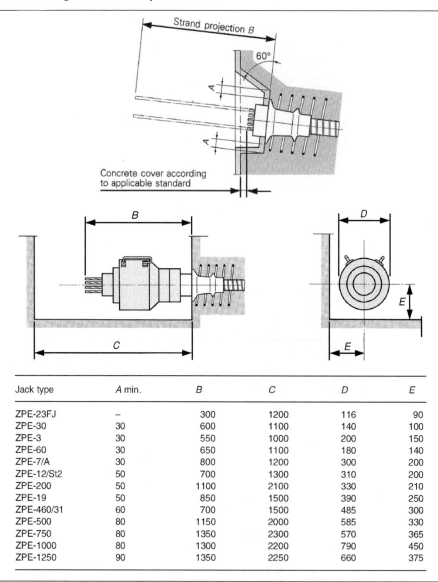

Jack type	A min.	B	C	D	E
ZPE-23FJ	–	300	1200	116	90
ZPE-30	30	600	1100	140	100
ZPE-3	30	550	1000	200	150
ZPE-60	30	650	1100	180	140
ZPE-7/A	30	800	1200	300	200
ZPE-12/St2	50	700	1300	310	200
ZPE-200	50	1100	2100	330	210
ZPE-19	50	850	1500	390	250
ZPE-460/31	60	700	1500	485	300
ZPE-500	80	1150	2000	585	330
ZPE-750	80	1350	2300	570	365
ZPE-1000	80	1300	2200	790	450
ZPE-1250	90	1350	2250	660	375

D.3. External tendons systems

The main differences between external and internal multi-strand tendons are the anchor arrangement and the use of HDPE ducts as described in sections 2.6 and 2.8. Again, all the major prestress suppliers have developed suitable arrangements for external tendons, with two typical VSL anchorage details given in Figures D8 and D9. The tendon and ducts are effectively debonded from the anchor trumpet and surrounding concrete to allow future replacement if necessary.

Figure D8 VSL External stressing anchorage type Ed

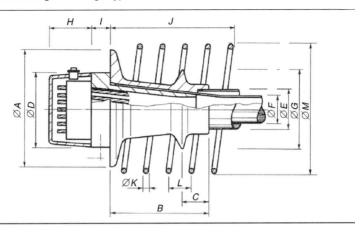

	Dimensions													
Tendon unit	A	B	C	D	E	F	G	H	I	J	K	L	M	N
Strand type 0.6″														
6-7	250	245	60	175	95	75	190	100	45	345	16	55	330	6
6-12	310	305	65	225	115	90	235	115	45	235	18	60	400	7
6-19	390	350	70	275	145	110	295	135	45	540	22	65	510	8
6-31	430	450	80	305	155	125	320	160	55	605	22	65	550	9
6-37	520	530	90	320	183	140	420	170	55	745	26	80	640	9

Figure D9 VSL External stressing anchorage type Edm

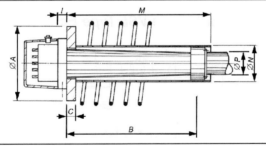

	Dimensions						
Tendon unit	A	B min	C	I	M	N	P
Strand type 0.6″							
6-1	80	35	10	35	According to the dimensions of the work at the anchorage	According to the dimensions of the opening at the anchorage	According to the material and dimensions of the duct
6-2	120	155	15	35			
6-3	150	155	25	35			
6-4	170	145	25	45			
6-7	220	145	35	45			
6-12	300	325	45	45			
6-19	370	485	55	45			
6-22	400	605	60	45			
6-31	470	635	75	55			
6-37	520	775	80	55			

D.4. Flat slab systems

Used in thin slabs, this multi-strand tendon arrangement utilises either 13 mm or 15 mm diameter strands. They are installed inside flat ducts to take up the minimum space and to be able to achieve the maximum eccentricity for the strand within the slab. Details of the Freyssinet flat slab system are shown in Figure D10. VSL, DYWIDAG, BBR and Mekano4 also provide similar flat slab systems.

Figure D10 Freyssinet flat slab tendons

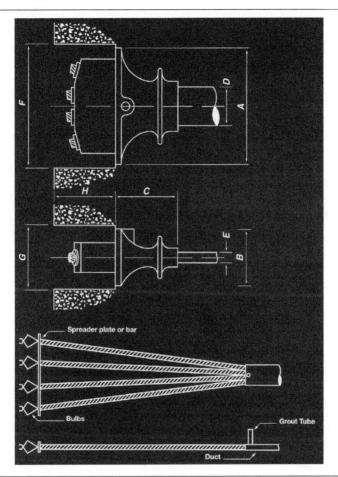

Tendon reference	No. of strands	Strand type	Total ultimate tensile strength kN	80% UTS	60% UTS	Transfer strength of concrete N/mm^2
4S13	4	12.9 Super	744	595	446	25
4S13D	4	12.7 Drawn	836	669	502	25
4S15	4	15.7 Super	1060	848	636	25
4S15E	4	15.7 Euro	1116	893	670	28
4S15D	4	15.2 Drawn	1200	960	720	33
5S13	5	12.9 Super	930	744	558	25
5S15	5	15.7 Super	1325	1060	795	25

	Anchor			Duct		Pocket		
Anchor ref. Dimension	Width A	Depth B	Length C	Width D	Depth E	Width F	Depth G	Length H
4S13	195	85	95	75	20	215	90	90
4S15	230	100	110	75	20	230	110	120
5S13	220	75	215	70	19	260	100	100
5S15	260	80	270	90	19	360	100	100

D.5. Bar systems

DYWIDAG, VSL, Freyssinet and McCalls provide bar prestressing systems, described in Chapter 2. Bars were used in some of the first prestressed concrete structures and most of the systems have been established for many years. Prestressing bars are either fully threaded along their complete length or are smooth with just the ends threaded.

DYWIDAG bar post-tensioning 'smooth bar' and 'threadbar' arrangements are shown in Figure D11. The bars are available in lengths of up to 30 m. The DYWIDAG smooth bar has threaded ends with a

Figure D11 DYWIDAG prestressing bars

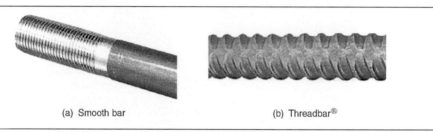

(a) Smooth bar

(b) Threadbar®

Figure D12 DYWIDAG post-tensioning bar details

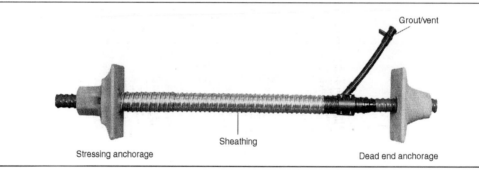

Stressing anchorage

Sheathing

Grout/vent

Dead end anchorage

Technical data

Type		Smooth bar			Threadbar®				
Steel grade $f_{p0\cdot1k}/f_{pk}$	N/mm²	835/1030	1080/1230		835/1030			1080/1230	
Diameter/type		32G	36C	26E	32E	36E	26D	32D	36D
Diameter	mm	32	36	26	32	36	26	32	36
Nom. cross section area	mm²	804	1018	551	804	1018	551	804	1018
Weight	kg/m	6.31	7.99	4.48	6.53	8.27	4.48	6.53	8.27
Pitch	mm	3	3	13.0	16.0	18.0	13.0	16.0	18.0
Yield load $f_{p0\cdot1k}A$	kN	671	1099	460	671	850	595	868	1099
Ultimate load $f_{pk}A$	kN	828	1252	568	828	1049	678	989	1252
Fatigue stress range									
Steel (0.55 f_{pk}/0.9$_{p0\cdot1k}$)	N/mm²	290/230	290/230	240/210	240/210	240/210	240/210	240/210	240/210
Nut/coupler 0.6 f_{pk}	N/mm²	98	98	98	78	78	78	78	78
Bending									
Elastic bending min. R	m	19.50	12.10	16.20	19.50	21.90	8.90	10.75	12.10
Cold bending min. R	m	4.80	5.40	5.30	6.40	7.20	5.30	6.40	7.20

Values based on German approval. Customisations according to different codes and concrete strengths applicable.

rolled-on thread manufactured in the factory to the specific length required. The rolled thread results in a characteristic low seating loss and high fatigue resistance. The DYWIDAG threadbar is hot-rolled and comes with thread-like deformations over its entire length. The threadbar is cut to length and is instantly threadable without further preparation. Figure D12 gives the data for the DYWIDAG bar systems. Details of the anchorage and coupler components are given in Figures D13 and D14 respectively. The steel duct details are given in Figure D15. The jacks used with the DYWIDAG bar systems are shown in Figure D16.

Figure D13 DYWIDAG bar anchor details

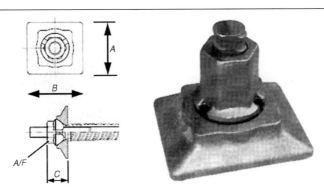

Technical data

Bar type		Smooth bar			Threadbar®				
Steel grade $f_{p0.1k}/f_{pk}$	N/mm²	835/1030	1080/1230		835/1030			1080/1230	
Nom. diameter	mm	32.0	36.0	26.5	32.0	36.0	26.5	32.0	36.0
Diameter/type	–	32G	36C	26E	32E	36E	26D	32D	36D
A	mm	–	160	–	–	140	120	140	160
B	mm	–	180	–	–	165	130	165	180
C	mm	–	75	–	–	107	90	107	115
Width across flats A/F	mm	–	65	–	–	60	50	60	65

Details of the anchorage zone

Bar type			Smooth bar			Threadbar®				
Steel grade $f_{p0.1k}/f_{pk}$		N/mm²	835/1030	1080/1230		835/1030			1080/1230	
Nom. diameter		mm	32.0	36.0	26.5	32.0	36.0	26.5	32.0	36.0
Diameter/type		–	32G	36C	26E	32E	36E	26D	32D	36D
Concrete strength class B25	Y	Centre distances	–	280	–	–	–	200	250	280
		Edge distances	–	160	–	–	–	120	145	160
	X	Centre distances	–	320	–	–	–	220	280	320
		Edge distances	–	180	–	–	–	130	160	180
Concrete strength class B35	Y	Centre distances	–	260 (380)	–	–	(340)	190	230	260 (380)
		Edge distances	–	150 (190)	–	–	(170)	115	135	150 (190)
	X	Centre distances	–	290 (440)	–	–	(400)	210	260	290 (440)
		Edge distances	–	165 (220)	–	–	(200)	125	150	165 (220)
Concrete strength class B45	Y	Centre distances	–	230	–	–	–	170	210	230
		Edge distances	–	135	–	–	–	105	125	135
	X	Centre distances	–	260	–	–	–	190	230	260
		Edge distances	–	150	–	–	–	115	135	150

Figure D14 DYWIDAG bar coupler details

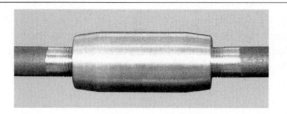

Figure D15 DYWIDAG bar duct sizes

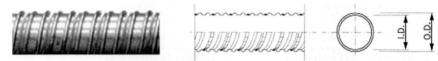

Type		Smooth bar			Threadbar®				
Steel grade $f_{p0.1k}/f_{pk}$	N/mm²	835/1030	1080/1230		835/1030			1080/1230	
Diameter/type		32G	36C	26E	32E	36E	26D	32D	36D
Diameter	mm	32	36	26	32	36	26	32	36
Internal diameter I.D.	mm	44	51	38	44	51	38	44	51
External diameter O.D.	mm	49	57	43	49	57	43	49	57
Min. centre distances	mm	79	92	68	79	92	68	79	92
Support distances	m	0.5–2.5			0.5–2.5				
Wobble angle β	°/mm	0.3			0.3				
Friction coefficient μ	–	0.25			0.5				

The values are based on the German Code (DIN). Adaptations to other code systems or concrete strengths (e.g. ASTM, BS, etc.) are possible.

Figure D16 DYWIDAG bar jack data

Jack 60 Mp series 04

Jack 110 Mp series 04

Bar type	Smooth bar		Threadbar®					
Steel grade	835/1030	1080/1230	835/1030			1080/1230		
Nom. diameter	32	36	26.5	32.0	36.0	26.5	32.0	36.0
Diameter/type	32G	36C	26E	32E	36E	26D	32D	36D
60 Mp	●		●	●	★	●	★	
110 Mp	○	●		○	●		●	●

★ Use 60 Mp if only stressed 55% to ultimate load.
○ Can be stressed with 110 Mp on certain exceptions.

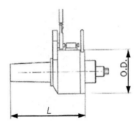

Technical data

Jack types	Length L (mm)	O.D. (mm)	Stroke (mm)	Piston area (cm²)	Capacity (kN)	(MPa)	Weight (kg)
60 Mp Series 04	401	190	50	132.5	625	50	36
60 Mp Series 05	456	190	100	132.5	625	50	44
110 Mp Series 01	494	267	50	235.6	1100	50	46
110 Mp Series 03	594	267	150	235.6	1100	50	54

Jack dimensions (for block-out design)

Jack types	A (mm)	B (mm)	C (mm)	D (mm)	E (mm)
60 Mp Series 04	225	176	106	106	300
60 Mp Series 05	225	231	106	106	300
110 Mp Series 01	275	219	125	120	375
110 Mp Series 03	275	319	125	120	375

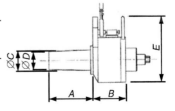

Figure D17 Freyssibar used in temporary works

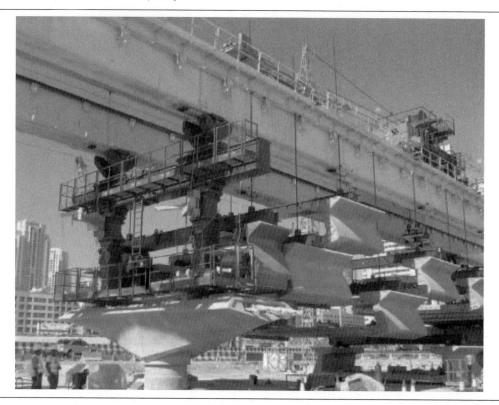

Freyssinet's Freyssibar prestressing system is used in bridgeworks, for both the permanent works and temporary works. They are threaded over their full length which simplifies installation and reuse. In Figure D17 the Freyssibars are being used as hangers for the erection of precast segments. Figures D18 and D19 show the Freyssinet bar system with the main dimensions of the bars, couplers and ducting.

Figure D18 Freyssinet's Fressibar prestressing bar characteristics and dimensions

Bar characteristic	Unit	Nominal diameter (mm)				
		26.5	32	36	40	50
Steel grade	MPa	1030	1030	1030	1030	1030
Cross section area	mm²	552	804	1018	1257	1964
Linear mass	kg/m	4.56	6.66	8.45	10.41	16.02
Characteristic breaking load : Fpk	kN	568	828	1048	1295	2022
0.1% proof load	kN	461	672	850	1049	1640
Tensionning force at 0.80 x Fpk	kN	454	662	838	1036	1618
Thread pitch	mm	6	6	6	8	8
Average Young's modulus	GPa	170	170	170	170	170
Minimum elongation at maximum force	%	3.5	3.5	3.5	3.5	3.5

Item	Sketch	Dimensions	Unit	Nominal bar diameter (mm)				
				26.5	32	36	40	50
Flat nut		Length	mm	37	41	46	55	71
		Width on flat surface	mm	50	56	62	65	90
Flat washer		External diameter	mm	65	70	75	80	105
		Thickness	mm	6	6	6	6	6
Flat plate		Dimensions	mm	110x125	125x125	140x160	160x160	200x200
		Thickness	mm	35	35	40	40	45
		Hole diameter	mm	34	40	44	50	60
Injection plate		Dimensions	mm	110x125	125x125	140x160	160x160	200x200
		Thickness	mm	35	35	40	40	45
		Hole diameter	mm	34	40	44	50	60
Threaded plate		Dimensions	mm	110x125	125x125	140x140	150x150	185x185
		Thickness	mm	40	50	50	60	70
Welded cap (option)		Length	mm	15	20	20	25	25
Spherical nut		Length	mm	45	51	56	60	71
		Width on flat surface	mm	50	56	62	65	90
Spherical plate		Dimensions	mm	160x115	160x125	160x140	160x160	190x190
		Thickness	mm	40	40	40	40	60

Figure D19 Freyssibar coupler and ducting dimensions

Coupler	Dimensions	Unit	Nominal bar diameter (mm)					Ref.
			26.5	32	36	40	50	
	External diameter	mm	45	50	60	65	76	C
	Length	mm	90	115	130	140	170	

Ducting	Dimensions	Unit	Nominal bar diameter (mm)					Ref.
			26.5	32	36	40	50	
Steel corrugated sheath	Internal diameter	mm	45	50	55	60	75	G1
	Thickness	mm	0.45	0.45	0.45	0.45	0.50	
	Volume of grout	L/m	1.0	1.2	1.4	1.6	2.5	
	Connection element (internal diameter)	mm	50	55	65	70	85	G'1
HDPE tube	External diameter	mm	63	63	75	75	90	G2
	Thickness	mm	5.8	5.8	6.8	6.8	8.2	
	Volume of grout	L/m	1.5	1.3	1.9	1.7	2.3	
For prolongation sleeve	External diameter	mm	70	76.2	88.9	95	114.3	GR
	Thickness	mm	2	2	2	2	2	
	Minimum length (L = sleeve)	mm	180 + L	205 + L	220 + L	230 + L	260 + L	
For coupling sleeve	External diameter	mm	88.9	88.9	101.6	114.3	152.4	GC
	Thickness	mm	2	2	2	2	2	
	Maximum length	mm	210	235	255	265	320	

D.6. Wire prestressing systems

Multi-wire stressing systems were used on some of the very first prestressed bridges and are still available from BBR. The BBRV system uses bundles of individual 7 mm diameter wires fixed into the anchorage with a BBRV button head. Each wire is mechanically fixed in the anchor head and can reach its full rupture load without any slippage. Before the tendons are grouted they are re-stressable and de-stressable.

The BBRV anchoring method allows the production of post-tensioning tendons with any number of single wires and therefore with any given magnitude of prestressing force. Standard tendon arrangements are illustrated in Figure D20 and tendons details given in Figure D21. Details of the BBRV anchorages, couplers and jacks are given in Figures D22, D23 and D24 respectively.

Figure D20 BBRV wire tendons

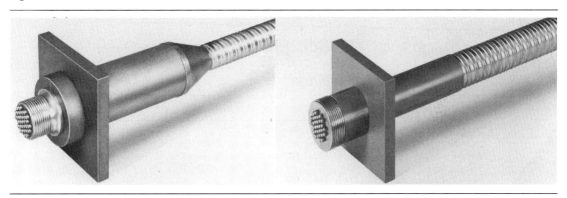

Figure D21 BBRV standard wire tendons

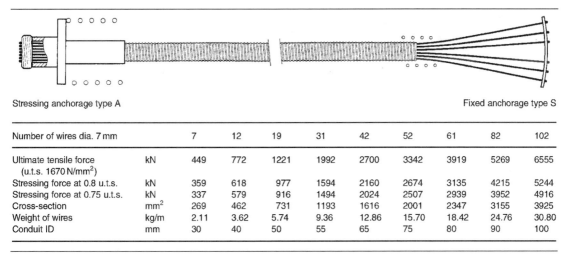

Stressing anchorage type A   Fixed anchorage type S

Number of wires dia. 7 mm		7	12	19	31	42	52	61	82	102
Ultimate tensile force (u.t.s. 1670 N/mm²)	kN	449	772	1221	1992	2700	3342	3919	5269	6555
Stressing force at 0.8 u.t.s.	kN	359	618	977	1594	2160	2674	3135	4215	5244
Stressing force at 0.75 u.t.s.	kN	337	579	916	1494	2024	2507	2939	3952	4916
Cross-section	mm²	269	462	731	1193	1616	2001	2347	3155	3925
Weight of wires	kg/m	2.11	3.62	5.74	9.36	12.86	15.70	18.42	24.76	30.80
Conduit ID	mm	30	40	50	55	65	75	80	90	100

Figure D22 BBRV wire anchorages

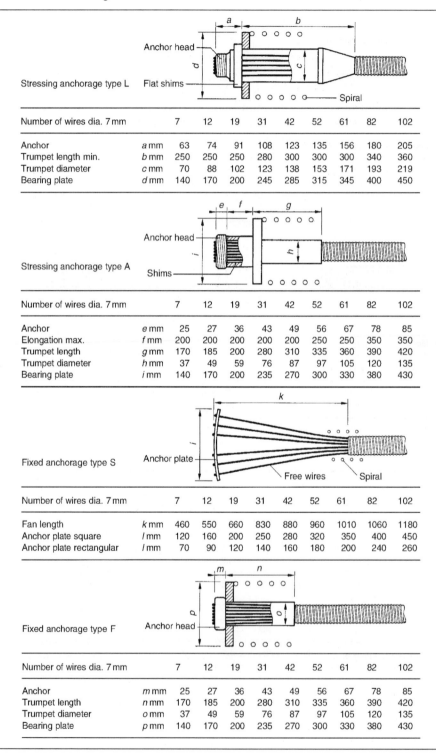

Stressing anchorage type L										
Number of wires dia. 7 mm		7	12	19	31	42	52	61	82	102
Anchor	*a* mm	63	74	91	108	123	135	156	180	205
Trumpet length min.	*b* mm	250	250	250	280	300	300	300	340	360
Trumpet diameter	*c* mm	70	88	102	123	138	153	171	193	219
Bearing plate	*d* mm	140	170	200	245	285	315	345	400	450

Stressing anchorage type A										
Number of wires dia. 7 mm		7	12	19	31	42	52	61	82	102
Anchor	*e* mm	25	27	36	43	49	56	67	78	85
Elongation max.	*f* mm	200	200	200	200	200	250	250	350	350
Trumpet length	*g* mm	170	185	200	280	310	335	360	390	420
Trumpet diameter	*h* mm	37	49	59	76	87	97	105	120	135
Bearing plate	*i* mm	140	170	200	235	270	300	330	380	430

Fixed anchorage type S										
Number of wires dia. 7 mm		7	12	19	31	42	52	61	82	102
Fan length	*k* mm	460	550	660	830	880	960	1010	1060	1180
Anchor plate square	*l* mm	120	160	200	250	280	320	350	400	450
Anchor plate rectangular	*l* mm	70	90	120	140	160	180	200	240	260

Fixed anchorage type F										
Number of wires dia. 7 mm		7	12	19	31	42	52	61	82	102
Anchor	*m* mm	25	27	36	43	49	56	67	78	85
Trumpet length	*n* mm	170	185	200	280	310	335	360	390	420
Trumpet diameter	*o* mm	37	49	59	76	87	97	105	120	135
Bearing plate	*p* mm	140	170	200	235	270	300	330	380	430

Figure D23 BBRV wire couplers

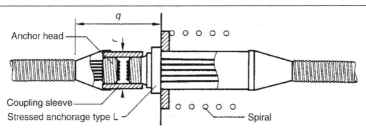

Fixed coupling type LK

Number of wires dia. 7 mm		7	12	19	31	42	52	61	82	102
Trumpet length	q mm	230	260	290	350	410	430	470	570	630
Trumpet diameter	r mm	70	88	102	123	138	153	171	193	219

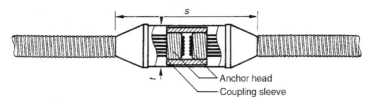

Movable coupling type LK1

Number of wires dia. 7 mm		7	12	19	31	42	52	61	82	102
Trumpet length min.	s mm	600	620	670	750	810	880	950	1080	1150
Trumpet diameter	t mm	70	88	102	123	138	153	171	193	219

Figure D24 BBRV wire jack details

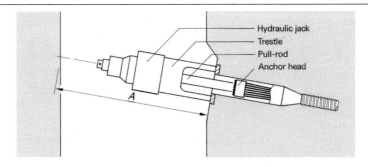

Type of stressing jack		NP 60	NP 100	NP 150	NP 200	NP 250	NP 300	GP 500
Max. jack force	kN	620	1030	1545	2060	2575	3090	5150
Jack diameter	mm	160	205	250	290	315	350	560
Jack stroke	mm	100	100	100	100	100	100	400
Jack weight	kg	28	50	83	117	147	196	1260
Space requirement A	mm	1700	1700	1700	2000	2000	2000	2500

D.7. Auxiliary equipment

The other equipment used for completing the installation of the tendons includes strand-pushing equipment, hydraulic pumps to operate the stressing jacks, and the grouting equipment. Details of the DYWIDAG equipment used are given in Figures D25 to D27.

Figure D25 DYWIDAG strand pushing equipment

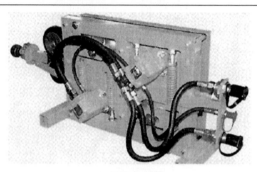

ESG 8-1

Type	Tensile or compression force (kN)	Pushing speed (m/s)	Weight (kg)	Dimensions $L \times W \times H$ (mm)	Hydraulic pumps
ESG 8-1	3.9	6.1	140	1400/350/510	ZP 57/28

Figure D26 DYWIDAG hydraulic pump data

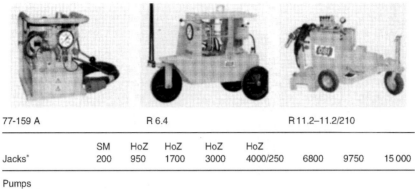

| 77-159 A | R 6.4 | R 11.2–11.2/210 |

Jacks*	SM 200	HoZ 950	HoZ 1700	HoZ 3000	HoZ 4000/250	6800	9750	15 000
Pumps								
77-159 A	•	•						
77-193 A	•	•	•					
R 3	•	•	•					
R 6.4	•	•	•	•	•			
R 11.2–11.2				•	•	•		
R 11.2–11.2/210					•		•	•	•
ZP 57/28				For all pushing devices				

* For pistons without power seating

Length *L*

W

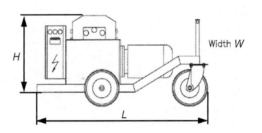

H

Width *W*

L

Technical data

Pumps	Operation pressure (MPa)	Capacity, V min (l/min)	Eff. oil amount (l)	Weight (kg)	Dimensions $L \times W \times H$ (mm)
77-159 A	70	3.0	10.0	60	420/380/480
77-193 A	70	3.0	10.0	63	420/380/480
R 3.0	70	3.0	13.0	98	600/390/750
R 6.4	60	6.4	70.0	310	1400/700/1100
R 11.2–11.2/210	55	11.2/22.4	170.0	720	2000/800/1300
ZP 57/58	16/22	53/80	175.0	610	1260/620/1330

Figure D27 DYWIDAG grouting equipment

MP 2000-5   MP 4000-2

Grouting equipment	Maximum injection pressure (MPa)	Capacity (l/h)	Weight (kg)	Dimensions $L \times W \times H$ (mm)
MP 2000-5	1.5	420	60	2000/950/1600
MP 4000-2	1.5	1500	63	2040/1040/1750
P 13 EMRT	8.0	3000	98	2150/1750/1500

Prestressed Concrete Bridges, 2nd edition
ISBN: 978-0-7277-4113-4

ICE Publishing: All rights reserved
doi: 10.1680/pcb.41134.401

Index